A Synopsis of North American Desmids

A SYNOPSIS OF
NORTH AMERICAN DESMIDS

Part II. DESMIDIACEAE: PLACODERMAE
Section 2

G. W. PRESCOTT
University of Montana Biological Station, Bigfork, Montana

HANNAH T. CROASDALE
Department of Biology, Dartmouth College, Hanover, New Hampshire

W. C. VINYARD
Department of Biology, California State University-Humboldt, Arcata, California

UNIVERSITY OF NEBRASKA PRESS • LINCOLN and LONDON

Preparation and publication of this synopsis was supported by the National
Science Foundation (Grants GB 15903, GB 42523).

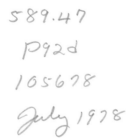

Library of Congress Cataloging in Publication Data

Prescott, Gerald Weber, 1899–
  Desmidiales.

  (North American flora. Series II, pt. 6)
  Pt. 2, published by University of Nebraska Press,
Lincoln under title: A synopsis of North American
desmids, lacks series statement.
  Cover title.
  Bibliography: p. 391
  CONTENTS: pt. 1. Sacҫodermae, Mesotaeniaceae.—
pt. 2. Desmidiceae: Placodermae.
  1. Desmidiales. 2. Algae—North America.
I. Croasdale, Hannah T., 1905–   joint author.
II. Vinyard, W. C., joint author.   III. Title.
IV. Title: A synopsis of North American desmids.
V. Series
QK569.D48P73   589'.47   70-183418
ISBN 0-8032-0899-5

Contents

Preface

Section 2 of the Cosmarieae for *A Synopsis of North American Desmids*[1] presents the genera *Euastrum* Ehrenberg 1832 and *Micrasterias* Agardh 1827. The sequence uses the traditional phylogenetic concept, which, however, is variously interpreted by different phycologists. One suggested evolutionary scheme is presented in Part I of the *Synopsis.*[2] Most students hold that the Placodermae, and probably the Saccodermae as well, arose from filamentous conjugalean ancestors. In not a few species there may be a reversion to a filamentous or pseudo-filamentous condition.

That the Placodermae have retrogressed in their sexual reproduction while they have evolved morphologically is borne out in *Euastrum* and *Micrasterias,* in which only a few species have been observed with zygospores. Taxonomy within these two genera invites critical studies because of the confusion at especially subspecific levels. Some taxa particularly show extreme variability, either genetically or as environmental responses, with many taxonomic epithets given to the latter incorrectly. Relatively few *Euastrum* and *Micrasterias* species are treated in Ralfs (1848)—used as the starting point for desmid nomenclature. We have found it necessary, and believe that it has been appropriate in many instances, to reassign names and to place others in synonymy. In some changes personal judgment has played a role in making decisions, with the realization that they are subject to revision as critical and monographic studies are undertaken.

The authors are grateful for additional contributed collections of desmids from Mr. Robert Frock (Pennsylvania), Dr. Patrick Gleason (Florida), Mr. W. D. Taylor (California), and Mr. G. W. Mathews (Kansas).

---

1. *A Synopsis of North American Desmids.* Part II. Desmidiaceae: Placodermae. Section 1. Lincoln: University of Nebraska Press, 1975.
2. Part I—Saccodermae, Mesotaeniaceae—was published in the North American Flora Series, Series II, Part 6, of the New York Botanical Garden (Bronx, New York, 1972).

A Synopsis of North American Desmids

EUASTRUM Ehrenberg ex Ralfs 1848, Brit. Desm., p. 78.

The genus *Euastrum* includes about 350 species, of which 116 (together with their varieties) have been reported thus far from North America. As in most of the Cosmarieae, cells are relatively short, being about 1.5 times longer than broad, oval in outline (outer boundary), and they are compressed as seen in vertical and lateral views. The chief differentiating characteristics are, first, the presence of a vertical notch in the apex and, second, of one or more facial protuberances or "tumors" which are more clearly evident in lateral than in "front" (facial) view. In not a few species the vertical notch is lacking or may be replaced by a concave, or broad V-shaped, apical margin. Rarely the apical margin is convex. In species without an apical notch, however, the facial protuberances and wall decorations characteristic of *Euastrum* are present.

The semicells are mostly 3-lobed, with 2 basal lobes and a polar lobe. In many species there are upper lateral lobules between the basal lobes and the polar lobe. The morphology of the polar lobe is regarded as having important taxonomic significance and to be more consistent than the forms taken by the basal and lateral lobes. It is thought that in this genus (as has been established for *Micrasterias*) the strongest influence of cytoplasmic DNA is within or toward the apex of the cell. Most of the *Euastrum* with entire margins or simple lobulations are small. The large species are variously lobed and have ornate wall features.

In lateral view the cells are oval in outline (rarely elliptic or subrectangular), with narrow, acute, spine-bearing poles, or with the poles truncate (convex or flat). In vertical view the cells are elliptic, oval, or quadrate in outline, mostly with rounded poles, and showing lateral tumors or protrusions.

The cell wall may be smooth, punctate, or scrobiculate, and often has granules and spines either on the face or on the margins of the lobes. There often are pores, either numerous, small, and incomplete (not penetrating entirely through the wall), or single or few in each semicell and complete, forming a conspicuous *pore organ* (Lütkemüller 1894, 1902).

The isthmus is relatively narrow in *Euastrum*, and the sinus between the semicells is mostly narrow and closed. In fact, the open sinus commonly possessed by other genera in the Cosmarieae is a rarity. Some species have a sinus which opens at the outer extremity because of the divergent angles of the basal lobes. There is 1 chloroplast in each semicell, with usually 1 pyrenoid; larger cells have 2 to several pyrenoids. The chloroplast may be a simple, axial mass, or may have flanges and ridges which expand against the wall.

Zygospores are known for a relatively few species, probably because species are predominantly heterothallic and there is little opportunity for individuals of different strains to become apposed for conjugation. Zygospores are globular, usually with spines, or with mammillate and hornlike protuberances. They are formed between gametangial cells.

Although *Euastrum* can be considered to be a distinct genus, there are some species (usually small) which approach *Cosmarium*, and a few have *Miscrasterias*-like features. For example, *Euastrum verrucosum* var. *moebii* Borge has been referred to *Micrasterias moebii* (Borge) by West & West (1897a) because of this intergradation, but since has been reassigned to *Euastrum* (*E. moebii* (Borge) Scott & Prescott). *Euastrum* is clearly differentiated from *Tetmemorus*, which also has a polar incision, by being relatively much shorter than the latter, and by its lobes and facial protuberances. Many taxa at the species level show considerable

1

morphological variation (*E. ansatum* Ehrenb. ex Ralfs, *E. binale* (Turp.) Ehrenb., *E. verrucosum* Ehrenb. ex Ralfs, *e.g.*). At least 42 varieties and forms of the latter species have been described, many of which are undoubtedly habitat variables. *Euastrum verrucosum* has been studied critically by Huber-Pestalozzi (1931), whereas *Euastrum ansatum* Ehrenb. ex Ralfs has been treated by Ducellier (1918). More of such examinations are needed for other species of *Euastrum* which are burdened with taxonomic epithets. The intergradation of these variables has made it difficult, and in some instances impossible, to draw clearly defined lines; accordingly there are numerous forms of uncertain position. The taxonomic status of such variables on a genetic basis is subject to interpretation of individual desmidologists, and is further complicated by the fact that little has been determined relative to whether variations are incidentally produced by environmental influences, or whether they are genetically induced. It is noteworthy that some variables within a local population are similar to the forms recognized in far-removed habitats where only 1 morphological type exists. Hence, we do not know whether the wide distribution of a form is related to dispersal, or whether 2 or more habitats have the same identical "determiner." Suffice to state, many taxonomic names are given to forms which are not genetic. In this synopsis epithets are retained in many instances with the philosophy that they serve useful purposes. They give an indication of the relative variability of a species and its genetic stability, and the names serve as designations for reference and discussion.

Species of *Euastrum* occur most frequently in soft-water habitats, more so it appears than do many other desmid genera in which there are species which are found in hard-water situations. The larger, more ornate species are restricted to highly acid habitats where there is a concentration of decaying organic matter. It is noteworthy perhaps that both *Euastrum* and *Micrasterias* are represented poorly in polar habitats (although there are many species of the former found in the Arctic), and are more profuse in the tropics and subtropics.

Key to Species of North American *Euastrum*

In the following key to species, differentiation is made primarily on the form of the polar lobe and the presence or absence of an apical incision. Of secondary importance is the shape of the basal lobes and the presence or absence of upper, lateral lobules. Reference is made in the key to the configuration of the lateral margins of the semicells, especially between the lateral lobules and the polar lobe: merely retuse or in the form of a sinus. An attempt has been made to differentiate taxa as they are seen in front or "face" view. In many instances in *Euastrum* species, as in several other desmid genera, final and/or more precise identification must include lateral and vertical views, and these are described in the species diagnoses, and those of most infraspecific taxa.

1. Apical margin of the polar lobe with a narrow, median, vertical incision, the margins of the notch parallel or nearly so.
  2. Apical angles, or margins of the cell with spines or long, sharp granules.
    3. Margins of semicells with upper lateral lobules, or prominent protrusions.
      4. Lobules with 2 or 3 long, sharp, spinelike teeth at their apices.
        5. A prominent U-shaped or V-shaped sinus between the polar lobes and the basal lobes or lobules.
          6. Spines on lobules solitary and simple, sometimes blunt and only slightly developed, the sinus below the polar lobe widely open.
            E. pseudocoralloides.
          6. Spines on lobules in the form of 2 or 3 long, sharp teeth or sharp tuberculations.
            7. Polar lobes nearly as wide at the apex as the basal lobes.
              8. Apical margin flat.   *E. abruptum.*
              8. Apical margin elevated.
                9. Apical margin undulate, the polar lobes inflated.   *E. pictum.*

9. Apical margin not undulate, the polar lobes not inflated.
  10. Basal lobes bilobed to form 2 approximately equal lobules which are trispinate.   *E. umbonatum.*
  10. Basal lobes with 3 lobules, the upper, lateral lobules with 2 blunt teeth.   *E. crameri.*
7. Polar lobes conspicuously narrower than the basal lobes.
  11. Apical margin flat, with a short submarginal spine at each angle. *E. incertum.*
  11. Apical margin slightly retuse, with a conspicuous subapical, bispinate process.   *E. evolutum.*
5. Without a prominent sinus between the polar lobe and the basal lobes.
  12. Apical margin retuse, but slightly elevated in the midregion; apical spines upwardly directed.   *E. oculatum.*
  12. Apical margin elevated to the median notch, apical spines horizontally projected.
    13. Margins of basal lobes converging, the upper lateral protrusions forming a shoulder below the subparallel margins of the polar lobes.   *E. quebecense.*
    13. Margins of basal lobes convex, without a shoulder below the polar lobes.   *E. pulchellum.*

4. Lobules without long, sharp teeth.
  14. Apex elevated, with a horizontally directed, short, blunt spine at the lateral angles.   *E. bidentatum.*
  14. Apical margin flat.
    15. Basal lobes with 3 equal lobules.   *E. kriegeri.*
    15. Basal lobes without 3 equal lobules.
      16. Basal lobes divided to form 2 equal lobules; semicells subquadrate.   *E. octogibbosum.*
      16. Basal lobes with 1 upper lateral, shoulderlike lobule; semicells pyramidal.   *E. personatum.*

3. Margins of semicells without upper lateral lobules.
  17. Basal lobes bearing a single, stout, horizontally directed spine.   *E. divaricatum.*
  17. Basal lobes without a horizontally directed spine.
    18. Polar lobes divided by the median incision to form 2 separate, prominent lobules.
      19. Apical margin flat, the median incision short and U-shaped; subapical sinus deep, creating 2 lobules in the polar lobe.   *E. trigibberum.*
      19. Apical margin not flat.
        20. Polar lobules bearing 3 stout spines.   *E. cuspidatum.*
        20. Polar lobules bearing a short, simple, blunt spine.   *E. quadrilobum.*
    18. Polar lobes not divided to form 2 prominent lobules.
      21. Apex broadly rounded.
        22. Basal lobes angularly and convexly rounded, the margins of the semicell retuse to a prominent subapical spine half the distance to the apex; apical incision deep.   *E. fissum.*
        22. Basal lobes emarginate and subrectangular, the margins of the semicell retuse to subapical spines which are near the apex; apical incision short.   *E. elegans.*
      21. Apex not broadly rounded.
        23. Cells quadrangular.
          24. Basal lobes convex; apical notch a shallow U, with marginal granules.   *E. denticulatum.*
          24. Basal lobes sharply or angularly rounded; apical notch a broad V.   *E. rectangulare.*
        23. Cells longer than wide.
          25. Basal lobes simple, the margins broadly rounded, smooth except for a pair of short spines; apical margin elevated.   *E. ciastonii.*
          25. Basal lobes bilobed, the margins crenate, apical margin flat.   *E. turneri.*
2. Apical angles and cell margins without spines (sometimes with mucros).
  26. Apical angles or cell margins bearing prominent granules or granular thickenings.
    27. Polar lobe inflated, larger than the basal lobes; apical lobules pincer-shaped.   *E. informe.*
    27. Polar lobes not inflated; lobules not pincer-shaped.

28. Polar lobes prominent, divided to form 2 rounded lobules; wall minutely
     granular at the margins (?).   *E. gemmatoides.*
28. Polar lobes not divided to form 2 prominent lobules.
   29. Apex elevated; the median incision relatively deep and open.   *E. protuberans.*
   29. Apex flat; median notch very shallow.   *E. lapponicum.*
26. Apical angles and cell margins smooth, without spines or granules.
   30. Cells 3 times longer than wide; the basal lobes and margins smooth, the margins
       subparallel to truncate apices.
     31. Cells 220–251 μm long.   *E. giganteum.*
     31. Cells 106–130 μm long.   *E. solum.*
   30. Cells less than 3 times longer than wide.
     32. Basal lobes extended horizontally to form simple, sharply rounded angles, the
       basal lobes cone-shaped; polar lobes expanded, anvil-shaped.   *E. wollei.*
     32. Basal lobes shaped otherwise.
       33. Polar lobes inflated or expanded, with their lobules horizontally extended.
         34. A narrow, closed or slightly open sinus between the polar lobes and the
           basal lobes or lobules.
           35. Polar lobe inflated or extended to equal the basal lobes or lobules.
             36. Basal lobes simple, not divided or bilobed.   *E. informe.*
             36. Basal lobes bilobed.   *E. floridense.*
           35. Polar lobes not extended as far as the basal lobes.
             37. Upper lateral lobules emarginate or bilobed.
               38. Upper lateral lobules extended farther than the basal lobes, which
                 are smaller.   *E. infernum.*
               38. Upper lateral lobules not extended as far as the basal lobes, which
                 are larger.   *E. oblongum.*
             37. Upper lateral lobules not emarginate.
               39. Cells subrectangular; margins of semicells subparallel, with shallow
                 retuse margins between the basal and the upper lateral lobules.

                                                 E. crassum.
               39. Cells oval; semicells truncate-pyramidal, the lateral margins deeply
                 retuse.   *E. ventricosum.*
         34. Without a narrow sinus between the polar lobes and the basal lobes.
           40. Semicells with upper lateral lobules; basal lobes distinct.
             41. Upper lateral lobules horizontally directed, extended farther than the
               polar lobules, which are narrowly rounded.   *E. pinnatum.*
             41. Upper lateral lobules somewhat upwardly directed, shorter than or
               equal to the polar lobules, which are broadly rounded.   *E. affine.*
           40. Semicells without lateral lobules; basal lobes not distinctly identified.
             42. Sinus of the cell (between semicells) narrow and closed, opening
               slightly outward; basal angles sharply rounded; lateral margins con-
               verging evenly except for a slight swelling.   *E. latipes.*
             42. Sinus open throughout.
               43. Polar lobules narrow, horizontally extended; face of semicell with
                 prominent, downwardly projected protrusions.   *E. insigne.*
               43. Polar lobules short, angularly rounded, not horizontally directed;
                 face of semicell without prominent protrusions.   *E. intermedium.*
       33. Polar lobes not inflated or expanded.
         44. Semicells with upper lateral lobules or prominent swellings on the margins
           of the basal lobes.
           45. Polar lobe scarcely evident, the basal lobes and the upper lateral lobules
             evenly incised throughout to form 7 lateral segments.   *E. tetralobum.*
           45. Polar lobe evident; basal lobes not so divided.
             46. Upper lateral lobules broad, emarginate.   *E. vinyardii.*
             46. Upper lateral lobules not emarginate.
               47. Polar lobe divided apically by a widely open incision forming 2
                 upwardly diverging lobules.   *E. gemmatoides.*
               47. Polar lobe not so divided.
                 48. Upper lateral lobules prominent, shoulderlike, forming upwardly
                   directed lobes, with a broad U-shaped sinus between the polar
                   lobe and the marginal lobules.   *E. humerosum.*
               48. Upper lateral lobules not upwardly directed.

49. Semicells quadrangular in outline; the polar lobes produced, with subparallel margins; basal lobes quadrate, the lateral margins retuse, subparallel to the shoulderlike upper, lateral lobules.
E. jenneri.

49. Semicells not quadrangular; polar lobes not prominently produced.
  50. Basal lobes prominently bilobed, with a sharp sinus between the upper basal lobules and prominent lateral lobules; the polar lobes very short and truncate, the apex flattened.
  E. allenii.
  50. Basal lobes not so divided; polar lobes not flat and truncate.
    51. Margins of polar lobe divergent to slightly inflated polar lobules; the semicells broadly triangular. *E. ampullaceum.*
    51. Margins of polar lobes not divergent.
      52. Margins of polar lobe converging.
        53. Upper lateral lobules prominent, the polar lobe clearly produced. *E. subinerme.*
        53. Upper lateral lobules merely low swellings; the polar lobe scarcely evident. *E. inerme.*
      52. Margins of polar lobe parallel.
        54. Upper lateral lobules prominent, the polar lobes decidedly produced; cell wall coarsely scrobiculate; face of semicell with 5 mucilage pores. *E. sinuosum.*
        54. Upper lateral lobules merely swellings on the margin of the basal lobe.
          55. Cells with elongate, pyramidal semicells; face of semicell with 3 prominent swellings across the base.
          E. didelta.
          55. Semicells broadly pyramidal, with 1 basal protrusion on the face. *E. everettense.*
44. Semicells without upper lateral lobules or prominent swellings.
  56. Lateral margins of cell equally crenulate throughout; vertical incision very deep; apical lobules pincer-shaped. *E. invaginatum.*
  56. Lateral margins not equally crenulate throughout.
    57. Basal lobes quadrangularly rounded, with subparallel margins; margins of semicell abruptly narrowed to form a long, produced polar lobe with subparallel margins. *E. longicolle.*
    57. Semicells shaped otherwise.
      58. Semicells broadly urn-shaped, the short polar lobes with laterally extended lobules that have sharp angles. *E. integrum.*
      58. Semicells shaped otherwise.
        59. Cells small, 26 μm maximum length; polar lobes broad, the apex flat. *E. rimula.*
        59. Cells larger, shaped otherwise.
          60. Basal lobes angularly swollen, the margins at first diverging and then abruptly retuse to a broad polar lobe with subparallel margins. *E. indicum.*
          60. Cells shaped otherwise.
            61. Semicells with broadly rounded basal lobes.
              62. Cells relatively small, maximum length 58 μm; face of semicell with a conspicuous excentric mucilage pore. *E. pingue.*
              62. Cells larger; face of semicell without an excentric mucilage pore.
                63. Cells stout, relatively broad, 111 μm maximum length; face of semicell without facial protrusions. *E. obesum.*
                63. Cells more elongate; face of semicell with 3 basal and 2 upper protuberances. *E. ansatum.*
            61. Semicells with narrowly rounded or angularly convex basal lobes.
              64. Base of semicells with subparallel margins, slightly retuse between basal angles and upper lateral swelling. *E. brasiliense.*
              64. Base of semicell differently shaped.

65. Cells relatively large, up to 157 μm long.
   66. Semicells with prominent upper lateral swellings on the
      margin.   *E. didelta.*
   66. Semicells symmetrically cuneate, without upper, lateral
      swellings on the margin.   *E. cuneatum.*
65. Cells smaller, 40–87 μm long.
   67. Basal margins at first diverging, the upper margins retuse
      above the basal lobes, up to 40 μm long.   *E. subhexalobum.*
   67. Basal margins convex, the upper margins mostly convex;
      up to 87 μm long.   *E. aboense.*
1. Apical margins without a narrow, vertical incision.
68. Apical margin retuse, invaginated, or with a broad U- or V-shaped notch (some
   varieties of species with a narrowed V-shaped notch). See *E. crispulum.*
  69. Margins, or polar angles with spines or with toothlike projections.
   70. Apex divided by a deep invagination to form 2 distinct polar lobules.
    71. Polar lobules furnished with 3 stout spines or with 3 stout, blunt, toothlike
      projections.
     72. Polar lobules with 3 spines.   *E. cuspidatum.*
     72. Polar lobules with 3 stout, blunt teeth.   *E. formosum.*
    71. Polar lobules with a single, blunt spine.   *E. quadrilobum.*
   70. Apex not divided to form 2 polar lobules.
    73. Basal lobes with a single, horizontally directed spine.   *E. sibiricum.*
    73. Basal lobes without a horizontal spine.
     74. Semicells pyramidal or truncate-pyramidal.
      75. Basal lobes bilobed, with a decided sinus between the basal lobes and the
        polar lobe; apical angles with a short, blunt spine. *E. pseudocoralloides.*
      75. Basal margins undulate to a very short polar lobe; apical angles with a
        long, sharp mucro.   *E. doliforme.*
     74. Semicells quadrate in outline.
      76. Basal lobes symmetrically convex at the margin, with 3 prominent gran-
        ules.   *E. boldtii.*
      76. Basal lobes not symmetrically convex; without prominent, marginal gran-
        ules.
       77. Basal lobes equally divided to form 3 lobules; polar lobe very short; all
         lobules tipped with 2 or 3 short spines.   *E. ovale.*
       77. Basal lobes not equally divided to form 3 lobules.
        78. Basal lobes angular, the margins usually somewhat divergent above the
          sinus; apex slightly wider than the base of the semicell; the lateral
          polar angles tipped with a blunt spine.   *E. rectangulare.*
        78. Basal lobes shaped differently.
         79. Apical notch narrowly V-shaped (usually a vertical incision).   *E. dubium.*
         79. Apical notch a broad, shallow V, or the apical margin retuse.
          80. Basal lobes simple, the margins retuse, diverging to the apical
            margin, the polar angles slightly outturned.   *E. prescottii.*
         80. Basal lobes equally bilobed.
          81. Face of semicell with a pair of granules below a low, median
            protrusion; the apical margin nearly flat; lateral view sub-
            rectangular.   *E. pseudoboldtii.*
          81. Face of semicell without granules below a median protrusion
            which is prominent; lateral view broadly oval, the semicells
            much extended in the midregion.   *E. dissimile.*
  69. Margins of semicells, or polar lobes without spines.
   82. Polar angles or margins with granules.
    83. Polar lobe elevated and distinct, subrectangular, and crownlike.
     84. Basal lobes bilobed.
      85. Apex flat, or very slightly retuse.   *E. spinulosum.*
      85. Apex strongly retuse.
       86. Basal lobes divided equally to form two lobules.   *E. gemmatum.*
       86. Basal lobes divided to form an upper, lateral smaller lobe.   *E. verrucosum.*
     84. Basal lobes not bilobed on the margin.
      87. Basal lobes conical, horizontally extended, with a slight swelling or pro-
        trusion on the polar side of the lobe.   *E. verrucosum.*

87. Basal lobes broadly and symmetrically rounded.
  88. Basal lobes with a slight granular protrusion on the polar side of the lobe.   *E. hypochondrum.*
  88. Basal lobes without a protrusion on the polar side.
    89. Basal lobes with a conspicuous spinelike process at the outer limits of the sinus.   *E. platycerum.*
    89. Basal lobe without a spinelike process at the outer limits of the sinus.   *E. ceylanicum.*
83. Polar lobes not elevated and subrectangular.
  90. Apex nearly flat, with a slight median invagination; basal lobes symmetrically convex, with granular margins.   *E. subornatum.*
  90. Apex and basal lobes otherwise.
    91. Apical lobules prominent, with a deep invagination; polar lobules bearing 3 sharp mucros.   *E. bipartitum.*
    91. Apical lobules only slightly produced; apical invagination a shallow notch; polar lobules with 3 sharp granules.   *E. gayanum.*
82. Polar angles or margins of semicells smooth, without granules.
  92. Basal lobes not divided or undulate.
    93. Sinus narrow and closed.
      94. Angles of basal lobes sharply and narrowly rounded.
        95. Margins of semicells deeply retuse from basal angles to sharply pointed angles of the polar lobe.
          96. Apical margin flat, with a slight median incision; polar lobes widely extended into sharp angles; cells 46 μm long.   *E. acmon.*
          96. Apical margin deeply and broadly retuse; cells 39–43 μm long.   *E. incavatum.*
        95. Margins of semicells and angles of polar lobes otherwise.
          97. Cells deeply lobed at the apex.   *E. tribulosum.*
          97. Cells with flat apical margins.   *E. luetkemuelleri.*
      94. Angles of basal lobes broadly rounded.
        98. Semicells with a median mucilage pore.
          99. Mucilage pore definitely excentric.   *E. pingue.*
          99. Mucilage pores centrally located.
            100. Semicells with a basal, median tubercular protrusion; lateral view narrowly oval.   *E. validum.*
            100. Semicells without a basal, median protrusion; lateral view broadly oval.   *E. croasdaleae.*
        98. Semicells without a median mucilage pore.
          101. Semicells subrectangular, without a facial protrusion.   *E. sublobatum.*
          101. Semicells subpyramidal or pyramidal, with facial protrusions.
            102. With with a slight undulation or swelling on the semicell margins between the basal lobes and the apical lobe.   *E. pyramidatum.*
            102. Without a marginal undulation between the basal lobes and the polar lobe.
              103. Semicells with 1 prominent, facial protrusion in the midregion and smaller swellings within the margin of the lobes; polar lobules rounded.   *E. crassangulatum.*
              103. Semicells with a facial protrusion at the base of the semicell; polar lobules sharply rounded.   *E. binale.*
    93. Sinus not narrow and closed.
      104. Sinus merely a V-shaped invagination.   *E. notatum.*
      104. Sinus deeper.
        105. Semicells urn-shaped, the basal angles laterally extended, the apical lobules expanded.   *E. intermedium.*
        105. Semicells rhomboidal.
          106. Face of semicell with a bigranular swelling; lateral and vertical views oval.   *E. binum.*
          106. Face of semicell with a prominent, lobular protrusion; lateral view of semicell broadly oval; quadrilobed in vertical view.   *E. prominens.*
  92. Basal lobes possessing lobules or with margins which are undulate.
    107. With a sinus between the polar lobes and the upper, lateral lobules.
      108. Basal lobes divided to form 6 spine-bearing lobules (3 on either side).   *E. ovale.*

108. Basal lobes not divided into 6 lobules.
    109. Sinus between the semicells narrow and nearly closed; upper, lateral lobules upwardly directed.   *E. marianopoliense.*
    109. Sinus between the semicells open, V-shaped; upper, lateral lobules horizontally extended or nearly so.   *E. urnaforme.*
107. Without a sinus between the the polar lobes and the upper, lateral lobules.
    110. Basal lobes subrectangular and bilobed; polar lobes decidedly produced.
        111. Polar lobes mostly larger than the basal part of the cell; upper margins diverging to wide, sharply pointed polar angles (often tipped with a sharp tuberculation).   *E. subalpinum.*
        111. Polar lobes not larger than the basal part of the cell; cell shaped otherwise.
            112. Basal lobes with the lower lobule smaller than the upper, which extends further than the basal lobule.   *E. pseudoberlinii.*
            112. Basal lobes with 2 equal lobules or protrusions on the margin.
                113. Face of semicell with 3 granular protrusions.   *E. gemmatum.*
                113. Face of semicell without 3 granular protrusions.
                    114. Lower lobules of the basal lobes emarginate; face of semicell smooth.   *E. pseudopectinatum.*
                    114. Lower lobules of the basal lobe not again bilobed; face of semicell with 3 protrusions.   *E. pectinatum.*
    110. Basal lobes not subrectangular.
        115. Semicells semicircular in outline, with undulate margins; apical margin slightly truncate, with a broad, V-shaped apical notch.   *E. incrassatum.*
        115. Semicells not semicircular in outline but quadrate, semirectangular, or subpyramidal.
            116. Lateral margins with 2 undulations between the basal angles and the polar lobe.   *E. montanum.*
            116. Lateral margins without 2 undulations between the basal angles and the polar lobe.
                117. Angles of basal lobes sharply rounded or acute, the margins diverging to sharp angles which extend further than the basal angles.   *E. solidum.*
                117. Angles of basal lobes convexly rounded; lower margins of semicell not diverging.
                    118. Apical margins with 2 undulations between the apical angles.   *E. crispulum.*
                    118. Apical margins without 2 undulations between the apical angles.
                        119. Face of semicell with a prominent, knoblike protrusion, median or supramedian.
                          120. Base of semicell with a conspicuous, supraisthmial tuberculation.   *E. bipapillatum.*
                          120. Base of semicell without a supraisthmial tuberculation.
                              121. Lateral margins of basal lobes subparallel, the upper, lateral lobules extending as far or nearly so as the basal lobes.   *E. insulare.*
                              121. Lateral margins of semicell converging to a slightly developed, upper, lateral protrusion.   *E. elobatum.*
                    119. Face of semicell without a prominent median protrusion.
                        122. Apical angles with a sharp mucro; poles of semicells in lateral view produced, with a rounded, median knob.   *E. dubium.*
                        122. Apical angles smoothly rounded.
                            123. Lateral margins of polar lobe diverging, the apex slightly inflated.   *E. crassicolle.*
                            123. Lateral margins of polar lobe not diverging; apex not inflated.
                              124. Margins of basal lobe above the basal angles with a prominent upper, lateral protrusion extending as far as the basal angles; apical margin in lateral view triundulate.   *E. erosum.*
                            124. Margins of semicell above basal angles converging to a short polar lobe with only a slight, upper, lateral pro-

 trusion; apical margin of semicell in lateral view not
  triundulate.   *E. pyramidatum.*
68. Apical margin undulate, flat, or convex.
 125. Cells with smooth margins.
  126. Polar lobe long, much produced, with subparallel margins (intramarginal granules at the apex); basal lobes forming a shoulder below the polar lobe.
                    E. attenuatum.
  126. Polar lobe short, only slightly produced.
   127. Apical margin smooth and flat.   *E. cornubiense.*
   127. Apical margin straight, but undulate.   *E. crispulum.*
 125. Cells with granular margins.
  128. Basal lobes much extended laterally, with downward bent rather than a
   straight horizontal projection.   *E. hypochondroides.*
  128. Basal lobes not laterally extended.
   129. Polar lobes much produced, with subparallel margins; basal lobes bilobed, with
    granular margins.   Varieties of *E. attenuatum.*
   129. Polar lobes not prominently produced.
    130. Basal lobes laterally extended; polar lobes crownlike.
     131. Apical angles granular, broadly rounded, and slightly inflated; face of semicell
      with 3 granular protrusions.   *E. ceylanicum.*
     131. Apical angles rounded but not inflated, apex with a circle of granules;
      face of semicell with 1 prominent granular protrusion. *E. sphyroides.*
    130. Basal lobes not laterally extended.
     132. Polar lobes sharply truncate, with 5 marginal undulations. *E. undulatum.*
     132. Polar lobes broadly rounded, dome-shaped, with a submarginal circle of
      granules.   *E. johnsonii.*

1. **Euastrum aboense** Elfving 1881, Acta Soc. Fauna Flora Fennica 2(2): 7. Pl. 1, Fig. 2.

Euastrum sinuosum Lenormand var. *aboense* (Elfv.) Cedergren 1932, Ark. f. Bot. 25A(4): 35.

 Cells medium-sized, 1.1 times longer than broad; semicells subpyramidal, the marginal outline slightly convex; apex truncate, plain, with a deep median notch; semicell margins slightly retuse below the apex, and with a secondary retuseness between the upper, lateral, slightly evident protrusion and the basal lobes; sinus deep and narrow; central protuberance on the face of the semicell low; in front view showing a transverse series of 3 pores in the lower part of the semicell, and with 2 supramedian pores, the wall also bearing a number of scattered, simple pits, lightly punctate throughout; lateral view oblong, the poles broadly rounded, with a low median protrusion on either side; vertical view elliptic, the poles broadly rounded, the lateral margins with 3 low swellings in the midregion. L. 52–70 μm. W. 30–42 μm. Isth. 10–13 μm. Th. 19–29 μm. Zygospore unknown.

 Cedergren (*l.c.*) transferred the species to *E. sinuosum*, and Krieger (1937, p. 500) recognizes this assignment. There are enough differences between the typical *E. sinuosum* and *E. aboense* var. *aboense* that we think it is appropriate to keep the two separate. *E. sinuosum* has a prominent and somewhat produced polar lobe with parallel margins. The poles are dilated; the wall coarsely scrobiculate, the pits set within 5 facial protrusions; the lateral view showing the base of the semicell somewhat extended or inflated.

 DISTRIBUTION: Idaho, Michigan, Montana. Newfoundland. Great Britain, Europe, Asia, South America, Arctic.

 PLATE LXI, fig. 9.

2. **Euastrum abruptum** Nordstedt 1869(1870), Vid. Medd. Naturh. Foren. Kjöbenhavn 1869, III(14/15): 217. Pl. 2, Fig. 3 var. **abruptum** f. **abruptum**.

Euastrum subglaziovii Borge 1899, Bih. Kongl. Svenska Vet.-Akad. Handl. 24: 25. Pl. 1, Fig. 29.

Cells medium-sized, subrectangular in outline, 1.5 times longer than broad; semicells quadrate; the basal lobes rather sharply angular at the margin, the sides slightly retuse to upper, lateral lobules with granular spines, then abruptly retuse to the short polar lobe, which is truncate and flat at the apex, the median notch prominent and open, the angles of the polar lobe furnished with an upwardly directed or divergent, stout spine; upper lateral lobules extended further than the basal lobes; face of semicell with a prominent median protrusion bearing 3 or more large tubercular granules, with a horizontal pair of wall pits above the median protrusion, a granule within the apical margin on either side of the median notch, an intramarginal granule at the lateral lobules and a toothlike granule within the margin of the basal angles; sinus narrow and closed throughout; lateral view vasiform, the poles broad and flat, bearing 2 blunt spines at each angle, the lateral margins retuse below the apex and extended into broadly rounded supraisthmial lateral swellings (the widest part of the semicell) which has a vertical pair of teeth at the margin; vertical view quadrate, the margins slightly retuse or convex, the 4 angles extended to form bi-or trispinate processes, and with a median swelling on either side. L. 37.5–52 μm. W. 27–35 μm. Isth. 7–9 μm. Zygospore unknown.

This species should be compared with *E. bidentatum* var. *speciosum* (Boldt) Schmidle, which has an elevated apical margin, and is oval rather than rectangular in vertical view.

DISTRIBUTION: Colorado, Kansas, Louisiana, Massachusetts, Michigan, Minnesota, New Jersey, North Carolina, New England States, Ohio, British Columbia. Asia, Africa, South America, Cuba, Panama Canal, East Indies.

PLATE LXXVIII, figs. 3–5b.

2a. Euastrum abruptum var. **lagoense** (Nordst.) Krieger 1937, Rabenhorst's Kryptogamen-Flora 13(1:3): 606. Pl. 83, Figs. 4–6.

Euastrum abruptum Nordstedt f. *minus* West & West 1898, Linn. Soc. Jour. Bot. 33: 289. Pl. 16, Fig. 10 (not including *E. abruptum* f. *minus* West & West in Taylor 1935, p. 199. Pl. 39, Fig. 1).

Euastrum binale var. *lagoensis* Nordstedt (1869)1870, Vid. Medd. Naturh. Foren. Kjöbenhavn 1869: 218. Pl. 2, Fig. 4.

A variety in general the same shape as the typical but one-half the size; margin of lobes without teeth and the face lacking the pair of small pores characteristic of the typical. L. 20–30 μm. W. 15–21 μm. Isth. 4.5 μm. Ap. 13.5 μm. Th. 15 μm.

This is a simpler, less ornate variety and in lateral view has smooth, broadly rounded poles rather than being flat, truncate, and quadrate as in the typical. This shape, the difference in ornamentation, in addition to the great difference in size seem to warrant a species status possibly. Taylor (1935, Pl. 39, Fig. 1) illustrates *E. abruptum* f. *minus*, which has been placed in synonymy with *E. denticulatum* (Kirch.) Gay by Krieger (1937, p. 583).

DISTRIBUTION: Alaska, Florida, Michigan, Minnesota, Mississippi, Montana, Ohio, Oklahoma. Newfoundland, Québec. Mozambique, South America, Jamaica.

PLATE LXXVIII, figs. 8–8b; PLATE LXXXV, fig. 4.

2b. Euastrum abruptum var. **rectangulare** Prescott var. nov. f. **rectangulare**.

Cellulae rectangulares, 1 2/5 plo longiores quam latae, a varietate typica differentes possessione spinae subapicalis parvae in marginibus lobi polaris, necnon marginibus lateralibus loborum basalium subparallelis; lobi basales inflatione bidenticulata in angulis superioribus lateralibus atque pari granulorum spiniformium in angulis inferioribus praediti; omnes lobi lateraliter subaeque extensi; spinae exteriores lobi polaris ad angulum sursum directae; pori mucosi non visibili. Long cum spinis 43–47.5 μm. Lat. 26–32.5 μm. Isth. 6–10 μm.

ORIGO: E fossa distante 9 milia passuum (14.5 km.) ad orientem a loco Brandenton, Florida dicto.

HOLOTYPUS: Scott Coll. Fla-121.

ICONOTYPUS: Plate LXXVIII, fig. 10.

Cells rectangular in outline, 1 2/5 times longer than broad, differing from the typical by having a small, subapical spine on the margins of the apical lobe, and by having the lateral margins of the basal lobe subparallel, the upper, lateral angles of the basal lobe with a bidenticulate swelling and a pair of spinelike granules on the lower basal angles of the basal lobe; all lobes about equally extended laterally; the outer spines of the apical lobe directed upward at an angle; mucilage pores not evident. L. 42–47.5 μm with spines. W. 26–32.5 μm. Isth. 6–10 μm.

DISTRIBUTION: Florida, Louisiana.

PLATE LXXVIII, fig. 10.

2c. **Euastrum abruptum** var. **rectangulare** f. **spinosum** Prescott f. nov.

Cellulae 1 1/4 plo longiores quam latae; a forma typica differentes possessione spinarum subapicalium longarum prominentum in marginibus lobi polaris, differentes necnon quod margines laterales lobi polaris atque loborum basalium subparalleli, et anguli inferiores loborum basalium unicum granulum parvum spiniforme praebent; pori mucosi infra incisuram apicalem nulli; sinus incisurae apicalis angustior quam in varietate typica. Long. cum spinis 47.5 μm, sine spinis 44 μm. Lat. cum spinis 32.5 μm. Ist. 10 μm.

ORIGO: E fossa distante 3 milia passuum (4.9 km.) ad orientem a loco Rigoleta, Louisiana dicto.

HOLOTYPUS: Scott Coll. La-84.

ICONOTYPUS: Plate LXXVIII, fig. 14.

Cells 1 1/4 times longer than wide; differing from the typical variety by having prominent, long, subapical spines on the margins of the polar lobe, and with the lateral margins of the polar lobe subparallel, the margins of the basal lobe also subparallel, the lower angles of the basal lobes with a single, small, spinelike granule visible; mucilage pores below the apical notch wanting; sinus of the apical notch narrower than in the typical variety. L. without spines 44 μm, with spines 47.5 μm. W. with spines 32.5 μm. Isth. 10 μm.

DISTRIBUTION: Louisiana, Mississippi.

PLATE LXXVIII, fig. 14.

2d. **Euastrum abruptum** var. **subglaziovii** Krieger 1937, Rabenhorst's Kryptogamen-Flora 13(1:3): 606. Pl. 83, Figs. 7, 8 f. **subglaziovii.**

Euastrum subglaziovii Borge var. *minor* Borge 1903, Ark. f. Bot. 1: 111. Pl. 4, Fig. 26.

Cells small, quadrangular in outline, about 1 1/5 times longer than broad; semicells rectangular in outline, slightly wider than long; polar lobes short with diverging margins, the apex truncate and flat, with a V-shaped median notch, wall at the angles extended into prominent, mostly horizontally directed spines; two small intramarginal granules on either side of the median notch; the upper, lateral lobes slightly produced and bearing 3 prominent wall thickenings or granules; margins of semicell retuse below the lateral lobules to reduced basal lobes, equal in extension to the lateral lobules; sinus deep and closed; central protrusion on the face of the semicell bigranulate; an intramarginal granule at each of the lateral and basal lobes; lateral view truncate-oval in outline, the apex broadly convex with a process at each angle, the lateral margins with a prominent, bigranulate protrusion in the midregion. L. 30–38 μm. W. 24.5–26 μm. Isth. 6.5 μm. Th. 16–17 μm.

Except for its smaller size this variety is similar to the typical in both front and lateral views. The angles at the apex bear simple mucros, however, rather than being bispinate.

DISTRIBUTION: Kansas, Louisiana. Africa, South America.

PLATE LXXVIII, fig. 7A, 7Aa.

2e. **Euastrum abruptum** var. **subglaziovii** f. **dichotypicum** Prescott f. nov.

Forma a forma typica differens quod lobuli superiores laterales loborum basalium prominentius evoluti, necnon lobi unius semicellulae forma a lobis alterius differentes, etiam leves, non granulis praediti; cellulae par pororum mucosorum, singulum utroque in latere et admodum super protuberationem super-ficialem centralem habentes, necnon protuberationem lateralem mediam 3 granula ferent praebentes; semicellulae a latere visae rhomboideae, marginibus lateralibus ab incisione media profunda cellulae ad protuberationes laterales prominentes abrupte divergentibus, polis dentem polarem prominentem atque 2 dentes sub-terminales uno infra alterum in agulis apicalibus praebentibus. Long. 37–38 μm. Lat. 27.5 μm. Isth. 9 μm. Crass. 19 μm.

ORIGO: E stagno Nelumbiano, distante 7 milia passuum (11.3 km.) ad sep-tentriones a loco Colfax, Louisiana dicto.

HOLOTYPUS: Scott Coll. La-38.

ICONOTYPUS: Plate LXXVIII, fig. 15.

A form differing from the typical variety by having the upper, lateral lobules of the basal lobes more prominently developed and with the lobes of one semicell differing in shape from those of the other, being smooth rather than furnished with granules; with a pair of mucilage pores, one on either side and immediately above the central, facial protuberance, with a lateral, median protuberance with 3 granules in view; in lateral view the semicell rhomboid, the lateral margins diverging sharply from the deep median incision of the cell to prominent lateral protuberances, then converging gently to rounded poles, with a prominent polar tooth and two subterminal teeth, one below the other at the apical angles. L. 37–38 μm W. 27.5 μm. Isth. 9 μm. Th. 19 μm.

DISTRIBUTION: Louisiana, New York.

PLATE LXXVIII, fig. 15.

2f. **Euastrum abruptum** var. **subglaziovii** f. **inflatum** Prescott f. nov.

Forma a forma typica differens ut anguli inferiores loborum basalium late rotundati et manifeste convexi, lobulis dente marginali praeditis; incisura apicalis semicellulae latior profundiorque quam in varietate typica. Long. 41.3–44 μm. Lat. 28.3–29 μm. Isth. 6.5–7.25 μm.

ORIGO: E stagno castorum ad Collegium in loco New Paltz, N.Y. dicto.

HOLOTYPUS: GWP Coll. Pyle NY-5 (22A-17).

ICONOTYPUS: Plate LXXVIII, figs. 9, 12.

A form differing from the typical variety by having the lower angles of the basal lobes of the semicell broadly rounded and prominently convex, the lobules furnished with a marginal tooth; apical notch of the semicell broader and deeper than in the typical. L. 41.3–44 μm. W. 28.3–29 μm. Isth. 6.5–7.25 μm.

DISTRIBUTION: New York.

PLATE LXXVIII, figs. 9, 12.

3. **Euastrum acmon** West & West 1897, Jour. Bot. 35: 81. Pl. 366, Figs 11, 12 var. **acmon.**

Cells relatively small, 1.5 times longer than broad; sinus narrow within and then opening broadly; semicells trapezoidal, the basal lobes rather sharply round-conical, somewhat upwardly turned, the margins retuse between the basal lobes and the polar lobe, the apex broad and flat, not quite equal to the basal lobes in width, the median notch shallow and broadly V-shaped; median facial protrusion of semicell lacking; wall smooth; vertical view elliptic with narrowly rounded poles and with a slight swelling on either side in the midregion. L. 58 μm. W. 44 μm. Isth. 10 μm. Ap. 26 μm. Th. 30 μm. Zygospore unknown.

The narrowly and sharply rounded basal lobes and the smooth, deeply retuse, and plane lateral margins extending to the broad, flat apex distinguish this species.

DISTRIBUTION: Typical form not reported from North America. Africa (Angola).

PLATE LXII, figs. 4, 4a.

3a. **Euastrum acmon** var. **turgidum** Scott & Prescott 1952, Hydrobiologia 4(4): 377. Pl. 1, Fig. 10.

Smaller than the typical with 2 prominent protrusions on the median, facial protuberance; polar incision broader and more concave than in the typical; lateral view of semicells transversely elliptic, lateral margins greatly distended so that the cell is approximately as thick as the semicell is high; vertical view broadly oval to almost circular in outline, the poles narrowly rounded and with two prominent marginal protrusions on each side in the midregion. L. 30–31 μm. W. 22 μm. Isth. 5–6 μm. Th. 18–19 μm.

DISTRIBUTION: Florida.

PLATE LXII, fig. 3.

4. **Euastrum affine** Ralfs 1848, Brit. Desm., p. 82. Pl. 18, Fig. 3 var. **affine** f. **affine.**

Euastrum humerosum Ralfs var. *affine* (Ralfs) Raciborski 1885, Pamiet. Wydz. III, Akad. Umiej. w Krakowie 10: 93.

Cells large, oval in outline, 2 times longer than broad; semicells truncate-pyramidal, the polar lobe relatively short, flat at the apex with laterally inflated angles, the median notch deep, narrow, and closed; upper, lateral lobules narrow (in height) and short, somewhat upwardly directed; basal lobes subrectangular, bilobed at the margin; facial protuberances of the semicell low, 5, with 3 above the isthmus and 1 on either side of the midregion in which there is a mucilage pore; sinus deep and narrow; lateral view oval in outline, truncate at the apex with rounded angles, the lateral margin retuse below the apex and diverging to the lateral, median swelling, then converging downward to rounded basal angles; vertical view broadly elliptic, the poles slightly produced and rounded at the apex, the lateral margins with 3 inflations in the midregion. L. 95–123.5 μm. W. 48–63 μm. Isth. 13.5–19.5 μm. Th. 29 μm. Zygospore unknown.

Krieger (1937, p. 521) has reduced several varieties of *E. didelta* and *E. humerosum* to synonymy with this species.

DISTRIBUTION: Alaska, California, Connecticut, Florida, Georgia, Louisiana, Minnesota, Mississippi, New Hampshire, New Jersey, North Carolina, Oregon, Pennsylvania, South Carolina, Utah, Virginia, Washington, Wisconsin, Wyoming. British Columbia, Labrador. Great Britain, Europe, Asia, South America, Arctic.

PLATE LXIV, fig. 10.

4a. **Euastrum affine** var. **affine** f. **equilaterale** Prescott & Scott 1945, Amer. Mid. Nat. 34(1): 232. Pl. 4, Fig. 2.

Semicells triangular, broad at the base, with slightly sloping or rounded basal lobes and the margins not retuse as in the typical, converging to the polar lobe, which has prominent apical lobules; sinus linear and closed throughout; the upper, lateral lobules of the semicell but very slightly produced; face of the semicell with 3 prominent protuberances, 1 central and 2 lateral, and with a conspicuous mucilage pore in the midregion. L. 82 μm. W. 46 μm. Isth. 15 μm.

DISTRIBUTION: Louisiana.

PLATE LXIV, figs. 11, 14, 14a.

4b. **Euastrum affine** var. **minus** Prescott var. nov.

Varietas a varietate typica differens quod minor, et lobi basales simplices (non bilobati), sinus non prorsus inaperto ut in varietate typica sed extrorsus late patens; lobi superiores laterales fere aeque longe ac lobi basales extendentes; margines superiores semicellulae a latere visae subparalleli et lobuli apicales plus inflati quam in varietate typica. Long. 79 μm. Lat. 37 μm. Ist. 10 μm. Lat. apic. 24 μm. Crass 33 μm.

ORIGO: E palude distante 9 milia passuum (14.5 km.) ad meridiem a loco Wilma, Florida dicto.

HOLOTYPUS: Scott Coll. Fla-251; 975.

ICONOTYPUS: Plate LXIV, figs. 12, 12a.

A variety differing from the typical by its smaller size and by having the basal lobes simple (not bilobed); the sinus not closed throughout as in the typical but opening widely outward; the upper, lateral lobes extended almost as far as the basal lobes; in lateral view the upper margins of the semicell subparallel and with the apical lobules more inflated than in the typical. L. 79 μm. W. 37 μm. Isth. 10 μm. Ap. 24 μm. Th. 33 μm.

DISTRIBUTION: Florida.

PLATE LXIV, figs. 12, 12a.

5. **Euastrum allenii** Cushman 1904, Bull. Torr. Bot. Club 31: 582. Pl. 26. Fig. 6.

Cells large, subrectangular in outline, 2 times longer than broad; polar lobe very short, truncately rounded and flattened at the apex, narrower than the body of the semicell, the median notch short, closed inwardly; margins of the semicell with 3 transverse, prominent basal protuberances or undulations, with a median mucilage pore; wall coarsely punctate; sinus narrow and closed throughout; lateral view subrectangular, the poles broadly truncate, lateral margins subparallel; vertical view oval, the poles rounded, the lateral margins with 3 swellings. L. 105–117 μm. W. 49–69 μm. Isth. 13–18 μm. Ap. 33 μm. Th. 24 μm. Zygospore unknown.

DISTRIBUTION: Newfoundland.

PLATE LXV, figs. 8–8b.

6. **Euastrum ampullaceum** Ralfs 1848, Brit. Desm., p. 83. Pl. 13, Fig. 4 f. **ampullaceum.**

Euastrum ampullaceum f. *scrobiculatum* Nordstedt 1873, Acta Univ. Lund 9: 8.

Cells medium-sized, elliptic in outline, 1.5 times longer than broad; the semicells broadly truncate-pyramidal, the polar lobe short and slightly inflated, the apex broadly rounded and with rounded angles, the median notch short and narrow; the lateral margins concave but diverging to slightly rounded upper, lateral angles, then concave to the relatively high, subquadrate basal lobes; facial protrusions 3 supraisthmial, the center one often bilobed, and with 2 protuberances, one on either side of the midregion, below which are 3 triangularly arranged mucilage pores; lateral view broadly oval in outline, the lower margins bilobed and subparallel, then converging to a narrowed, broadly truncate pole, with 2 mucilage pores at the margin on either side of the semicell between the 2 lobules; vertical view broadly elliptic, the poles slightly produced and rounded, the lateral margins slightly triundulate and with 3 mucilage pores showing in the midregion. L. 66–102 μm. W. 32–63 μm. Isth. 12–15.5 μm. Ap. 24–42 μm. Th. 35 μm. Zygospore subspherical, the wall with numerous bluntly pointed papillae, 54–72 μm in diameter.

DISTRIBUTION: Alaska, Connecticut, Florida, Louisiana, Massachusetts, Michigan, Mississippi, New Hampshire, New Jersey, North Carolina, Oregon, South Carolina. British Columbia, Newfoundland. Great Britain, Europe, Asia, South America, Faeroes, Arctic.

PLATE LXII, figs. 1, 1a; PLATE LXIII, fig. 3.

6a. **Euastrum ampullaceum** formae Prescott & Scott 1945, Amer. Mid. Nat. 34(1): 233. Pl. 4, Fig. 6.

Forms with the polar lobules more prominently developed than in the typical, approaching *Euastrum affine* Ralfs in shape. L. 82 μm. W. 46 μm. Isth. 15 μm.

DISTRUBUTION: Alaska, Louisiana, Mississippi. Ontario.

PLATE LXIII, figs. 1, 4.

6b. **Euastrum ampullaceum** f. **lata** Irénée-Marie 1956, Rev. Algol. 2(1/2): 113. Fig. 2.

A form larger than the typical, with shorter polar lobes; the wall coarsely granular. L. 93–95.5 μm. W. 55–57 μm. Isth. 20 μm. Ap. 25.5–26 μm.

DISTRIBUTION: Mississippi. Québec.

PLATE LXIII, figs. 2, 2a, 6, 6a.

7. **Euastrum ansatum** Ehrenb. ex Ralfs 1848, Brit. Desm., p. 85. Pl. 14, Figs. 2a–2f var. ansatum f. ansatum.

Euastrum ansatum Ehrenberg 1832, Entwickl. Infusor., p. 82; 1838, Die Infusionsthierchen, p. 162. Pl. 21, Fig. 11.
Euastrum ansatum var. *valleculatum* Schaarschmidt 1882, Mag. Tudom. Akad. Math. e Termész. Köz. 18: 263. Fig. 1.
Euastrum didelta f. *longicolle* Irénée-Marie 1958, Nat. Canadien 85(5): 123. Fig. 13.
Euastrum ralfsii Rabenhorst 1868, Flor. europ. algar. dulcis et submar., III, p. 184.

Cells medium-sized, narrowly oval in outline, 2 times longer than broad; semicells broadly rounded in the basal angles, then with retuse converging margins to a truncate, flat apex with rounded angles, the median notch short and closed; the sinus deep and narrow; face of semicell with 3 protrusions at the base and 1 on either side of the midregion; the wall coarsely punctate; lateral view narrowly oval, the lateral margins of the semicell at the base with 2 undulations, narrowed above and the margins retuse to the broadly rounded poles; vertical view oval, the poles slightly produced, the lateral margins with 3 undulations. L. 61–(110)115 μm. W. 32–51 μm. Isth. 8–15.5 μm. Ap. 16–24 μm. Th. 23–(30)36 μm. Zygospore spherical, the wall with numerous spines, 35–55.5 μm in diameter without spines.

DISTRIBUTION: Widely distributed in the U.S. and Canada. Cosmopolitan.

PLATE LVIII, figs. 6, 6a, 7f, 8.

7a. Euastrum ansatum Ehrenb. ex Ralfs formae Scott & Prescott 1945, Amer. Mid. Nat. 34(1): 233. Pl. 3, Fig. 2 var. **ansatum** forma.

A form with a central mucilage pore and a median swelling of the semicell just above the isthmus; membrane coarsely punctate. L. 110 μm. W. 33 μm.

DISTRIBUTION: Colorado, Connecticut, Georgia, Louisiana, Utah, Wisconsin. Ontario.

PLATE LVIII, fig. 8.

7b. Euastrum ansatum var. **ansatum** f. **angustum** Prescott f. nov.

Forma figura formae typicae similis, sed relative angustior, lobis basalibus angustius rotundatis. Long. 100 μm. Lat. 39 μm. Ist. 13.5 μm.

ORIGO: E stagno iuxta viam veterem, distante 6 milia passuum (9.7 km.) ad orientem a loco Pearlington, Mississippi dicto.

HOLOTYPUS: Scott Coll. Miss-46.

ICONOTYPUS: Plate LVIII, figs. 12, 14, 14a.

A form similar in shape to the typical but relatively narrower, the basal lobes more narrowly rounded. L. 100 μm. W. 39 μm. Isth. 13.5 μm.

DISTRIBUTION: Mississippi.

PLATE LVIII, figs. 12, 14, 14a: PLATE LIX, figs. 9, 9a.

7c. Euastrum ansatum var. **ansatum** f. **brevius** Prescott f. nov.

Forma brevior crassiorque quam forma typica; semicellulae ad basim latae, lobis basalibus latius rotundatis, marginibus concavis usque ad lobum polarem breviorem atque relative latiorem. Long. 61.6 μm. Lat. 40.8 μm. Ist. 13.6 μm.

ORIGO: E lacu in loco He Devil Mountain, Idaho (Seven Devils Range) dicto.

HOLOTYPUS: GWP Coll. OI-9.

ICONOTYPUS: Plate LIX, fig. 10.

A form shorter and stouter than the typical; the semicells broad at the base, the basal lobes more broadly rounded and the margins concave to the shorter, relatively broader polar lobe. L. 61.6 μm. W. 40.8 μm. Isth. 13.6 μm.

DISTRIBUTION: Idaho.

PLATE LIX, fig. 10.

7d. Euastrum ansatum var. **ansatum** f. **scottii** Wade 1952, Dissert., p. 147. Pl. 5, Fig. 3; 1957a, Rev. Algol., 2(4): 252. Pl. 1, Fig. 1.

A form with an elongate polar lobe and without lateral swellings above the basal lobes; face of semicell with a single protuberance in the midregion immediately above the isthmus, and with a single mucilage pore above the protuberance; wall coarsely punctate. L. 105 μm. W. 47 μm.

DISTRIBUTION: Michigan.

PLATE LIX, fig. 8.

7e. Euastrum ansatum var. **ansatum** f. **simplicius** Croasdale f. nov.

Forma firme minor quam forma typica, brevior, margines laterales semicellulae fere recti ab angulis basalibus, ad apicem late truncatam convergentes; superficies semicellularum unicam protrusionem admodum super sinum in regione isthmiali, atque unicum porum mucosum media in parte habens; membrana tenuiter punctata; semicellulae a latere visae late ovatae, inflationem manifestam media in parte praebens. Long. 66 μm. Lat. 33 μm.

ORIGO: Alaska centralis.
HOLOTYPUS: HC Coll.
ICONOTYPUS: Plate LIX, fig. 1.

A form averaging smaller than the typical, shorter in proportions, the lateral margins of the semicell almost straight from the basal angles to the broadly truncate apex; face of the semicell with a single protrusion immediately above the sinus in the region of the isthmus, and with a single mucilage pore in the midregion; wall finely punctate; lateral view broadly oval, the semicell with a pronounced swelling in the midregion. L. 66 μm. W. 33 μm.

This form approaches *Euastrum cuneatum* Jenn., which also has nearly straight lateral margins of the semicell. The proportions are more those of *E. ansatum*, however, as is the presence of a median mucilage pore. In lateral view the semicell is relatively thicker and with a median protrusion which is unlike *E. ansatum*. *E. cuneatum* is a larger species.

DISTRIBUTION: Alaska.
PLATE LIX, fig. 1.

7f. Euastrum ansatum var. **attenuatum** Schmidle 1896a, Flora 82(3): 309. Pl. 9, Fig. 16.

A variety somewhat larger than the typical and relatively broader in the base of the semicell, the basal angles sharply rounded; lateral margins of the semicell decidedly retuse to an apex relatively narrower than in the typical; face of semicell with a low, median protrusion, above which is a semicircle of mucilage pores. L. 120 μm. W. 52 μm.

DISTRIBUTION: Indiana. Australia.
PLATE LVIII, fig. 10; PLATE LIX, fig. 11.

7g. Euastrum ansatum var. **concavum** Krieger 1937, Rabenhorst's Kryptogamen-Flora 13(1:3): 487. Pl. 58, Fig. 12 f. **concavum**.

Euastrum ansatum Ehrenb. (p.p.) Hylander 1928, Algae of Connecticut, p. 80. Pl. 10, Fig. 11.

A variety with the basal lobes more sharply rounded than in the typical, the lateral margins of the semicell decidedly retuse (concave) to the broadly truncate, flat pole, which is slightly inflated, the angles sharply rounded, the median notch very short and closed. L. 69-82 μm. W. 35-44 μm. Isth. 10-13 μm. Ap. 17-21 μm. Th. 25-26 μm.

DISTRIBUTION: Connecticut (as *E. ansatum*), Montana, New Hampshire, Vermont. Labrador. Europe, South America, Arctic.
PLATE LIX, fig. 12.

7h. Euastrum ansatum var. **dideltiforme** Ducellier 1918, Bull. Soc. Bot. Genève II, 9: 42. Textfigs. 16, 17 f. **dideltiforme**.

A variety with basal angles more sharply rounded than the typical and with a slight enlargement forming low, rounded, lateral lobes, the margins of the polar lobe subparallel to the flattened apex, the median notch relatively deep and narrow; face of semicell with 3 swellings supraisthmial and one on either side of the midregion where there is a mucilage pore (sometimes wanting ?); wall coarsely punctate. L. 73-90 μm. W. 36-45 μm. Isth. 12-15 μm. Ap. 20-22 μm. Th. 26-28 μm.

DISTRIBUTION: Alaska, Montana, North Carolina. Ontario. Europe, Asia, Australia, Arctic, Africa.
PLATE LIX, figs. 15, 15a.

7i. **Euastrum ansatum** var. **dideltiforme** f. **elongatum** Růžička 1955, Biologia 10(5): 594. Pl. 2, Fig. 19.

A form relatively longer than the typical, 2.5 times longer than broad, the polar lobe more produced. L. 111–126 μm. W. 47–52.5 μm. Isth. 17–20 μm.
DISTRIBUTION: Ontario. Europe.
PLATE LIX, fig. 13.

7j. **Euastrum ansatum** var. **dideltiforme** f. **major** Irénée-Marie & Hilliard 1963, Hydrobiologia 21(1/2): 104. Pl. 2, Fig. 12.

A form somewhat larger throughout than the typical; otherwise similar. L. 122–123 μm. W. 61–61.5 μm. Isth. 20–20.5 μm. Ap. 23–25 μm.
It seems reasonable to regard this as an incidental growth form of var. *dideltiforme*.
DISTRIBUTION: Alaska.
PLATE LX, figs. 3–3b.

7k. **Euastrum ansatum** var. **javanicum** (Gutw.) Krieger 1937, Rabenhorst's Kryptogamen-Flora 13(1:3): 488. Pl. 58, Fig. 9.

Euastrum dideltoides (Racib.) West & West f. *javanica* Gutwinski 1902, Bull. Inter. Acad. Sci. Cracovie, Cl. Sci. Math. et Nat. 1902(9): 602. Pl. 39, Fig. 55.

A variety more slender than the typical; face of semicell with 2 vertically arranged mucilage pores in the midregion. L. 88–100 μm. W. 46 μm. Isth. 13–15 μm. Th. 26 μm.
DISTRIBUTION: Louisiana. East Indies.
PLATE LIX, fig. 2.

7l. **Euastrum ansatum** var. **longum** Grönblad 1945, Acta Soc. Sci. Fennica, n. s; B, II(6): 12. Pl. III, Fig. 38.

Cells very slender; the basal angles narrowly rounded, the margins of the semicells converging to a very slight swelling and then tapering to an elongate polar lobe, the upper part of which has subparallel margins; face of semicell with 3 low swellings across the base immediately above the isthmus. L. 118 μm. W. 47 μm. Isth. 17 μm. Th. 31 μm.
DISTRIBUTION: Massachusetts, New Hampshire, Vermont. South America.
PLATE LIX, figs. 14–14b(f.).

7m. **Euastrum ansatum** var. **pyxidatum** Delponte 1873, Desm. Subalp., p. 103. Pl. 6, Figs. 32–34.

A variety similar to var. *dideltiforme* in the shape of the basal lobes, with a pronounced apical lobe with parallel margins; 1 facial protuberance in the midregion immediately above the isthmus but not prominent; a slight swelling within the margin of the basal lobes and at each of the upper, lateral marginal swellings as seen in face view, with a small median mucilage pore. L. 57–100 μm. W. 33–48 μm. Isth. 11–15.6 μm. Ap. 17–21 μm. Th. 22–29 μm.
DISTRIBUTION: California, Mississippi, North Carolina, South Carolina, Virginia. Québec. Great Britain, Europe, Asia, Africa, South America, Arctic.
PLATE LVIII, figs. 11–11b.

7n. **Euastrum ansatum** var. **rhomboidiale** Ducellier 1918, Bull. Soc. Bot. Genève, II, 9: 44. Textfigs. 26–28.

Cells with the same proportions as the typical; semicells pyramidal, the lateral margins practically straight from sharply rounded basal angles, converging to the flat apex, the polar lobe scarcely evident; lateral view elliptic, the semicells with 2 undulations near the base, then with retuse margins to a rounded apex. L. 70–107 μm. W. 39–53 μm. Isth. 10–16 μm. Ap. 18–21 μm.

DISTRIBUTION: British Columbia. Europe, Iceland, Greenland.

Pl ATE LIX, figs. 17, 17a.

7o. **Euastrum ansatum** var. **simplex** Ducellier 1918, Bull. Soc. Bot. Genève, II, 9: 37.

A variety smaller than the typical, with broadly convex and high basal lobes, the margins of the semicell abruptly concave to the truncate apex, the margins of the polar lobe subparallel; face of semicell with 3 basal protuberances, and 1 on either side of the midregion where there is a mucilage pore. L. 59–82 μm. W. 29–49.5 μm. Isth. 10–13 μm. Ap. 16–19 μm. Th. 19–24 μm.

DISTRIBUTION: Alaska. Labrador. Europe, Asia, Indonesia, Australia, Africa, Arctic.

PLATE LIX, figs. 3, 4; Pl. LX, figs. 2, 2a.

7p. **Euastrum ansatum** var. **subconcavum** Prescott nom. nov.

Euastrum didelta var. *intermedium* Ducellier, in Irénée-Marie 1958, Nat. Canadien 85(5): 122. Pl. I, Fig. 12.

A variety with lateral margins of the semicell decidedly concave above sharply rounded basal angles, becoming subparallel at the polar lobe, the angles of the apex broadly rounded; facial protuberances as in the typical but without a median mucilage pore; in lateral view narrowly oval in outline, the semicells oval, the lateral margins with 2 protrusions, and the apex broadly truncate-rounded. L. 75.5 μm. W. 41.8 μm. Isth. 11.3 μm.

Euastrum didelta (Turp.) Ralfs var. *intermedium* Ducellier has been placed in synonymy with *E. didelta* by Krieger (1937, p. 517). This species is differentiated from *E. ansatum* mostly on the basis of the swollen apical region of the former. The plant described by Irénée-Marie (*l.c.*) seems to belong to *E. ansatum*, in which the upper margins of the semicell are subparallel and the apex not enlarged.

DISTRIBUTION: Québec.

PLATE LIX, figs. 5, 5a, 16, 16a.

7q. **Euastrum ansatum** var. **submaximum** Borge 1899, Bih. Kongl. Svenska Vet.-Akad. Handl. 24, Afd. III(12): 25. Pl. 2, Fig. 53.

A variety in which the semicells are very broad at the base, the lateral margins slightly undulate (with a single obscure inflation above the sharply rounded basal angles); facial view with a broad, median swelling; lateral view of semicells broadly oval with a prominent protrusion in the midregion of the lateral margins, the poles broadly rounded. L. 54–117 μm. W. 57–64 μm. Isth. 13–16 μm. Ap. 23.5–26 μm.

DISTRIBUTION: Michigan. Cuba.

PLATE LVIII, figs. 15–15b.

7r. **Euastrum ansatum** var. **triporum** Krieger 1937, Rabenhorst's Kryptogamen-Flora 13(1:3): 492. Pl. 59. Fig. 8.

A stout variety, with a broad and distinct polar lobe; margins of the semicells with a scarcely discernible swelling above the basal angles, the margins lightly retuse between; face of semicell showing 2 mucilage pores, 1 on either side and immediately above the central, basal protrusion and 1 median in the semicell between the 2 upper, lateral protrusions. L. 65–85 μm. W. 35–45 μm. Isth, 12 μm. Th. 27 μm.

DISTRIBUTION: California, Montana. Europe, Asia, Australia.

PLATE LIX, figs. 6–7a.

8. Euastrum attenuatum Wolle 1881, Bull. Torr. Bot. Club 8: 3. Pl. 6, Fig. 20 var. attenuatum.

Cells of medium size, trapeziform, 1 1/3 times longer than broad; basal lobes of the semicell subquadrate with slightly retuse margins; sharply narrowed above the basal lobes to form a narrow polar lobe with subparallel margins, the apex truncate and without a median notch, the wall at the apex with a circle of intramarginal plications; sinus narrow and closed; face of the semicell with 2 vertically arranged protrusions in the midregion and with a swelling within the margin of the basal lobes; lateral view elliptic in outline, the base of the semicells with subparallel but slightly undulate margins, the polar lobe produced and truncate at the apex, showing a ring of intramarginal plications; vertical view oval in outline, the poles rounded and somewhat produced, the lateral margins with 3 protrusions. L. 53–67 μm. W. 32–45 μm. Isth. 11–13 μm. Ap. 11–13 μm. Th. 9.7–24 μm. Zygospore unknown.

DISTRIBUTION: Alaska, Florida, Maine, Michigan, Montana, Nebraska, New Hampshire, New Jersey, Pennsylvania, Washington. Québec. Europe, Asia, South America, Puerto Rico.

PLATE LXVIII, figs. 10–10b.

8a. Euastrum attenuatum var. hastingsii (Wolle) Prescott comb. nov.

Euastrum hastingsii Wolle 1892, Desm. U.S., p. 113. Pl. 42, Figs. 16, 17.
Euastrum attenuatum var. *brasiliense* Grönblad 1945, Acta Soc. Sci. Fennica, n.s., B, II(6): 12. Pl. 3, Fig. 39.

A variety with the polar lobe broader and shorter than the typical, the poles crenate at the margin, with a circle of granules immediately above the plications; vertical view quadrate, and with the poles not produced as prominently as in the typical. L. 59 μm. W. 32 μm. Isth. 10 μm. Th. 23 μm.

Krieger (1937, p. 541) has placed *E. hastingsii* in synonymy with *E. attenuatum*. Wolle's description and illustrations of *E. hastingsii* are sufficiently different from his *E. attenuatum* to warrant a separation in our judgment. (cf. Pl. LXVIII, figs. 11–11b, and Pl. LXVIII, fig. 13). Although Grönblad's var. *brasiliense* includes some facial features not illustrated by Wolle (*l.c.*), it seems impossible to separate this variety from Wolle's *hastingsii*; hence we are placing it in synonymy with the latter. Bourrelly (1957, p. 1059. Pl. 1, fig. 12) has described a form of var. *brasiliense* from Africa which seems to merit a new epithet. His plant is illustrated on Pl. LXVIII, figs. 11–11b, for comparison.

DISTRIBUTION: Florida, New Hampshire. South America.

PLATE LXVIII, figs. 13, 13a; Pl. LXIX, figs. 1, 1a.

8b. Euastrum attenuatum var. lithuanicum Wołoszynska 1919a, Rozpr. Wydz. matem.-przyr. Akad. Umiej. Krakowie, B, 57: 48. Pl. III, Figs 8–10 f. lithuanicum.

A variety relatively broader and stouter than the typical; the basal lobes more distinctly bilobed at the margins and with the lobules bearing sharp granules, margins of the semicell concave to form a broad polar lobe, rounded and dome-shaped at the apex, with prominent intramarginal plications; facial protuberances apparently similar to the typical; lateral view with the basal lobes much more extended than in the typical and with the polar lobe approximately as broad as in the facial view. L. 65 µm. W. 40 µm. Isth. 14 µm. Th. 30 µm.

DISTRIBUTION: Michigan. Europe, South America.

PLATE LXIX, figs. 2, 2a.

8c. Euastrum attenuatum var. lithuanicum fa. pulchellum Prescott & Scott 1945, Amer. Mid. Nat. 34(1): 233. Pl. 1, Fig. 3; Pl. 4, Fig. 10.

Euastrum attenuatum var. *pulchellum* (Presc. & Scott) Bourrelly 1966, Inter. Rev. Ges. Hydrobiol. 51(1): 88. Pl. 1, Fig. 3.

Cells about 1.5 times longer than broad; semicells trapezoidal, the basal angles bilobed, margins of the semicells diverging at first and then deeply emarginate, the lobules with 1 to 4 granules; upper lateral margins deeply concave and then diverging slightly to a broad truncate apex with rounded angles; polar lobe with about 12 longitudinal folds and a ring of small granules on the apical surface immediately above the folds; sinus sharply pointed but not closed inwardly, gradually opening outwardly; face of semicell with 2 median protuberances, one above the other, and 2 lateral protuberances, one on either side near the margins of the basal lobes. L. 51–55 µm. W. 33–37 µm. Isth. 11–12 µm. Th. 22 µm.

This form differs from the typical mostly by the open sinus and the prominence of the longitudinal folds. The granules around the apex are quite conspicuous. It is slightly smaller throughout than the typical.

DISTRIBUTION: Louisiana, Minnesota, Mississippi. Ontario.

PLATE LXIX, figs. 3–3b.

8d. Euastrum attenuatum var. splendens (Fritsch & Rich) Grönblad & Scott, in Grönblad, Prowse & Scott 1958, Acta Bot. Fennica 58: 12. Figs. 33, 34. Photo 354 f. splendens.

Euastrum splendens Fritsch & Rich 1937, Trans. Roy. Soc. South Africa 25(2): 175. Figs. 9a–c.

A variety differing from the typical by having extended and broadly rounded basal lobes from a widely open sinus, the margins deeply concave to form a columnlike polar lobe with subparallel margins; the apex slightly inflated; face of semicell with a median protrusion encircled by large granules, and with a granular protrusion on either side in the basal lobes; vertical view broadly oval, equally 8-lobed, the lateral lobes being slightly broader than the polar lobes (the poles of the cell in this view scarcely distinguishable, however). L. 61–66 µm. W. 39 µm. Isth. 11–12 µm. Th. 30 µm.

DISTRIBUTION: Montana. Africa, South America.

PLATE LXIX, figs. 6, 6a.

8e. Euastrum attenuatum var. splendens f. foersteri Prescott nom. nov.

Euastrum attenuatum var. *brasiliense* Grönblad (p.p.), in Förster 1972, Nova Hedwigia 23(2/3): 533. Pl. 6, Fig. 12.

A form similar to the typical in respect to the basal lobes, but with the polar lobe definitely more produced, with subparallel margins; apex with a circle of fewer tubercles; basal lobes bilobed at the apices, horizontally directed. L. 64–65 μm. W. 29–34 μm. Isth. 9–9.5 μm. Ap. 14–15 μm. Th. 21.5 μm.

DISTRIBUTION: Florida.

PLATE LXIX, figs. 4, 4a.

9. **Euastrum bidentatum** Nägeli 1849, Gatt. Einz. Algen, p. 122. Pl. 7-D, Figs. 1a–1f var. **bidentatum** f. **bidentatum.**

Euastrum pictum Börg. var. *subrectangulare* West & West 1898, Linn. Soc. Jour. Bot. 33: 293. Textfig. 1–f.

Euastrum rostratum Ralfs 1844, Ann. & Mag. Nat. Hist. 14: 192.

Cells below medium size, broadly oval in outline, 1.5 times longer than broad, the margins of the semicell undulate, the basal lobes bilobed and with a slightly developed lateral lobule which may have a toothlike projection; the polar lobe short, broad, the angles extended into a toothlike process and with the apical margin broadly rounded and slightly undulate, with a deep, closed median notch; polar lobe nearly as wide as the basal lobes; face of semicell with a prominent median protrusion bearing several large granules, above which are 2 horizontally arranged mucilage pores, with several granules on the face of the cell, especially within the margin of the lobes; sinus narrow and closed; lateral view elliptic, the margins of the semicell undulate, with granulations, the poles narrowly rounded; vertical view elliptic, with poles bearing several granules and with a median lateral protrusion bearing prominent granules. L. 42–61 μm. W. 27–39 μm. Isth. 6.3–13 μm. Ap. 13–29 μm. Th. 20–26 μm. Zygospore globose, the wall furnished with blunt spines.

DISTRIBUTION: Widely distributed throughout the United States. British Columbia, Newfoundland, Québec. Great Britain, Europe, Asia, Africa, South America, Azores, Arctic.

PLATE LXXVII, figs. 5–5b (f.), 6.

9a. **Euastrum bidentatum** var. **bidentatum** f. **lata** Irénée-Marie 1958, Nat. Canadien 85(5): 113. Pl. 1, Fig. 3.

A variety slightly larger than the typical, the basal lobes simple, with diverging margins, the upper lateral lobes more produced than in the typical; polar lobe very short, with a subapical spine and with the apical margin elevated and with rounded lobules at the margin of the deep vertical notch; face of semicell with a prominent median protuberance bearing 3 vertical rows of granules, and with granules within the margins of the lobes; sinus narrow and closed; lateral view narrowly elliptic in outline, the poles slightly produced, with a prominent median protrusion on each side, bearing prominent granules. L. 61.2–61.8 μm. W. 45.1–45.3 μm. Isth. 10.5–10.6 μm.

DISTRIBUTION: Québec.

PLATE LXXVII, figs. 7, 7a.

9b. **Euastrum bidentatum** var. **oculatum** (Istv.) Krieger 1937, Rabenhorst's Kryptogamen-Flora 13(1:3): 603. Pl. 85, Fig. 2 f. **oculatum.**

Euastrum elegans (Bréb.) Kützing var. *oculatum* Istvanffi 1887, Notarisia 1887: 236.

Cells broadly oval in outline, the lateral margins of the semicell in general similar to the typical, but with the polar lobe more elevated and more convex; face of semicell with a protrusion bearing granules in circular arrangement, above which is a prominent pair of mucilage pores horizontally disposed. L. 50–70 μm. W. 32–50 μm. Isth. 7–11 μm.

DISTRIBUTION: Alaska, Minnesota. Europe, Siberia.

PLATE LXXVII, fig. 9.

9c. Euastrum bidentatum var. **oculatum** f. **glabrum** Scott & Prescott 1952, Hydrobiologia 4(4): 378. Pl. 3, Fig. 1.

Differing from the variety by having the supraisthmial protuberances smooth and evenly rounded, and by having more or less prominent depressions in the face of the semicell immediately within the margin of each lateral invagination of the wall; in lateral view semicell inversely cone-shaped from a widely open median incision of the cell, with a smoothly rounded protuberance near the base of the semicell and a slight bulge about halfway between the basal protuberance and the somewhat narrowly rounded apex; a mucilage pore showing between the basal protuberance and the intermediate protuberance. L. 60–61 μm. W. 36–37 μm. Isth. 9–10 μm. Th. 25 μm.

DISTRIBUTION: Florida.

PLATE LXXVII, figs. 10, 10a.

9d. Euastrum bidentatum var. **quadrioculatum** Scott & Prescott 1952, Hydrobiologia 4(4): 379. Pl. 1, Fig. 4.

Differing from the typical by having 2 pores within the lateral margins on each side of the semicell in the mid-zone, in addition to the 2 pores in the face of the semicell; and also differing by having a reduction in the number and size of granules on the wall; central facial protuberance reduced to a low swelling; lateral view showing 6 mucilage pores, one on either side in the midregion and 4 pores quadrately placed around the center of the ovate semicell; vertical view quadrangularly oval, the lateral margins parallel or slightly convex and with a lateral, median protuberance which is without decorations, the poles abruptly rounded, 4 mucilage pores showing, 2 on either side immediately within the margin of the central protuberance. L. 56 μm. W. 36 μm. Isth. 8 μm. Th. 22 μm.

This variety should be compared with *E. oculatum* Boerges., which it resembles in shape of cell as seen in side view. The form of the basal lobes and the deep polar incision identify this variety with *E. bidentatum*.

DISTRIBUTION: Florida.

PLATE LXXVII, figs. 11–11b.

9e. Euastrum bidentatum var. **rotundatum** Irénée-Marie 1958, Nat. Canadien 85(5): 113. Fig. 4.

A variety resembling both *Euastrum bidentatum* and *E. dubium* Näg., broadly oval in outline, 1 2/3 times longer than broad; basal lobes rounded and low (narrow), the margins retuse between the basal lobes and the upper, lateral lobules, which are sharply distinct from the polar lobe, which has diverging lateral margins to the subapical spines, giving the semicell a capitate appearance; the apical margin more elevated than in the typical; lateral view elongate-oval in outline, with only a very slight median protrusion evident in the midregion of the semicells. L. 42–45.6 μm. W. 25.8–28.2 μm. Isth. 7–8 μm.

DISTRIBUTION: Québec.

PLATE LXXVII, figs. 14, 14a.

9f. **Euastrum bidentatum** var. **speciosum** (Boldt) Schmidle 1892, Bih. Kongl. Svenska Vet.-Akad. Handl., Afd. III (8), 24: 47.

Euastrum elegans (Bréb.) Kützing var. *speciosum* Boldt 1888, Bih. Kongl. Svenska Vet.-Akad. Handl., Afd. 3, 13(5): 9. Pl. 1, Figs. 10, 11.
Euastrum papilio Raciborski 1885, Pamiet. Wydz. Akad. Umiej. Krakowie 10: 95. Pl. 13, Fig. 9 (Reprint, p. 39. Pl. 4).

Cells relatively shorter than the typical, the basal lobes with a double process at the angles; the upper, lateral lobules bigranulate and somewhat more produced; polar lobe with an elevated apex, the median notch widely open; median protrusion of the semicell with 3 vertical rows of large granules. L. 43–70 μm. W. 31–43 μm. Isth. 5–10 μm. Ap. 30–34 μm. Th. 23–24 μm.

DISTRIBUTION: Alaska, California, Florida, Iowa, Mississippi, Nevada, New Hampshire. Labrador, Québec. Europe, Asia, Africa, South America, Arctic, Azores.

PLATE LXXVII, fig. 8.

10. **Euastrum binale** (Turp.) Ehrenb. in Ralfs 1848, Brit. Desm., p. 90. Pl. XIV, Figs. 8a–8e, 8g, 8h var. **binale** f. **binale**.

Heterocarpella binalis Turpin 1820. Fig. 14.
Euastrum binale (Turp.) Ehrenberg (1840)1841, Berlin Monatsb. Akad. 1840: 208.

Cells small, quadrangular in outline, the basal lobes broadly rounded, with semicells quadrangular in outline; lateral margins retuse above the basal lobes to very short, broad polar lobes, with a somewhat inflated apex, the apical margin biconvex with a slight invagination as a median notch; facial protuberance low and simple, without granules; sinus narrow and closed throughout; lateral view oval in outline, the semicell with a slight swelling at the base, narrowing to a broadly rounded apex; vertical view oval, the poles rounded, with a low swelling on either side in the midregion. L. 12–36 μm. W. 10–26 μm. Isth. 3–8.5 μm. Ap. 9–16.8 μm. Th. 8.5–17 μm. Zygospore globose with many blunt spines, diameter without spines 19.6–26 μm.

DISTRIBUTION: Alaska, Connecticut, Florida, Iowa, Kansas, Kentucky, Maine, Massachusetts, Michigan, Mississippi, Nebraska, New Hampshire, New Jersey, North Carolina, Ohio, Oklahoma, Pennsylvania, Rhode Island, Virginia, Utah. British Columbia, Québec. Great Britain, Europe, Asia, Africa, Australia, South America, Arctic, Azores, Hawaii.

PLATE LXX, figs. 7, 7a.

10a. **Euastrum binale** var. **binale** f. **minor** West 1888, Jour. Bot. 26: 340; 1889, Jour. Roy. Microsc. Soc. 5: 19. Pl. 3, Fig. 24.

Euastrum binale (Turp.) Ehrenb. var. *minus* (West) Krieger 1937, Rabenhorst's Kryptogamen-Flora 13(1:3): 552 Pl. 75, Fig. 17.

A form similar to the typical but decidedly smaller. L. 10–12 μm. W. 7.5–11.3 μm. Isth. 2–3.5 μm. Ap. 7–8.5 μm. Th. 6–6.3 μm.

We do not believe that the difference in size justifies the recognition of this taxon as a variety rather than a form of the typical as recommended by Krieger (*l.c.*).

DISTRIBUTION: Alaska, Maine, Massachusetts. Ontario, Québec. Great Britain, Europe, Asia, Africa, Arctic.

PLATE LXX, fig. 5.

10b. **Euastrum binale** var. **cosmarioides** (West & West) Krieger 1937, Rabenhorst's Kryptogamen-Flora 13 (1:3): 550. Pl. 75, Figs. 6, 8.

Euastrum cosmarioides West & West 1895, Trans. Linn. Soc. London, Bot. 5: 54. Pl. 6, Fig. 23.

A variety similar in shape to the typical but with a slightly wider polar lobe, the apex angles sharp; facial protuberance wanting or scarcely evident; lateral view of semicell nearly circular and without a median lateral protrusion; vertical view oval and showing a slight protrusion on the sides. L. 22–27 μm. W. 15–19 μm. Isth. 4.6–6.6 μm. Th. 12.5 μm.

DISTRIBUTION: Florida. Asia, Africa.

PLATE LXX, figs. 6–6b.

10c. **Euastrum binale** var. **groenbladii** Messikommer 1927, Inaug. Dissert., Zürich, p. 98. Pl. 1, Fig. 17.

A variety with a narrower polar lobe and with the angles less curved than in the typical, margin of apex with 2 pairs of small granules on either side of the median depression; vertical view oval with a slight protrusion on either side. L. 13–21 μm. W. 12–16.5 μm. Isth. 4 μm. Th. 8 μm.

DISTRIBUTION: Florida, Massachusetts. Europe, Africa, Arctic.

PLATE LXX, figs. 8–8b.

10d. **Euastrum binale** var. **gutwinskii** (Schmidle) Krieger 1937, Rabenhorst's Kryptogamen-Flora 13(1:3): 551. Pl. 75, Figs. 13–15 f. **gutwinskii.**

Euastrum binale f. *gutwinskii* Schmidle 1893, Ber. d. Deutsch. Bot. Ges. 11(10): 552.

A variety somewhat larger than the typical, the basal lobes produced, and with a marginal undulation between the basal lobes and the polar lobe, which is more prominent but narrower than in the typical; lateral view oval in outline, the semicells with a basal inflation, the poles broadly rounded. L. 16–30 μm. W. 12.5–30 μm. Isth. 4.3–8 μm. Ap. 11–13.2 μm. Th. 11–13 μm. Zygospore globose with long spines, 19–28 μm in diameter without spines, 35–49 μm in diameter with spines.

DISTRIBUTION: Alaska, Colorado, Connecticut, Georgia, Kentucky, Louisiana, Massachusetts, Michigan, Mississippi, New Hampshire, New Jersey, North Carolina, Utah, Vermont, Wyoming. Labrador, Newfoundland, Québec. Great Britain, Europe, Asia, Africa, South America, Arctic.

PLATE LXX, fig. 9.

10e. **Euastrum binale** var. **gutwinskii** f. **apertum** Croasdale 1964, in Croasdale & Grönblad 1964, Trans. Amer. Microsc. Soc. 83(2): 164. Pl. VII, Figs. 16–19.

A form with a widely open sinus, the basal angles broadly rounded, the margins of the semicell with a prominence above the basal lobes, then sharply retuse to form a short polar lobe with a flat apical margin, with a pronounced median invagination. L. 26–28 μm. W. 18.5–22 μm. Isth. 3.5–8 μm. Th. 13 μm.

DISTRIBUTION: Labrador.

PLATE LXX, figs. 10–10b.

10f. **Euastrum binale** var. **hians** (West) Krieger 1937, Rabenhorst's Kryptogamen-Flora 13(1:3): 551. Pl. 75, Fig. 16.

Euastrum binale f. *hians* West 1892, Linn. Soc. Jour. Bot. 29: 140. Pl. 20, Fig. 14.

A variety in which the margins of the basal lobes diverge sharply from a closed sinus, retuse to the relatively wide polar lobe, the apical margin only slightly retuse in the midregion. L. 11–16 μm. W. 10–12.5 μm. Isth. 2.5–3.5 μm. Th. 6–7 μm.

DISTRIBUTION: Iowa, Michigan, Mississippi, North Carolina, North Dakota, Ohio. Great Britain, Europe, Asia, Africa, South America, Arctic.

PLATE LXX, figs. 11–11b, 12.

10g. **Euastrum binale** var. **obtusiusculum** Schaarschmidt (1882) 1883, Mag. Tudom. Akad. Math. e Termész. Köz. 18: 263. Pl. 1, Fig. 3.

A variety narrower than the typical, 1.25 times longer than broad; basal lobes angular at their apices and upturned; margins of semicell with an undulation between the basal lobes and the polar lobe, which is acute at the angles, winglike lobules at the apex; lateral view truncate-oval the poles flat and the polar lobe somewhat produced, the body of the semicell broadly rounded. L. 20–24 μm. W. 16–20 μm.

DISTRIBUTION: Massachusetts. Europe.

PLATE LXX, figs. 16, 16a.

10h. **Euastrum binale** var. **papilliferum** Gutwinski 1909, Bull. Acad. Sci. Cracovie, Sci. Math. Nat. 1909: 467. Pl. 8, Fig. 41.

A variety shorter and stouter than the typical, the basal lobes broadly rounded but bearing a short, spinelike projection, with slightly undulate margins; the polar lobe pronounced with subparallel margins, the apical margin bearing 2 teeth at the angles and with 2 marginal teeth on either side of the median notch; lateral view with the semicells much swollen at the base, with a broadly rounded polar lobe; vertical view oval, the poles broadly rounded and with a median swelling on either side. L. 18–24 μm. W. 13–19 μm. Isth. 6.6–8 μm. Ap. 14 μm.

DISTRIBUTION: Alaska. Europe, Asia, South America, Arctic.

PLATE LXX, figs. 13–13b.

10i. **Euastrum binale** var. **poriferum** Prescott var. nov.

Cellulae 1 1/3 plo longiores quam latae, var. obtusiusculo magnitudine similes, sinu, autem, profundiore; margines semicellularum lobulum manifestum inter lobes basales polaremque praebentes; semicellulae truncato-pyramidales; superficies par pororum mucosorum conspicuorum media in parte habens; semicellulae a latere visae late ovatae, membrana mediae partis perspicue incrassata et poro mucoso praedita; semicellulae a vertice visae ovatae, membrana utroque in latere mediae partis incrassata et par pororum mucosorum praedita. Long. 32 μm. Lat. 24 μm. Ist. 7 μm. Crass. 15 μm.

ORIGO: E palude distante 11.5 milia passuum (18.5 km.) ad occidentem a loco Torrey, Florida dicto.

HOLOTYPUS: Scott Coll. Fla-113.

ICONOTYPUS: Plate LXX, figs. 14–14b.

Cells 1 1/3 times longer than broad, similar to var. *obtusiusculum* Schaar. in size but with a deeper sinus; margins of semicell with a distinct lobule between the polar lobe and the basal lobes, the outline truncate-pyramidal; face with a pair of conspicuous mucilage pores in the midregion; lateral view of semicell broadly oval with a distinct thickening of the wall in the midregion where there is a mucilage pore; vertical view oval, the wall thickened on either side of the midregion where there is a pair of mucilage pores. L. 32 μm. W. 24 μm. Isth. 7 μm. Th. 15 μm.

DISTRIBUTION: Florida.

PLATE LXX, figs. 14–14b.

10j. **Euastrum binale** var. **pseudogutwinskii** Grönblad 1921, Acta Soc. Fauna Flora Fennica 49(7): 14. Pl. 3, Figs. 24, 25.

A variety with the basal lobes broadly rounded, the margins of the semicells undulate above, and then retuse to form a pronounced polar lobe with subparallel margins; the apical margin flat with a slight median incision; cell wall with granules intramarginal along the apex and within the lateral undulations of the basal lobes; a slight median protrusion bearing a mucilage pore; vertical view oval, the poles rounded and with a slight lateral swelling showing a mucilage pore. L. 19–23 μm. W. 13–16.6 μm. Isth. 3–3.5 μm. Th. 10.5 μm.

DISTRIBUTION: Florida, Minnesota. Europe, Arctic.

PLATE LXX, fig. 15.

10k. **Euastrum binale** var. **retusum** West 1892a, Jour. Roy. Microsc. Soc. 1892: 723. Pl. 9, Fig. 6.

A variety with the margins of the basal lobes diverging to an angle, then retuse to a prominent polar lobe with sharp angles, the apical margin decidedly retuse; facial protuberance with a circular patch of granules, the wall definitely punctate throughout; vertical view oval, the poles rounded, with a lateral protuberance bearing granules on either side, the wall granular along all margins. L. 27 μm. W. 21 μm. Isth. 8 μm. Th. 12 μm.

DISTRIBUTION: Connecticut. Great Britain, Europe, Asia.

PLATE LXX, fig. 17.

10l. **Euastrum binale** var. **sectum** Turner 1892, Kongl. Svenska Vet.-Akad. Handl. 25(5): 81. Pl. 10, Figs. 35, 39, 47.

A variety similar to var. *gutwinskii* and probably not to be differentiated; basal lobes with an undulation above the basal angles; polar lobe as in var. *gutwinskii*; wall smooth. L. 17–31.2 μm. W. 13.4–22 μm. Isth. 4.8–7 μm. Ap. 8.5–14 μm.

DISTRIBUTION: Arizona, Florida, Massachusetts, Oklahoma. British Columbia, Labrador. Europe, Asia, South America.

PLATE LXX, fig. 18.

10m. **Euastrum binale** var. **tumoriferum** Kossinskaja 1936a, Acta Inst. Bot. Acad. Sci. URSS, II, 3: 417. Pl. 1, Fig. 10.

A variety similar to var. *gutwinskii* but differs in having a more pronounced polar lobe and with slightly more extended apical angles; facial view with an intramarginal lobule immediately beyond the upper, lateral angles. L. 25–28.8 μm. W. 18–19.2 μm. Isth. 4.8–5 μm. Ap. 9.6–10 μm. Th. 13–13.2 μm.

It is doubtful whether this variety is worthy of a separate taxon status. It seems to us that it perhaps should be placed as a form of var. *gutwinskii*.

DISTRIBUTION: Alaska. Europe, North Siberia.

PLATE LXX, figs. 19–19b.

11. **Euastrum binum** Prescott sp. nov.

Cellulae parvae, 1–1.5 plo longiores quam latae; semicellulae transverse elongatae, angulariter ovatae, sinus extrorsum late patens, in apicem acutus; margines laterales lobi basalis subparalleli utra sinum, angulis rotundatis, deinde retusi in lobum polarem brevem truncatum, incisura media haud profunda; 3 granula minuta vix distinguibilia intra marginem apicalem utroque in latere incisurae mediae; inflatio superficialis 2 tubercula praebens; sinus relative profundus; semicellulae a latere visae ovatae, polis late rotundatis, inflationem lateralem mediam prominentem utroque in latere praebentes; semicellulae a vertice visae ovatae, polis late rotundatis, inflationem bipapillatam mediam utroque in latere praebentes. Long. 23 μm. Lat. 18 μm. Ist. 5 μm. Crass. 12 μm.

ORIGO: E palude distante 9 milia passuum (14.5 km.) ad meridiem a loco Wilma, Liberty Co., Florida dicto.

HOLOTYPUS: Scott Coll. Fla-251.

ICONOTYPUS: Plate LXII, figs. 14–14b; LXXII, figs. 14–14b.

Cells small, 1–1 1/5 times longer than broad; the semicells transversely elongate, angularly oval, the sinus widely open outwardly, acute at the apex; lateral margins of the basal lobe subparallel beyond the sinus, the angles rounded, the margins retuse in the short polar lobe, which is truncate with a relatively shallow median notch; 3(?) minute, scarcely discernible granules within the polar margin on either side of the median notch; facial swelling with 2 tubercles; lateral view oval in outline, the poles broadly rounded, with prominent, lateral, median swellings, the sinus relatively deep; vertical view oval, the poles broadly rounded, with a bipapillate swelling median on either side. L. 23 μm. W. 18 μm. Isth. 5 μm. Th. 12 μm.

This species has some *Cosmarium*-like features. It should be compared with *E. berlingii* Boldt, one of the few species of *Euastrum* in which the sinus is open.

DISTRIBUTION: Florida.

PLATE LXII, figs. 14–14b; PLATE LXXII, figs. 14–14b.

12. **Euastrum bipapillatum** Grönblad 1920, Acta Soc. Fauna Flora Fennica 47(4): 29. Pl. 2, Figs. 35, 36.

A small species with truncate-pyramidal semicells that are slightly wider than long; cell 1.75 times longer than broad; basal lobes sharply rounded, the margins converging toward the apex with an upper, lateral lobule about halfway, the polar lobe slightly produced with subparallel margins, the apical margin broad, the angles sharply rounded, with a slight median invagination; face of semicell with a suprasisthmial and a median tubercle; sinus narrow and closed; lateral view narrowly oval, the poles rounded, with a prominent tubercular swelling on either side below the apex. L. 28–40 μm. W. 21 μm. Isth. 12.2 μm. Th. 13–15 μm. Zygospore unknown.

DISTRIBUTION: Michigan, New Hampshire, Vermont. Europe, South America, Arctic.

PLATE LXXII, figs. 11, 11a.

13. **Euastrum bipartitum** Krieger 1932, Arch. f. Hydrobiol. Suppl. 11: 211. Pl. 20, Fig. 16 var. **bipartitum.**

A small *Euastrum*, subrectangular in outline 1 1/3 to 1 1/4 times longer than broad; semicells 4-lobed; the basal lobes truncately rounded at the apices, bearing 3 short spines, the margins deeply retuse to form a polar lobe with 2 lobules, the angles rounded, with 3 sharp mucros, the apical margin deeply concave; face of semicell with a median protrusion furnished with a circle of large granules; wall with granules within the margins of all lobes; lateral view narrowly oval, the poles rounded and granular, the margins concave and diverging to the basal protuberance which is truncate and emarginate; vertical view oval, the poles broadly rounded but with an apiculate spine, 2 spines subapical on either side, the margins with a truncate protuberance in the midregion. L. 23–28 μm. W. 19–20 μm. Isth. 6 μm. Th. 13 μm. Zygospore unknown.

DISTRIBUTION: Typical form not reported from North America. East Indies.

PLATE LXXIII, figs. 11, 11a.

13a. **Euastrum bipartitum** var. **pseudotuddalense** (Messikommer) Prescott comb. nov.

Euastrum pseudotuddalense Messikommer 1942, Beitr. Geobot. Landes. der Schweiz 24: 141. Pl. 4, Fig. 5.

A small *Euastrum* about the same size as the typical; the semicells subquadrate in outline; basal lobes biundulate and rectangular at the angles; the polar lobe broad, nearly as wide as the basal lobes, the apex divided to form 2 lobules, the margin broadly invaginated, V-shaped, the apical lobules tipped with a granule; face of semicell with a circular protrusion in the midregion; wall with a cluster of granules within all the lobes; lateral view oval, the poles broadly rounded and granular, the margins diverging to a pronounced, truncate protuberance; vertical view oval, the poles rounded and granular, the lateral margins with a stout, truncate protuberance. L. 20 μm. W. 21 μm. Isth. 6 μm. Th. 14μm.

This variety should be compared with *Euastrum tuddalense* Ström, from which it differs in the shape of the basal lobes and in vertical view. It seems to be more nearly related to *E. bipartitum* Krieger than to *E. tuddalense*.

DISTRIBUTION: Labrador. Europe.

PLATE LXXIII, figs. 10–10b; PLATE LXXX, figs. 7–7b.

14. **Euastrum boldtii** Schmidle 1895(1896), Österr. Bot. Zeit. 45: 27. Pl. 16, Fig. 5 var. **boldtii**.

Euastrum boldtii Schmidle var. *isthmochondrum* Grönblad 1921, Acta Soc. Fauna Flora Fennica 49(7): 10. Pl. 3, Figs. 1–4.

Cells small, quadrangular in outline, 1.5 times longer than broad; basal lobes rounded, with 3 spinelike granules; the polar lobe produced, with subparallel margins, the apical margin broad and flat, with the angles extended into horizontal spines, the median notch a shallow invagination; facial protuberance with 3 large granules and with a single tubercle below it and immediately above the isthmus; cell wall with intramarginal granules; sinus narrow and closed; lateral view broadly oval, the apex convex, sometimes with a toothlike granule, a prominent protuberance on either side in the midregion of the semicell, and with a tubercle at the base; vertical view oval, the poles rounded, with a bituberculate swelling on either side in the midregion. L. 23–42 μm. W. 16–25 μm. Isth. 4–8 μm. Ap. 18 μm. Th. 14.5 μm.

DISTRIBUTION: Alaska, Idaho, Oregon. Newfoundland, Québec. Europe, South America, Greenland and the Arctic.

PLATE LXXII, figs. 15–15b, 22–22b (f.)

14a. **Euastrum boldtii** var. **groenbladii** Krieger 1937, Rabenhorst's Kryptogamen-Flora 13(1:3): 564. Pl. 77, Fig. 8.

Euastrum denticulatum (Kirch.) Gay f. Boldt 1888, Studier Öfver Sötvattens Alger och deras utbredning II–III: 6. Pl. 1, Fig. 9 Helsingfors.

Cells slightly larger than the typical but the same shape, with the margins less sharply granular and without the supraisthmial tubercle as found in the typical. L. 30–32 μm. W. 22–25 μm. Isth. 6–9 μm. Ap. 18–19 μm. Th. 15–17 μm.

The form of Boldt (*l.c.*) which Krieger cites as synonymous with this variety is similar in shape and size but has 2 tubercles below the median swelling. Grönblad shows a cell with 2 tubercular granules above the median protrusion, but neither he nor Krieger notes this.

DISTRIBUTION: Labrador. Europe, Greenland, Azores.

PLATE LXXII, figs. 16–17a.

15. Euastrum brasiliense Borge 1903, Ark. f. Bot. 1: 112. Pl. 5, Fig. 1 var. **brasiliense.**

Euastrum ansatum var. *turgidum* Börgesen (1890)1891, Vid. Medd. Naturh. Foren. Kjöbenhavn 1890: 938. Pl. 3, Fig. 15.

Cells medium-sized, broadly oval in outline, 2 times (or more) longer than broad; basal lobes abruptly rounded, the margins of the semicells retuse or undulate to a slightly developed shoulder or upper, lateral lobule, then retuse to a broad polar lobe that is convex or flattened at the apex, the median notch deep and narrow; with a low facial protuberance immediately above the isthmus; the wall coarsely punctate; sinus relatively short and closed; lateral view elongate-oval, the poles broadly rounded, the lower margins of the semicell subparallel; vertical view elliptic, with the margins lightly swollen in the midregion. L. 64–89 μm. W. 32–43 μm. Isth. 12–16 μm. Th. 27–31 μm. Zygospore unknown.

DISTRIBUTION: North Carolina. Africa, South America.

PLATE LVIII, figs. 3, 3a.

15a. Euastrum brasiliense var. **convergens** Krieger 1937. Rabenhorst's Kryptogamen-Flora 13(1:3): 484. Pl. 57, Fig. 17.

Euastrum brasiliense f. Fritsch & Rich 1924, Trans. Roy. Soc. South Africa 11: 330. Textfigs, 8E, 8F.

A variety in which the basal lobes are slightly produced, broadly rounded, with a retuse margin above to a slight inflation between the basal lobes and the polar lobe; cell wall coarsely punctate. L. 96–102 μm. W. 48–52 μm. Isth. 15–16 μm.

DISTRIBUTION: Florida. Africa, South America.

PLATE LVIII, figs. 4, 9.

16. Euastrum ceylanicum (West & West) Krieger 1937, Rabenhorst's Kryptogamen-Flora 13(1:3): 627. Pl. 90, Figs. 16, 17.

Euastrum spicatum West & West var. *ceylanicum* (West & West) Krieger 1932, Arch. f. Hydrobiol. Suppl. 11: 215. Pl. 21, Fig. 5.

Cells less than medium-sized, quadrate in outline, 1 1/6 times longer than broad; basal lobes broadly rounded and horizontally extended, narrowed abruptly to a narrow polar lobe, the margins slightly diverging or nearly parallel, the apical margin slightly retuse or flat, the angles broadly rounded or slightly swollen, the margins of the apical angles and the basal lobes with sharp granules, and with a few granules within the margins of all lobes; facial swellings 3, the median prominent with a circle of granules, the 2 lateral swellings less pronounced and with smaller granules; sinus narrow and closed, opening slightly outward; vertical view narrowly elliptic, the poles sharply rounded and granular, with 3 lateral swellings, the median being more prominent and granular; lateral view elliptic with a prominent median swelling on either side of the semicell. L. 46–78 μm. W. 39–63 μm. Isth. 11–16 μm. Ap. 21–28 μm. Th. 29 μm. Zygospore unknown.

It is clear that *Euastrum ceylanicum, E. substellatum* Nordst., *E. sphyroides* Nordst. and *E. spicatum* Turner are closely related. *E. ceylanicum* is characterized by having horizontally directed basal lobes, with a narrow and closed sinus; the face of the semicell has 3 protrusions, which are best seen in vertical view. *E. sphyroides* has relatively shorter basal lobes and a single facial protuberance. *E. substellatum* has a widely open sinus and basal lobes which are usually slightly upturned. *E. spicatum* has been reduced to synonymy with *E. ceylanicum*.

DISTRIBUTION: Wisconsin. Asia, East Indies.

PLATE LXXX, figs. 17, 17a, 18 (f.).

17. Euastrum ciastonii Raciborski 1892, Rozpr. Akad. Umiej. w Krakowie 22: 387. Pl. 7, Fig. 28 var. **ciastonii f. ciastonii.**

Cells under medium size, rectangular in outline, 1.75 times longer than broad; basal lobes of the semicell broadly rounded, often with a tooth at the angles, the margins retuse to the polar lobe, which is slightly inflated, the apical margin elevated to a deep and narrow median incision, the angles beset with a short, sharp spine; median protrusion on the face of the semicell low, with a circle of 4 large granules; wall with a submarginal granule on either side of the apical notch, with a pair of sharp spines at the margins of the basal lobes and a pair intramarginal; sinus narrow and closed; lateral view narrowly rectangular in outline, with a prominent median swelling with granules on either side of the semicell, the poles flat, truncate, with a tubercle at each angle and also a tubercle immediately below (supbolar), the margins retuse between the apex and the median swelling; vertical view elliptic, the poles with 1 or 2 spines, with a granular, median swelling on either side. L. 29–49 μm. W. 21–29 μm. Isth. 4.5–10 μm. Ap. 17–31 μm. Th. 16–21 μm. Zygospore unknown.

DISTRIBUTION: Alaska, Florida, Georgia, Massachusetts, Mississippi, New Hampshire, North Carolina. South America.

PLATE LXXVIII, figs. 6, 6a.

17a. Euastrum ciastonii var. **ciastonii f. hians** Irénée-Marie 1952, Hydrobiologia 4(1/2): 163. Pl. 18, Fig. 10.

Cells about 1.75 times longer than broad, but differing from the typical by the opening of the sinus. L. 44–50 μm. W. 25–26 μm. Isth. 7.5–8 μm. Ap. 23–25 μm.

DISTRIBUTION: Québec.

PLATE LXXVIII, fig. 11.

17b. Euastrum ciastonii var. **ciastonii f. minor** Irénée-Marie 1958, Nat. Canadien 85(5): 118. No. fig.

Cells very small, similar in all respects to the typical but about 1/3 the size. L. 24.8–25.8 μm. W. 17.5–20 μm. Isth. 4.8–5.6 μm.

DISTRIBUTION: Québec.

17c. Euastrum ciastonii var. **apertisinuatum** Scott & Prescott 1952, Hydrobiologia 4(4): 379. Pl. 1, Fig. 5.

A variety differing from the typical by having a sinus that opens outwardly from a sharply rounded apex; also differing in its possession of additional granules immediately within the apical margins of the polar lobules. L. 40–44 μm. W. 24–27 μm. Isth. 6–7 μm.

DISTRIBUTION: Florida. Ontario.

PLATE LXXVIII, figs. 16, 16a.

17d. **Euastrum ciastonii** var. **asymmetricum** Scott & Prescott 1952, Hydrobiologia 4(4): 380. Pl. 1, Fig. 6.

Differing from the typical by having the lower lateral margins of the basal lobes distinctly diverging to rather abruptly rounded lateral angles, the upper, lateral margins more deeply retuse, and in having a crest of 3 granules immediately within the apical margin of the polar lobules on either side of the median notch; in face view semicells differ by having 2 protuberances in the midregion which bear 2 to 4 granules, one protuberance being distinctly larger than the other; vertical view oval, the poles with a marginal series of 4 granules, with 2 protuberances in the midregion on each side, one of which is larger and bidentate rather than simple, and in position diagonally opposite the bidentate protuberance on the other side of the cell. L. 45 μm. W. 29–30 μm. Isth. 8 μm. Th. 21 μm.

DISTRIBUTION: Mississippi.

PLATE LXXVIII, figs. 19, 19a.

18. **Euastrum cornubiense** West & West 1905a, Monogr. II: 70. Pl. 40, Fig. 8 var. **cornubiense. f. cornubiense.**

Cells small, rectangular in outline, 1.5 times longer than broad; basal lobes slightly bilobed on the margins; polar lobe broad, truncate and slightly swollen, the apical margin broadly rounded or flat, sometimes slightly retuse; facial protrusion prominent but simple, without granules; sinus narrow and closed; lateral view oval, the poles broadly rounded, semicell with a prominent protrusion on either side in the midregion; vertical view oval, the poles rounded and with a prominent protrusion on either side in the midregion. L. 24–30 μm. W. 15–31 μm. Isth. 3.5–4 μm. Th. 8.5–11.5 μm. Zygospore unknown.

DISTRIBUTION: California. Newfoundland. Great Britain, Swedish Lappland.

PLATE LXXIV, figs. 14–14b, 23 (aberrant form).

18a. **Euastrum cornubiense** var. **cornubiense** f. **granulatum** Prescott & Scott 1945, Amer. Mid. Nat. 34(1): 235. Pl. 1, Fig. 6.

A form slightly larger than the typical and with rhomboid semicells, the upper lobule of the bilobed basal lobes more prominent, and with a less prominent, granule-bearing undulation between the basal lobes and the polar lobe, which is short, with a more decided retuse apical margin; with a prominent tubercle at the base of the semicell immediately above the isthmus. L. 36 μm. W. 25 μm. Isth. 7.5 μm. Th. 18 μm.

DISTRIBUTION: Mississippi.

PLATE LXXIV, figs. 15–15b.

18b. **Euastrum cornubiense** var. **cornubiense** f. **hypochondrum** Scott & Grönblad 1957, Acta Soc. Sci. Fennica, n.s., B, II(8): 12. Pl. 3, Figs. 13, 14.

A form about the same size as the typical, subrectangular in outline, but with truncate-pyramidal semicells; the basal lobes sharply rounded with the margins slightly converging upward to a low swelling and then diverging slightly to a somewhat swollen polar lobe, the angles of the polar lobe sharply rounded and bearing a single granule at the margin, the apical margin decidedly retuse in the midregion; median facial protrusion less prominent than with the typical. L. 29–31 μm. W. 18–20 μm. Isth. 4.5–6 μm. Zygospore globose with many broadly rounded protrusions, 20 μm in diameter.

DISTRIBUTION: Florida. Africa.

PLATE LXXIV, figs. 12–12b.

18c. Euastrum cornubiense var. **medianum** (Nordst.) Krieger 1937, Rabenhorst's Kryptogamen-Flora 13(1:3): 574. Pl. 78, Figs. 7–9 f. **medianum.**

Euastrum elegans (Bréb.) Kützing var. *medianum* Nordstedt 1888a, Kongl. Svenska Vet.-Akad. Handl. 22(8): 34. No fig.

A variety slightly larger than the typical, with the basal lobes of the semicell bilobed as in the typical, the polar lobe more inflated and with a granule at the angles as in f. *hypochondrum*; lacking an undulation between the basal lobes and the polar lobe. L. 36–40 μm. W. 36–40 μm. Isth. 5 μm. Th. 16 μm.

DISTRIBUTION: Florida. New Zealand, Australia, Java.

PLATE LXXIV, figs. 13–13b.

18d. Euastrum cornubiense var. **medianum** f. **subgranulatum** Prescott f. nov.

Forma magnitudine formae typicae similis, lobus polaris, autem, acute exsertus, marginibus lateralibus lobi polaris divergentibus, angulis apicalibus granulum acutum atque granulum submarginale ferentibus; granulum isthmiale tuberculare prominens ad basim semicellularum; membrana minute punctata. Long 28–30 μm. Lat. 17–19 μm. Ist. 4 μm. Crass. 10 μm.

ORIGO: E fossa distante 8 milia passuum (12.9 km.) ad meridiem a loco Punta Gorda, Florida dicto.

HOLOTYPUS: Scott Coll. Fla-77.

ICONOTYPUS: Plate LXXIV, figs. 16, 16a.

A form similar in size to the typical variety, but with the polar lobe sharply exserted, the lateral margins of the polar lobe diverging, the apical angles bearing a sharp granule and with a submarginal granule at the apical angles; with a prominent isthmial tubercular granule at the base of the semicell; wall minutely punctate. L. 28–30 μm. W. 17–19 μm. Isth. 4 μm. Th. 10 μm.

DISTRIBUTION: Florida.

PLATE LXXIV, figs. 16, 16a.

19. Euastrum crameri Raciborski 1889, Pam. Wydz. Mat.-Przyr. Akad. Umiej. 17: 104. Pl. 2, Fig. 5.

Cells average size, broadly oval in outline, 1.5 times longer than broad; the basal lobes broad and triundulate at the margins, with an emarginate protrusion on the upper, lateral angles, the margins retuse to form the polar lobe which is broad, the apical margin elevated from prominent spines on the angles to a deep, closed, median notch; an intramarginal spine opposite each apical angle, a more prominent spine within the margin of the sinus which forms the polar lobe, an intramarginal spine opposite the upper, lateral lobules and the uppermost of the lobules of the basal lobes; 3 prominent protrusions across the base of the semicell immediately above the isthmus; face of the semicell with 2 median mucilage pores transversely disposed; sinus narrow and closed; lateral view broadly elliptic, the poles produced and narrowly rounded, the semicells subtriangular, the sides broadly convex and with 2 prominent spinelike protrusions at the base; vertical view quadrate, the poles rectangular and with 3 prominent protrusions, and with a protrusion on either side in the midregion, a pair of mucilage pores evident within the margins. L. 42–75 μm. W. 29–52 μm. Isth. 7–15 μm. Ap. 29–32 μm. Th. 24–31 μm. Zygospore unknown.

DISTRIBUTION: Florida. Europe, Asia.

PLATE LXXX, fig. 9.

20. **Euastrum crassangulatum** Börgesen 1890, Vid. Medd. Naturh. Foren. Kjöbenhavn 1890: 942. Pl. 3, Fig. 25.

A small, *Cosmarium*-like species, trapezoidal in outline, 1 1/5 times longer than broad; the semicells truncate-pyramidal, the basal lobes broadly rounded and smooth, the margins retuse beyond to the polar lobe, which is broadly truncate, with a slightly retuse apical margin; face of semicell with a smooth, median protrusion; with a low swelling within the 2 lobules of the polar lobe and 1 within each basal lobe; the wall otherwise smooth and thickened at the angles of all lobes; sinus narrow and closed; lateral view broadly oval, the poles truncate and flattened, with a prominent protrusion on each side of the semicell in the midregion; vertical view elliptic, the poles rounded and smooth, with a very slight inflation on either side in the midregion. L. 25-28 μm. W. 17-19 μm. Isth. 3.5-6 μm. Ap. 12 μm.

DISTRIBUTION: Michigan, Minnesota. British Columbia. Europe, South America.

PLATE LXIX, figs. 12-12b.

21. **Euastrum crassicolle** Lundell 1871, Nova Acta Reg. Soc. Sci. Upsal., III, 8(2): 25. Pl. 2, Fig. 8 var. **crassicolle**.

A small *Euastrum*, trapezoidal in outline, 1.75 times longer than wide; the semicells truncate-pyramidal, the basal lobes extended and narrowly rounded, with a suprabasal inflation on the lateral margins, sharply retuse to the polar lobe, which is somewhat extended, with subparallel margins, the angles of the polar lobe rounded, the apical margin slightly retuse in the midregion; face of semicell with a slight, almost indiscernible median swelling; the wall smooth and uniform in thickness throughout; sinus narrow and closed; lateral view narrowly oval, the poles broadly rounded, the apical lobe somewhat exserted, with a low swelling in the midregion on either side; vertical view oval, the poles broadly rounded, with a low median swelling on either side. L. (16)20-32 μm. W. 13.5-18.5 μm. Isth. (3.5)5.2-8.8 μm. Ap. 6.5-11 μm. Th. 9.5-12 μm. Zygospore unknown.

DISTRIBUTION: Alaska, Colorado, Iowa, Louisiana, Maine, Massachusetts, New Jersey, New York. British Columbia, Labrador, Québec. Great Britain, Europe, Asia, Africa, South America, Cuba, Arctic.

PLATE LXXIV, fig. 17.

21a. **Euastrum crassicolle** var. **bicrenatum** DeToni 1889, Sylloge algarum, p. 1075; Krieger 1937, Rabenhorst's Kryptogamen-Flora 13(1:3): 575. Pl. 78, Figs. 15, 16.

Euastrum chickeringii Prescott, In: Prescott & Magnotta 1935, Pap. Michigan Acad. Sci. Arts & Lettr. 20: 164. Pl. 25. Figs. 15, 16.

A variety differing from the typical by the upper, lateral lobules of the semicell being extended equally to the basal lobes so that the lower margins of the semicells do not converge toward the polar lobe; lateral view with the median swellings of the semicell more pronounced. L. 26-31 μm. W. 13.8 -17 μm. Isth. 4.5-9 μm. Th. 13 μm.

DISTRIBUTION: Iowa, Michigan. Europe, Africa, Arctic.

PLATE LXXIV, figs. 18, 18a.

21b. **Euastrum crassicolle** var. **dentiferum** Nordstedt 1875, Öfv. Kongl. Vet.-Akad. Förhandl. 1875(6): 31. Pl. 8, Fig. 32.

A variety differing from the typical by having the polar lobe more inflated, the apical angles extended into horizontally directed spines, and by a sharper depression in the midregion of the apical margin; lateral view similar to the typical. L. 26–34 µm. W. 14–18 µm. Isth. 5–8.8 µm. Th. 11–12 µm.

DISTRIBUTION: Alaska, Colorado. Europe, Greenland, Arctic.

PLATE LXXIV, figs. 19–19b.

22. **Euastrum crassum** (Bréb.) Ralfs 1848, Brit. Desm., p. 81. Pl. XI, Figs. 3a–3f var. **crassum**.

Cosmarium crassum Brébisson, in Meneghini 1840, Synop. Desm., p. 222.
Euastrum crassum var. *michiganense* Prescott (p.p.), in Prescott & Scott 1845, Amer. Mid. Nat. 34(1): 236. Pl. 1, Fig. 14.
Euastrum crassum var. *ornatum* (Wood) Hansgirg 1888, Prodromus Algenflora von Böhmen, Erster Theil, p. 205.
Euastrum magnificum var. *crassioides* Hastings, in Wolle 1892, Desm. U.S., p. 108. No fig.
Euastrum ornatum Wood 1872, Smithson. Contrib. to Knowledge 241: 137. Pl. 21, Fig. 12; 1870, Proc. Philadelphia Acad. Nat. Sci. 1869: 17.

A large *Euastrum*, rectangular in outline, 1.75–2 times longer than broad; semicells quadrate, basal angles inflated, sharply rounded, the lateral margins retuse to upper, lateral lobules which extend nearly as far as the basal lobes and are slightly upwardly directed; with a narrow, deep sinus between the upper, lateral lobules and the polar lobe, which is short and inflated, the lateral angles sharply rounded, the apical margin flat or broadly convex with a deep, narrow median incision; the face of the semicell with 3 prominent protrusions immediately above the sinus, the protrusions of the 2 semicells in contact; wall coarsely scrobiculate or deeply punctate; lateral view subrectangular, the semicells with lower, lateral margins subparallel, converging, and retuse to the broadly truncate poles, the sinus between the 2 semicells short, narrow, and usually closed outwardly by the projecting basal protrusions; vertical view oval, the poles broadly rounded and often bilobed, with a slight protrusion in the midregion on either side. L. 145–230 µm. W. 77–106 µm. Isth. 21–32 µm. Ap. 51–63.5 µm. Th. 56–75 µm. Zygospore unknown.

Prescott & Scott (*l.c.*) list and illustrate a form of *Euastrum crassum* which cannot be referred to var. *michiganense* Prescott 1935, in Prescott & Magnotta, Pap. Michigan Acad. Sci., Arts & Lettr. 20: 165. Pl. 26, Figs. 1, 2. It appears to be a form of the typical. See var. *michiganense* below.

DISTRIBUTION: Widely distributed in the United States and Canada. Great Britain, Europe, Asia, South America, Cuba, Faeroes, Arctic.

PLATE LXI, figs. 14, 14a.

22a. **Euastrum crassum** var. **bigemmiferum** Grönblad 1945a, Mem. Soc. Fauna Flora Fennica 21: 56. Pl. 1, Figs. 3, 4.

A variety differing from the typical by having 2 median horizontally placed protuberances on the face of the semicell on either side of a median triangle of mucilage pores, with a central median pore (4 in all). L. 163 µm. W. 87 µm. Isth. 27 µm. Th. 61 µm.

DISTRIBUTION. Massachusetts. Finland.

PLATE LXI, figs. 15, 15a.

22b. **Euastrum crassum** var. **glabrum** Prescott var. nov.

Varietas a varietate typica differens quod minor, et quod inflationes laterales marginis super lobos basales habet; differt quoque quod membrana levis, non scrobiculata, et quod 1 tumorem supraisthmialem habet atque porum mucosum medium non habet. Long. 139.5 µm. Lat. 74.8 µm. Ist. 20.4 µm.

ORIGO: E lacu Horseshoe nominato in loco Mt. Jefferson Park, Cascade Mts., Oregon dicto.

HOLOTYPUS: GWP Coll. OI-24.

ICONOTYPUS: Plate LXI, fig. 12.

A variety differing from the typical by its smaller size, by having lateral swellings of the margin above the basal angles, by its smooth rather than scrobiculate wall and by having only 1 supraisthmial tumor; also by its lack of a median mucilage pore. L. 139.5 µm. W. 74.8 µm. Isth. 20.4 µm.

DISTRIBUTION: Oregon.

PLATE LXI, fig. 12.

22c. Euastrum crassum var. latum Prescott var. nov.

Varietas a varietate typica differens quod cellula latior, 1.5 plo longior quam lata, et lobi laterales superiores ad marginem trilobati, marginibus lateralibus mediis loborum basalium inflationem praebentibus, non retusis ut in specie typica; superficies semicellulae tumore admodum super isthmum praedita; membrana profunde scrobiculata; semicellulae a latere visae ad apicem perspicue angustae. Long. 175 µm. Lat. 98–99 µm. Ist. 28 µm. Crass. ad basim 51 µm.

ORIGO: E lacu Horseshoe nominato in loco Mt. Jefferson Park, Cascade Mts., Oregon dicto.

HOLOTYPUS: GWP Coll. OI-24.

ICONOTYPUS: Plate LXI, figs. 13, 13a.

A variety differing from the typical by its broader cell, 1.5 times longer than broad, and by the upper, lateral lobules being trilobed at the margin, the median, lateral margins of the basal lobes having a swelling rather than being retuse as in the typical. Face of the semicell with a single tumor immediately above the isthmus; wall deeply scrobiculate; lateral view of the semicell decidedly narrow toward the apex. L. 175 µm. W. 98–99 µm. Isth. 28 µm. Th. 51 µm at base.

DISTRIBUTION: Oregon.

PLATE LXI, figs. 13, 13a.

22d. Euastrum crassum var. michiganense Prescott, in Prescott & Magnotta 1935, Pap. Michigan Acad. Sci., Arts & Lettr. 20: 165. Pl. 26, Figs. 1, 2.

A variety relatively narrower and longer than the typical; semicells with a single, supraisthmial facial protuberance; lateral view of semicell uniformly tapering from a broadly rounded base, the poles broadly rounded. L. 156–159.5 µm. W. 87–89.7 µm. Isth. 23–25.5 µm.

DISTRIBUTION: Massachusetts, Michigan, Mississippi, Montana. Québec.

PLATE LXI, fig. 7.

22e. Euastrum crassum var. microcephalum Krieger 1937, Rabenhorst's Kryptogamen-Flora 13(1:3): 512. Pl. 65, Fig. 6.

Euastrum crassum (Bréb.) Kützing (p.p.), Smith 1924, Wisconsin Geol. & Nat. Hist. Surv. Bull. 57(II): 22. Pl. 56, Fig. 1.

A variety smaller than the typical, 2 times longer than broad, having a decidedly short apical lobe, the upper lateral lobules of the semicell high, the sinus between the lateral lobules and the polar lobe very short; face of semicell with a group of 4 mucilage pores in the midregion. L. 100–230 µm. W. 45–80 µm. Isth. 16–30 µm. Ap. 25–62.5 µm.

DISTRIBUTION: Louisiana, Wisconsin. Newfoundland, Québec. Europe, Arctic.

PLATE LXI, fig. 8.

22f. Euastrum crassum var. **scrobiculatum** Lundell 1871, Nova Acta Reg. Soc. Sci. Upsal., III, 8(2): 18. Pl. 2, Fig. 1.

Euastrum crassum var. *pulchrum* Cushman 1905, Rhodora 7: 251. Pl. 64, Fig. 1.
Euastrum crassum var. *taturnii* West & West 1905a, Monogr. II: 8. Pl. 55, Fig. 4.
Euastrum crassum var. *taturniii* f. *allorgei* Laporte 1931, Encyclop. Biol. 9: 82. Pl. 10, Figs. 107, 108.

A variety differing from the typical by the possession of a pattern of mucilage pores in the midregion of the semicell. L. 132-170 μm. W. 64.5-90 μm. Isth. 20-26 μm. Ap. 43-54 μm. Th. 51-78 μm.

DISTRIBUTION: Alaska, Florida, Louisiana, Massachusetts, Michigan, Mississippi, New Hampshire, New Jersey, North Carolina. Newfoundland, Nova Scotia, Québec. Great Britain, Europe, South America, Cuba, Faeroes.

PLATE LXII, fig. 15.

22g. Euastrum crassum var. **suboblongum** (Jackson) Prescott comb. nov.

Euastrum crassum var. *tumidum* f. *suboblongum* Jackson 1971, Dissert., p. 164. Pl. 32, Fig. 3.

A variety stouter and larger than the typical, cells trapezoidal in outline, about 1 2/3 times longer than broad; basal lobes of the semicell biundulate, the margins then sharply retuse to the much exserted lateral lobules, which are bilobed at their extremities; apex flat or broadly convex with a deep, narrow median incision; face of the semicell with a single facial, suprasthmial protrusion, and without a central mucilage pore as in var. *tumidum* f. *undulatum* Presc. L. 170-177 μm. W. 100-104 μm. Isth, 24-26 μm. Pol. lob. 52-54 μm. Th. 56.5-58 μm.

DISTRIBUTION: Montana.

PLATE LXI, figs. 11, 11a.

22h. Euastrum crassum var. **tumidum** Okada 1953, Mem. Fac. Fish. Kagoshima Univ. 3(1): 210. Pl. 2, Figs. 1, 2 f. **tumidum.**

A variety stouter and relatively broader than the typical, the basal lobes broad with a very slight concavity on the nearly parallel lateral margins, extending to very broad and inflated upper, lateral lobules, the sinus between the lobules and the small polar lobe narrow and closed; polar lobe cuneate, the apical margin convex with a short and narrow median incision; face of semicell with a supraisthmial protrusion, a mucilage pore in the center of the semicell (in ours); wall coarsely punctate; vertical view triangular, on 2 sides with a slight convex swelling, 2 angles broadly rounded and somewhat produced, the third broadly truncate. L. (154)160-205 μm. W. 81-106 μm. Isth. 25-30 μm. Th. 96-102 μm.

DISTRIBUTION: Montana. Asia (Japan).

PLATE LXII, figs. 13, 13a.

22i. Euastrum crassum var. **tumidum** f. **undulatum** Prescott nom. nov.

Euastrum crassum var. *tumidum* Okada (p.p.), in Jackson 1971, Dissert., p. 164. Pl. 32, Fig. 2.

A variety and form stouter than the typical, the form differing from the typical variety by having the lateral margins of the semicells 3- to 4-undulate, in having a mucilage pore in the midregion, and in having the polar lobe larger and more expanded. L. 154–162 μm. W. 96–102 μm. Isth. 26–27 μm.

The typical variety (Okada 1953, Mem. Fac. Fish. Kagoshima Univ. 3(1): 210. Pl. 2, Figs. 1, 2) is a short, stout plant, the lateral margins of the semicell evenly convex from base to apex of the lateral lobules, which are sharply rounded at their extremities.

DISTRIBUTION: Montana.

PLATE LXII, fig. 16.

23. Euastrum crispulum (Nordst.) West & West 1905a, Monogr. II: 72. Pl. 40, Figs. 15–18.

Euastrum sublobatum Brébisson var. *crispulum* Nordstedt 1873, Acta Univ. Lund 9: 10. Pl. 1, Fig. 9.

A small *Euastrum*, subquadrangular in outline, 1.5 times longer than broad; semicells truncate-pyramidal, the basal lobes extended and narrowly rounded, with a distinct undulation of the margins between the basal lobes ansd the polar lobe which is flat at the apex, with the angles slightly produced, and with the apical margin undulate, the median invagination being very slight and broad; face of semicell without protuberances; sinus narrow and closed; lateral view elliptic, the poles rounded, the lateral margins smooth, slightly inflated at the base of the semicell; vertical view elliptic, the poles rounded, lateral margins with a slight protrusion in the midregion: L. 21.6–28(35) μm. W. 15–20.5(26) μm. Isth. 3.7–7.5 μm. Ap. 9–11(20) μm. Th. 11–15 μm. Zygospore unknown.

West & West (*l.c.*) place this species in synonymy with *E. pyramidatum* West. We prefer to keep these two separate at present.

DISTRIBUTION: Alaska. Québec. Great Britain, Europe, Africa, South America.

PLATE LXXI, figs. 7–7b.

24. Euastrum croasdaleae Grönblad 1956a, Soc. Sci. Fennica Comm. Biol. 15(12): 25. Figs. 30–32.

A small *Euastrum*, subquadrangular in outline, 1.5 times longer than broad; semicells truncate-pyramidal, the basal lobes extended and broadly rounded, the margins retuse to the polar lobe, which is broad at the apex, with 2 lobules, the apical margin retuse between the lobules; face of semicell with a slight swelling in which there is a mucilage pore; sinus narrow and closed; lateral view broadly elliptic, the poles narrowly rounded and smooth, with a mucilage pore showing in the midregion on either side. L. 17–17.3 μm. W. 11–12.5 μm.

DISTRIBUTION: Massachusetts.

PLATE LXIX, figs. 13, 13a.

25. Euastrum cuneatum Jenner ex Ralfs 1848, Brit. Desm. p. 90, Pl. 32, Fig. 3 var. cuneatum.

Euastrum cuneatum var. *granulatum* Cushman 1905, Rhodora 7: 251.

Cells medium-sized, trapezoidal in outline, 2 times or more longer than broad; the semicells cuneate, the basal angles narrowly rounded, the lateral margins converging evenly to a narrowed apex which is flat, the angles broadly rounded, the median notch narrow and relatively shallow; face of semicell with 3 supraisthmial protuberances at the base, the central one often being slightly bilobed; sinus narrow and closed; wall distinctly punctate; lateral view narrowly oval in outline, the poles broadly rounded, the lateral margins symmetrically convex without a median protrusion; vertical view oval, the poles slightly produced and narrowly rounded, the lateral margins with 3 undulations. L. 87-157 μm. W. 40-69 μm. Isth. 13-22 μm. Ap. 28-34 μm. Th. 40-42 μm. Zygospore unknown.

This species shows decided variations in cell proportions and in the number and distribution of facial protuberances. Only 1 of these has been found in North America, but it may be that Cushman's var. *granulatum* should be recognized. We have not been able to see his plant and he published no figure. Krieger (1937, p. 480) recognizes it as synonymous with the typical. It should be looked for by field students, especially in New England.

Euastrum cuneatum is a fairly large, simple species with entire margins. It is not common in North America but occurs in widely distributed habitats. In Montana and Wisconsin it is restricted to acid situations and *Sphagnum* bogs.

Grönblad & Scott in Grönblad, Prowse & Scott (1958, p. 16) have transferred *E. cuneatum* var. *solum* Nordstedt to *E. solum* var. *angustum* var. nov. The intergradations of these species merit a critical analysis.

DISTRIBUTION: Alaska, Iowa, Montana, New Hampshire, Washington, Wisconsin. Newfoundland. Great Britain, Europe, Asia, Australia, New Zealand, Africa, Cuba, Arctic.

PLATE LVIII, figs. 1-1b.

25a. **Euastrum cuneatum** var. **subansatum** Kossinskaja 1936, Acta Inst. Bot. Akad. Nauk SSR, II, 3: 416. Pl. 1, Fig. 5.

A variety smaller than the typical; basal lobes broadly rounded, with a second marginal swelling immediately beyond, then retuse to a polar lobe, which is more abruptly truncate at the apex (resembling *E. ansatum* Ehrenb. ex Ralfs). L. 72-82 μm. W. 37-43.2 μm. Isth. 9.6-14.4 μm. Ap. 18-20 μm.

DISTRIBUTION: Michigan. Europe, Arctic.

PLATE LVIII, fig. 2.

26. **Euastrum cuspidatum** Wolle 1883, Bull. Torr. Bot. Club 10: 18. Pl. 27, Fig. 18.

A smaller than average *Euastrum*, subquadrate in outline, 1.25 times longer than broad, the semicells almost equally 4-lobed, the basal lobes horizontally extended with 3 prominent spines at the margin, the incision between the basal lobes and the polar lobe deep and broad, the polar lobe divided to form 2 spine-bearing lobules; face of semicell with a slight median protrusion, bearing tubercles; wall with a small tubercle within the margin of each lobe; sinus narrow and closed. L. 30-34 μm (with spines). W. 24-27 μm with spines. Isth. 4.5-5.6 μm. Th. 14.5 μm. Zygospore unknown.

DISTRIBUTION: Louisiana, New Jersey, North Carolina. Newfoundland.

PLATE LXXVII, fig. 4.

27. **Euastrum denticulatum** (Kirch.) Gay 1884, Bull. Soc. Bot. France 31: 335 var. **denticulatum** f. **denticulatum**.

Euastrum coralloides var. *subintegrum* West & West f. Scott & Prescott 1961, Hydrobiologia 17(1/2): 24. Pl. 13, Fig. 3, in Sieminska 1965, Trans. Amer. Microsc. Soc. 84(1): 121. Pl. 8, Figs. 23, 24.

Cells small, 1 1/5 times longer than broad, subrectangular in outline, the semicells truncate-pyramidal, the apical lobe short, with subparallel margins, the apex truncate and flat, the angles furnished with a stout spine, the median notch small, V-shaped, bordered outwardly by knoblike wall thickenings, with 2 intra-marginal granules on either side of the median notch; basal lobes broadly rounded, furnished with 3 marginal and 3 intramarginal granules; sinus deep and narrow; facial protrusions with 3 large granules and a pair of small granules below near the isthmus; lateral view oval in outline, the semicell truncate-pyramidal, the poles truncate with a short spine at each of the angles, the lateral margins with a subapical spine, the base broadly rounded or swollen, with several marginal granules; vertical view oval, the poles with 4 marginal granules and with a large tubercle on the lateral margins in the midregion. L. (13)19–35 μm. W. (11.5)14–25 μm. Isth. 3.8 μm. Ap. (10)12–16(21) μm. Th. (7.5)9.5–15 μm. Zygospore globose, the wall with slender spines, 32 μm in diameter including spines.

This species may be confused with some of the forms of *E. trigibberum* West & West, which has a very prominent median swelling that gives the vertical and lateral views a different appearance than *E. denticulatum*. In these views the lateral margins are decidedly concave between the poles and the median protuber-ance of the semicells. Krieger (1937, p. 583) places *Euastrum abruptum* f. *minus* West & West, in Taylor 1935, p. 199, Pl. 39, Fig. 1 with *Euastrum denticulatum*. West & West 1905a, Monogr. II: 56. Pl. 39, Figs. 1–4 illustrate a few forms of *E. denticulatum*, one of which (Fig. 4) is, in our opinion, not representative of this species, and is one which has some features in common with *Euastrum abruptum*.

DISTRIBUTION: Widely distributed throughout the United States. New-foundland. Great Britain, Europe, Asia, Africa, Australia, New Zealand, South America, Azores.

PLATE LXXV, figs. 1, 2(f.); PLATE LXXVIII, fig. 7.

27a. Euastrum denticulatum var. denticulatum f. glabrum Prescott f. nov.

Forma lobo polari minus prominenter producto quam in forma typica; lobi basales undulationem mediam parvam in marginibus habentes; membrana levis, dentibus ad angulos superiores multo reductis, inflationem levem mediam in superficie semicellulae praebens. Long. 33 μm. Lat. 24.5 μm. Ist. 8.5 μm.

ORIGO: E loco Muskelunge Lake, Wisconsin dicto.
HOLOTYPUS: GWP Coll. W-180.
ICONOTYPUS: Plate LXXV, fig. 7(f.).

A form with the polar lobe less prominently produced than in the typical; the basal lobes with a slight median undulation; wall smooth, the teeth at the upper angles much reduced, and with a smooth median swelling in the midregion of the semicell. L. 33 μm. W. 24.5 μm. Isth. 8.5 μm.

DISTRIBUTION: Wisconsin.
PLATE LXXV, fig. 7f.

27b. Euastrum denticulatum var. angusticeps Grönblad 1921, Acta Soc. Fauna Flora Fennica 49(7): 13. Pl. 3, Figs. 10, 11.

Euastrum tuddalense Ström var. *novae-terrae* Taylor 1935, Pap. Michigan Acad. Sci., Arts & Lettr. 20: 209. Pl. 41, Fig. 3.

A variety similar in size to the typical, but with a narrower polar lobe; basal angles broadly rounded with 3 sharp granules on the margin, the margins retuse to the polar lobe, which has subparallel margins, the apex flat and the angles extended into short spines, the median notch shallow and open; face of semicell with a median protrusion, bearing a circle of 4 granules; wall with 2 pairs of sharp granules within the margin of the 2 polar lobules and the basal lobes; sinus narrow and closed; vertical view oval, the poles rounded, bearing 4 marginal granules, with a distinct protrusion in the midregion on either side, bearing 2 nodules. L. 19–25.2 μm. W. 15–20.8 μm. Isth. 4.5–5.4 μm. Ap. 9.6–12 μm. Th. 12.2–12.6 μm.

The plant in Newfoundland (Taylor, *l.c.*) bears an intramarginal spine on each apical lobule rather than granules as in the typical, and there are minor variations in the apex of the basal lobes. We believe that *Euastrum abruptum* f. *minus* West & West (Taylor, *l.c.*, p. 199, Pl. 39, Fig. 1) belongs with *Euastrum denticulatum*. This is illustrated Pl. LXXVIII, fig. 7 and should be compared with Grönblad's variety. The margins of the apical notch are spine-bearing, unlike the typical, and marginal spines are somewhat more prominent; otherwise similar.

DISTRIBUTION: California, Massachusetts, Michigan. Newfoundland, Nova Scotia, Québec. Europe, Asia, South America.

PLATE LXXV, figs. 3, 4, 4a., 6, 6a, 10, 10a.

27c. Euastrum denticulatum var. bidentatum (Irénée-Marie) Prescott comb. nov.

Euastrum sinuosum var. *bidentatum* Irénée-Marie 1958, Nat. Canadien 85(5): 140. Fig. 22.

A variety differing from the typical by having a deeper and narrower apical incision and in having a marginal granule on either side of the notch; facial view with 3 protrusions at the base of the semicell, and 2 oblique pairs of smaller swellings on either side of the apex of the vertical incision. L. 33–35 μm. W. 22.5–23.8 μm. Isth. 7.2–7.5 μm.

This somewhat incompletely described form seems to us to be related to *E. sinuosum* (Irénée-Marie, *l.c.*). The deep apical incision is not typical of *E. denticulatum*, but the form of the cells, especially in the basal lobes, seems to warrant its inclusion with that taxon. Irénée-Marie mentions the lateral view but does not illustrate it.

DISTRIBUTION: Québec.

PLATE LXXVI, fig. 1.

27d. Euastrum denticulatum var. dangeardii Laporte 1931, Encyclop. Biol. 9, p. 88. Figs. 188–193.

A variety larger than the typical, subrectangular in outline, the semicells quadrate in outline, the basal lobes with bicrenate, nearly vertical margins, the margins retuse to a shoulderlike lateral lobule, and then retuse to the short polar lobe, which is similar to the typical; face of semicell with a median, granular protrusion, and with several granules across the face; lateral view elliptic, the poles rounded but furnished with a spinelike projection and with a prominent protrusion on either side in the midregion; vertical view broadly elliptic, the poles sharply rounded, the lateral margins slightly crenate or undulate, with a bilobed median protrusion on either side. L. 30.5–33 μm. W. 23.5–25.5 μm. Isth. 6–7.5 μm.

Although Krieger (1937, p. 583) places this variety in synonymy with the typical species, we agree with Jackson (1971, p. 166) that it should be maintained as a separate taxon. The differences are especially apparent in the side and end views of the cell. The upper, lateral lobules or undulations are different from the typical. Jackson (*l.c.*) shows a mucilage pore immediately below the apical notch. Different authors have interpreted the varieties of this species variously and there is some confusion. Until population studies can be made it seems best to recognize the varieties as originally described.

DISTRIBUTION: Montana. Europe.
PLATE LXXV, figs. 5–5b.

27e. Euastrum denticulatum var. **nordstedtianum** Irénée-Marie 1947, Nat. Canadien 74(3/4): 111. Pl. 1, Fig. 6.

Cells very small, the semicells trapezoidal, with a very deep constriction, the sinus closed, linear, but rounded at the apex; basal angles rectangular, with 6 granules, 4 borne within the margin and 2 on the margin; basal lobes with 2 lobules, separated from the polar lobe by a deep, rounded sinus; polar lobe with 3 granules in a triangle on either side of the apical incision, with 2 spines on each lateral margin of the lobe, and with 1 isolated granule at the middle of each half of the apical margin; center of each semicell with a swelling with a ring of 7 or 8 granules surrounding 2 elongate, longitudinally disposed granular tuberculations. L. 40–43 μm. W. 29.8–30.5 μm. Isth. 8.5–9.7 μm. Ap. 21.5–22.5 μm.

DISTRIBUTION: Québec.
PLATE LXXV, figs. 13, 13a.

27f. Euastrum denticulatum var. **quadrifarium** Krieger 1937, Rabenhorst's Kryptogamen-Flora 13(1:3): 585. Pl. 80, Figs. 20, 21 f. **quadrifarium.**

Euastrum denticulatum (Kirch.) Gay f. Krieger 1932, Arch. f. Hydrobiol. Suppl. 11: 212. Pl. 20, Fig. 14.

A variety differing from the typical by having the polar lobe more sharply produced, with a narrower sinus between it and the basal lobes; facial protrusion of the semicell with 4 granules arranged in a circle, with 2 mucilage pores on either side and above the midregion. L. 20–28(3) μm. W. 14–23 μm. Isth. 5–6 μm. Th. 11–14 μm.

DISTRIBUTION: Illinois, Michigan, Montana. Québec. East Indies, Africa, Arctic.
PLATE LXXV, fig. 9.

27g. Euastrum denticulatum var. **quadrifarium** f. **simplex** Prescott f. nov.

Forma figura magnitudineque formae typicae similis, granulis, autem, paucioribus et membrana sine poris mucosis; granula in angulis loborum vix manifesta; inflatio media semicellulae unicum tuberculum superficiale habens. Long. 20.5 μm. Lat. 15.5 μm. Ist. 5.5 μm.

ORIGO: E fossa iuxta iter prope locum Minocqua, Wisconsin dictum.
HOLOTYPUS: GWP Coll. Wis-160.
ICONOTYPUS: Plate LXXV, fig. 8.

A form similar in shape and size to the typical variety but with fewer granules and without mucilage pores in the wall; marginal granules on the angles of the lobes scarcely evident; with a single facial tubercle in the median swelling of the semicell. L. 20.5 μm. W. 15.5 μm. Isth. 5.5 μm.

DISTRIBUTION: Wisconsin.
PLATE LXXV, fig. 8.

27h. Euastrum denticulatum var. **rectangulare** West & West 1895, Trans. Linn. Soc. London, Bot., II, 5(2): 53. Pl. 6, Figs. 20, 21.

A quadrangular variety, differing from the typical by not having a distinct polar lobe, the basal angles narrowly rounded, the lateral margins undulate to the apex, which is as wide as the base of the semicell; with a pair of prominent granules on either side of the apical notch; facial protrusion blunt but with 2 spiny granules evident as seen in vertical view, which is broadly elliptic and has a spinelike extension of the poles. L. 19-22 μm. W. 15-17.5 μm. Isth. 5.5 μm. Ap. 12-17 μm. Th. 10-11 μm.

DISTRIBUTION: Florida, Madagascar, Panama Canal.

PLATE LXXV, figs. 12, 12a.

28. Euastrum didelta (Turp.) Ralfs 1848, Brit. Desm. p. 84. Pl. 14, Figs. 1a-1d var. **didelta** f. **didelta.**

Heterocarpella didelta Turpin 1828, p. 315. Pl. 13, Fig. 16.

Euastrum didelta f. *ansatiforme* Schmidle 1898, Bih. Kongl. Svenska Vet.-Akad. Handl. 24: 46. Pl. II, Fig. 28.

Euastrum didelta var. *ansatiforme* (Schmidle) Ducellier 1915, Bull. Soc. Bot. Genève 7: 84. Fig. 9.

Euastrum didelta var. *ansatiforme* Schmidle (p.p.), in Irénée-Marie 1938, p. 124. Pl. 16, Fig. 4.

A medium-sized *Euastrum*, elliptic in outline, 1.8-2 times longer than broad; the semicells pyramidal, the basal lobes broadly rounded, the margins converging to upper, lateral inflations, then retuse to a produced and much narrowed polar lobe, which has somewhat parallel margins, but slightly inflated at the apex, the apical margin flat and truncate, with a deep, narrow median notch; face of semicell with 3 basal protrusions, a median mucilage pore between 2 secondary protrusions; sinus narrow and closed; lateral view narrowly oval in outline, the poles broadly truncate, with a lateral, median swelling, narrowed to slightly protruding basal angles; vertical view oval, the poles broadly rounded, the lateral margins with undulations and a mucilage pore in the midregion of the margin. L. 109-195 μm. W. 52-91 μm. Isth. 11-25.7 μm. Ap. 20-35.5 μm. Th. 37-45 μm. Zygospore globose with numerous papillalike protrusions, 74-86 μm. in diameter (without papillae).

DISTRIBUTION: Widely and generally distributed in the United States. British Columbia, Québec. Great Britain, Europe, Asia, Australia, Africa, South America, Arctic, Azores.

PLATE LXIV, figs. 1, 1a.

28a. Euastrum didelta var. **didelta** f. **latior** Grönblad 1942, Acta Soc. Fennica, II, B, II:36. Pl. 1, Fig. 32.

A form similar to the typical in most respects, but with the semicells decidedly wider at the base and with the polar lobe pronounced. L. 160 μm. W. 99 μm.

DISTRIBUTION: Oregon. Finnish Lappland.

PLATE LXIV, figs. 3, 3a.

28b. Euastrum didelta var. **didelta** f. **scrobiculata** Nordstedt 1873, Acta Univ. Lund 9: 9.

Euastrum didelta var. *scrobiculata* (Nordst.) Irénée-Marie 1956, Rev. Algol. 2(1/2): 114. Fig. 13.

A form similar to the typical but with a coarsely scrobiculate wall.

This form is not recognized by Krieger (1937, p. 517) but placed in synonymy with the typical, which normally has a sparsely punctate wall. This does not seem to warrant the assignment to the variety taxonomic level by Irénée-Marie (*l.c.*).

DISTRIBUTION: Québec. Europe.

PLATE LXIV, fig. 8.

28c. **Euastrum didelta** var. **alaskense** Irénée-Marie & Hilliard 1963, Hydrobiologia 21(1/2): 105. Pl. 2, Fig. 11.

A variety similar to the typical in size and proportions, but with the basal angles quadrangular and slightly bilobed, the margins of the semicell deeply retuse and the upper, lateral inflations slightly more prominent so that the semicell is less pyramidal than in the typical. L. 142-145 μm. W. 68.5-70 μm. Isth. 19-19.5 μm.

DISTRIBUTION: Alaska.
PLATE LXIV, fig. 4.

28d. **Euastrum didelta** var. **crassum** (Presc. & Scott) Förster 1972, Nova Hedwigia 23: 535. Pl. 6, Fig. 7.

Euastrum obesum var. *crassum* Prescott & Scott 1942, Trans. Amer. Microsc. Soc. 61(1): 9. Pl. 1, Fig. 23.
Euastrum everettense var. *crassum* (Presc. & Scott) Jackson 1971, Dissert., p. 170. Pl. 34, Fig. 5.

A variety relatively more stout than the typical; semicells more pyramidal, with the polar lobe shorter and relatively broader; basal angles more narrowly rounded than the typical; face view of semicell with a submedian prominence, and without a mucilage pore; wall coarsely scrobiculate. L. 98-103 μm. W. 56-57 μm. Isth. 14.5-16 μm. Th. 39-41 μm.

Distribution; Alaska, Florida, Georgia, Mississippi, North Carolina, Virginia. Ontario.
PLATE LXIV, figs. 5-5b.

28e. **Euastrum didelta** var. **everettensiforme** (Wolle) Ducellier 1915, Bull. Soc. Bot. Genève 7: 87. Figs. 16, 17.

Euastrum everettense Wolle 1892, Desm. U.S., p. 112. Pl. 32, Figs. 5-7.

A variety differing from the typical by having the upper, lateral lobules higher on the margin of the semicells, the margins between the basal lobes and the upper, lateral protrusions concave, the polar lobe very short with slightly diverging margins; face of semicell without a mucilage pore. L. 75-95 μm. W. 40-59 μm. Isth. 13-19.5 μm.

Krieger (1937, p. 425) has placed this variety in synonymy with *Euastrum humerosum* Ralfs. The characteristics of that species with the bilobed basal lobes, the lateral protrusions at the base of the semicell, and the configuration of the upper, lateral lobules seem to us to exclude the type of cell found in *Euastrum didelta* and its varieties.

DISTRIBUTION: Massachusetts. Québec.
PLATE LXIV, fig. 7.

28f. **Euastrum didelta** var. **nasutum** (Scott & Grönblad) Förster 1972, Nova Hedwigia 23(2/3): 536. Pl. 6, Fig. 6.

Euastrum obesum var. *crassum* f. *nasutum* Scott & Grönbl. 1957, Acta Soc. Sci. Fennicae, n.s., B, II(8): 14. Pl. 3, Figs. 4, 5.

A variety differing from the typical by having narrowly rounded basal angles and with the face of the semicell showing a single inframedian swelling and a large tubercular swelling median and just above the isthmus, which shows prominently in the lateral view. L. 105 μm. W. 55 μm. Isth. 16 μm. Th. 36 μm.

DISTRIBUTION: Florida, Georgia.
PLATE LXIV, figs. 6-6b.

28g. **Euastrum didelta** var. **quadriceps** (Nordstedt) Krieger 1937, Rabenhorst's Kryptogamen-Flora 13(1:3): 520. Pl. 67, Figs. 8-10.

Euastrum quadriceps Nordstedt (1869)1870, Medd. Naturh. Foren. Kjöbenhavn 1869(14/15): 216. Pl. 2, Fig. 5.

A variety with pyramidal semicells, the basal angles narrowly rounded, the upper, lateral swellings less prominent than in the typical; face of semicell with 2 protrusions immediately below the polar lobe, and with 3 prominent tubercular swellings across the base of the semicell immediately beyond the isthmus; no median mucilage pore. L. 114-132 μm. W. 51-66 μm. Isth. 15-22 μm. Ap. 24-29 μm. Th. 30-42 μm.

DISTRIBUTION: Michigan. South America.

PLATE LXIV, fig. 2.

28h. **Euastrum didelta** var. **truncatum** Krieger 1937, Rabenhorst's Kryptogamen-Flora 13(1:3): 520. Pl. 68, Fig. 1.

A variety differing from the typical by having the basal angles biundulate, furnished with a tumorlike extension; facial tumors similar to the typical in number and arrangement, and with a median mucilage pore. L. 148 μm. W. 84 μm. Isth. 19 μm.

DISTRIBUTION: Montana. Finland, Portugal.

PLATE LXIV, fig. 9.

29. **Euastrum dissimile** (Nordst.) Schmidle 1898, Bih. Kongl. Svenska Vet.-Akad. Handl. 24, Afd. III(8): 45.

Euastrum binale var. *dissimile* Nordstedt 1875, Öfv. Kongl. Vet.-Akad. Förhandl. 1875(6): 31. Pl. 8, Fig. 21.

A small *Euastrum*, subquadrangular in outline; the semicells quadrangular, the basal angles narrowly rounded, the lateral margins diverging slightly to prominent upper, lateral swellings, and then sharply retuse to the very short polar lobe; apex relatively broad, retuse with a broad median notch, the angles furnished with short, horizontally directed spines; face of the semicell with 2 median protrusions and a slight swelling within the margin of the upper, lateral angles and 1 within the basal angles; lateral view narrowly oval in outline, the poles broadly truncate and slightly produced, with an apical, median tubercle; the semicell broadly rounded and extended in the midregion; vertical view quadrangular, the poles quadrately truncate with a spinelike protrusion at each angle, the lateral margins with a broad, flat protrusion in the midregion. L. 26-29 μm. W. 18-22 μm. Isth. 7-8 μm. Th. 13-16 μm.

DISTRIBUTION: Alaska. Europe, Arctic.

PLATE LXX, figs. 23, 23a.

30. **Euastrum divaricatum** Lundell 1871, Nova Acta Reg. Soc. Sci. Upsal., III, 8(2): 21. Pl. 2, Fig. 5 var. **divaricatum**.

Euastrum candianum Delponte 1876, Mem. Real. Accad. Sc. Torino 28: 107. Pl. 6, Figs. 11, 12.

A small *Euastrum*, broadly oval in outline, 1.25 times longer than broad; cells short-pyramidal, the basal angles broadly rounded and furnished with a prominent, horizontally directed spine, the lateral margins converging with a slight undulation immediately beyond the basal angles, the polar lobe short, flat at the apex, with a short spine at the angles, the median notch relatively deep and somewhat opened, with an elevation of the apical margin on either side of the opening; face of semicell with a granule-bearing, median protrusion, granules within the margin of the apex of the median notch, and with intramarginal granules at the apical margin; lateral view elliptic, the poles sharp, furnished with a spine, and with spiny granules immediately below the apex, the semicells much swollen and granular on the margin; vertical view elliptic, the poles sharply rounded and furnished with a spine, the lateral margins convex, with spiny granules on either side of the apex, and with a granular swelling in the midregion. L. 35–46.8 μm. W. 27–37 μm. Isth. 5–10 μm. Ap. 14–20 μm. Th. 14–19 μm. Zygospore unknown.

DISTRIBUTION: Georgia, Louisiana, Massachusetts, Michigan, New Hampshire, New Jersey, North Carolina, Pennsylvania, Wisconsin. Newfoundland. Great Britain, Europe, Asia, Africa, Costa Rica, South America, Arctic.

PLATE LXXVII, fig. 1.

30a. Euastrum divaricatum var. elevatum (Irénée-Marie) Prescott nom. nov.

Euastrum divaricatum var. *inerme* Irénée-Marie 1958, Nat. Canadien 85(5): 123. Fig. 13 (Not *E. divaricatum* f. *inermis* Eichler & Gutwinski 1894, Rozpr. Wydz. matem.-przy. Akad. Umiej. w Krakowie 28: 173. Pl. 5, Fig. 40.)

A variety relatively broader at the base of the semicell than the typical; the basal lobes rectangular, the polar lobe elevated on either side of the deep median notch, the spines at the angles of the polar lobe more prominent; vertical view oval rather than elliptic. L. 48.3–48.5 μm. W. 35.4–35.8 μm. Isth. 9.8–10 μm. Ap. 20.5 μm.

This variety is somewhat similar to f. *inermis* Eichl. & Gutw. but in the latter the basal lobes are not angular and subparallel margined, but are spiniferous.

DISTRIBUTION: Québec.

PLATE LXXVII, figs. 2, 2a.

30b. Euastrum divaricatum var. spinosum (Irénée-Marie) Prescott nom. nov.

Euastrum candianum var. *munitum* f. *candiana* Irénée-Marie 1958, Nat. Canadien 85(5): 116. Fig. 7.

A variety smaller than the typical, 1.5 times longer than broad (including spines); differing from the typical by having the margins of the basal lobes furnished with 5 sharp spines, with 3, upwardly directed spines on either side of the apical notch, which is deeper and narrower than in the typical; face of semicell with a smooth, median protrusion; lateral view narrowly oval, the poles broadly rounded and furnished with a toothlike spine, the semicells slightly swollen and smooth in the midregion of the lateral margins. L. 29 μm. W. 20 μm. Isth. 4.8 μm.

Irénée-Marie (*l.c.*) described a plant from Québec which resembles Turner's var. *munitum* of *Euastrum candianum*. The latter has been placed in synonymy with *E. divaricatum* Lund., but Krieger (1937, p. 654) does not accept Turner's variety of *E. candianum*.

DISTRIBUTION: Québec.

PLATE LXXVII, figs. 3. 3a.

31. Euastrum doliforme West & West 1898, Linn. Soc. Jour. Bot. 33: 289. Pl. 16, Fig. 12 var. doliforme.

A small *Euastrum*, truncate-oval in outline, 1.5 times longer than broad; semicells truncate-pyramidal; the basal angles subquadrate and bilobed on the margin, lateral margins of semicell converging abruptly and slightly retuse to an upper, lateral undulation, concave to the polar lobe, which is relatively broad and flat, with a broad V-shaped, retuse invagination, the angles extended into a toothlike spine; face of semicell with a prominent supraisthmial, median tubercle in the midregion; cell wall punctate; lateral view narrowly oval, the poles broadly rounded and slightly produced above a median, marginal swelling; vertical view oval, the poles broadly rounded, lateral margins with a prominent protrusion in the midregion. L. 41-44 μm. W. 27-28 μm. Isth. 7.5-10 μm. Th. 17 μm. Zygospore unknown.

DISTRIBUTION: Florida, Newfoundland.

PLATE LXXI, figs. 14, 14a.

31a. **Euastrum doliforme** var. **groenbladii** Croasdale 1956, Trans. Amer. Microsc. Soc. 75(1): 4. Pl. 1, Fig. 6.

A small variety, somewhat more stout than the typical, the basal angles narrowly rounded and not biundulate as in the typical, and with a sharp tubercle; the spines at the apical angles of the polar lobe reduced to tubercles, or lacking; face of semicell with 2 or 3 irregularly placed mucilage pores in the midregion. L. 32-34 μm. W. 24-26 μm. Isth. 8-9 μm. Ap. 15-17 μm. Th. 16 μm.

DISTRIBUTION: Alaska. Labrador, Europe.

PLATE LXXII, figs. 10, 10a.

32. **Euastrum dubium** Nägeli 1849, Gattungen Einz. Algen, p. 122. Pl. 70, Fig. 2 var. **dubium** f. **dubium**.

Euastrum dubium var. *anglicum* (Turner) West & West 1905a, Monogr. II: 44. Pl. 38, Fig. 9.

A small *Euastrum*, subrectangular in outline, 1.5 times longer than broad; the semicells subquadrate, the basal angles broadly rounded, and with an upper, lateral rounded lobule equal to the basal lobes, the margins sharply retuse to the short polar lobe, which is broad and flat but with a broad, V-shaped invagination in the midregion, the angles of the polar lobe sharply pointed but scarcely spiniferous; face of semicell with a low median protrusion bearing 1 to 3 granules, with small granules intramarginal at all lobes; lateral view narrowly oval, the poles narrowly rounded and slightly produced above 2 subapical granulations at the margin, no lateral median swelling of the semicells in evidence; vertical view subrectangular, the angles sharply rounded and with a slight protrusion in the midregion of the apical margin, lateral margins with a slight convex protrusion in the midregion. L. 25-39 μm. W. 17-26 μm. Isth. 4-7.5 μm. Ap. 10-16.8 μm. Zygospore unknown.

DISTRIBUTION: Alaska, Iowa, Kansas, Kentucky, Massachusetts, Michigan, Mississippi, Nebraska, New York, Ohio, Pennsylvania, South Carolina, Virginia, Washington. British Columbia, Ontario. Great Britain, Europe, Kerguelen, Asia, Africa, South America, Arctic, Azores.

PLATE LXXIII, fig. 12; PLATE LXXIV, figs. 3-3b.

32a. **Euastrum dubium** var. **cambrense** (Turner) West & West 1905a, Monogr. II: 45. Pl. 38, Fig. 10.

A variety in which the upper, lateral lobules are more prominent than in the typical and bear a sharp, upwardly directed spine; the facial protuberance with a circle of 6 granules. L. 32 μm. W. 20 μm. Isth. 5 μm.

This variety might be regarded as having no more difference from the typical than to designate it only as a form. Although it does present a different aspect from the typical, Krieger (1937, p. 655) designates it as a questionable taxon.

DISTRIBUTION: Mississippi. Wales.
PLATE LXXIV, fig. 4.

32b. Euastrum dubium var. **canadensis** Irénée-Marie 1958b, Rev. Algol. 4(2): 109.
Fig. 10.

A variety differing from the typical by its larger size, by the deeper incision
in the polar lobe, and by the pronounced beak on the polar lobules; margins of
the semicell above the basal lobes more concave.

This form bears a strong resemblance in facial view to Irénée-Marie's *E.
dubium* f. *mauritiana*, which herein has been transferred to *E. lapponicum* var.
mauritiana Prescott comb. nov.
DISTRIBUTION: Québec.
PLATE LXXIII, figs. 16–16b.

32c. Euastrum dubium var. **latum** Krieger 1937, Rabenhorst's Kryptogamen-Flora
13(1:3): 572. Pl. 79, Figs. 7–9.

A variety with stouter cells, 1.25 or less times longer than broad; the
semicells short, truncate-pyramidal; spines at the angles of the polar lobe more
prominent than in the typical; vertical view oval rather than subrectangular. L.
29–33 μm. W. 23–25 μm. Isth. 6–7.5 μm. Th. 12–13 μm.
DISTRIBUTION: Québec. Europe.
PLATE LXXIV, figs. 8, 8a.

32d. Euastrum dubium var. **maius** Croasdale 1965, Trans. Amer. Microsc. Soc.
84(3): 312. Pl. 2, Figs. 6, 7.

A variety somewhat larger than the typical, and relatively broader; the apical
notch a shallow, broad, invagination scarcely evident; lateral view narrowly oval in
outline, the poles broadly rounded and simple, the lateral margins with a promi-
nent protrusion in the midregion similar to var. *pseudocambrense* Grönblad (1920,
Pl. 6, Figs. 32, 33). L. 40–44 μm. W. 25–28 μm. Isth. 6–8 μm. Ap. 16–18 μm.
Th. 9 μm.
DISTRIBUTION: Devon Island.
PLATE LXXIV, fig. 9.

32e. Euastrum dubium var. **ornatum** Wołoszynska 1919, Rozpr. Wydz. Mat.-
Przyr. Akad. Umiej. w Krakow, Ser. B, 57: 49. Pl. 3, Fig. 32.

A variety in which the median protrusion of the semicells bears 3 prominent
granulations surrounding a mucilage pore; granules on the face of the semicell
larger and more numerous; basal angles bilobed, the upper, lateral lobules more
pronounced than in the typical and somewhat angular; apical margin of the polar
lobe with a deeper and narrower incision. L. 28–31 μm. W. 20–22 μm. Isth.
3.5–7.5 μm. Ap. 12–14.5 μm.
DISTRIBUTION: Alaska, Oklahoma. Poland.
PLATE LXXIV, fig. 7.

32f. Euastrum dubim var. **poriferum** Prescott & Scott 1945, Amer. Mid. Nat.
34(1): 237. Pl. 2, Fig. 6.

A variety in which the semicells are less pyramidal than the typical, nearly
subrectangular in outline with the upper, lateral lobules extended as much as the
rounded basal lobes; apical margin with a broad invagination, giving the polar lobe

the appearance of being bilobed, the angles of the polar lobe and the margins of the basal lobes furnished with a small tubercle; median protrusion of the semicell simple but with a mucilage pore, face of the semicell with an intramarginal granule at each of the 6 lobes; wall distinctly punctate; lateral view broadly oval, the poles truncate but triundulate, the margins with prominent swellings, showing a mucilage pore. L. 34–38 μm. W. 25–26 μm. Isth. 8–9 μm. Th. 17.5–18 μm.

DISTRIBUTION: Alaska, Louisiana, Mississippi.

PLATE LXXIII, figs. 14, 15, 15a, 17–17b, 18–18b; Pl LXXIV, figs. 1, 1b.

32g. **Euastrum dubium** var. **scrobiculatum** (Lütk.) Krieger 1937, Rabenhorst's Kryptogamen-Flora 13(1:3): 573. Pl. 79, Fig. 13.

Euastrum dubium f. *scrobiculata* Luetkemueller 1910, Verh. k. k. Zool.-Bot. Ges. in Wien 60: 482. Pl. 2, Fig. 1.

A variety similar in outline to var. *ornatum* and somewhat similar to it in wall decorations; median protrusion of the semicell with 2 mucilage pores and a granule, the median swelling bounded on either side by a semicircle of 3 granules. L. 29–30 μm. W. 20–21 μm. Isth. 6 μm. Th. 13 μm.

DISTRIBUTION: Montana. Bohemia.

PLATE LXXIV, figs. 6, 6a.

32h. **Euastrum dubium** var. **snowdoniense** (Turner) West & West 1905a, Monogr. II: 45. Pl. 38, Fig. 11.

Euastrum snowdoniense Turner 1895, The Naturalist 18: 343. Fig. 2.

A variety rectangular-oval in outline, the semicells quadrangular, the basal lobes broadly rounded, with the upper, lateral lobules extended equally with the basal lobes; the polar lobe short and slightly inflated, the median notch a broad V, giving the apex a bilobed appearance; face of semicell with a triangle of small protuberances in the midregion. L. 31 μm. W. 20 μm. Isth. 6.5 μm.

Krieger (1937, p. 571) includes this variety under the typical, but it has features which are as different, or more so, from the typical as several other recognized varieties.

DISTRIBUTION: Michigan. Great Britain.

PLATE LXXIV, fig. 5.

32i. **Euastrum dubium** var. **spiniferum** Prescott var. nov.

Varietas relative longior quam varietas typica, 1 2/3 plo longior quam lata; cellulae late ovatae; semicellulae semi-pyramidales, lobis basalibus late rotundatis et pari dentium acutorum praeditis; protrusio lateralis superior parva, etiam spina intramarginali praedita; lobus polaris brevis, angulis spinas manifestas sursum directas praebentibus, margine apicali incisionem acutam forma litterae "V" habente, protuberatio superficialis granula magna in triangulo ordinata habens; membrana spinam dentiformem ad basim utriusque spinae polaris ferente. Long. 42.5 μm. Lat. 27.8 μm. Ist. 8 μm.

ORIGO: E palude Little Dead Horse dicto, Minnesota.

HOLOTYPUS: GWP Coll. Minn-XXXI (84).

ICONOTYPUS: Plate LXXIV, fig. 10.

A variety relatively longer than the typical, 1 2/3 times longer than broad; cells broadly oval in outline; semicells semipyramidal, the basal lobes broadly rounded and furnished with a pair of sharp teeth, the upper, lateral protrusions slight and also furnished with an intramarginal spine; polar lobe short, the angles with prominent, upwardly directed spines, the apical margin with a sharp, V-shaped incision; facial protuberance with a triangle of large granules, wall with a toothlike spine at the base of each polar spine. L. 42.5 μm. W. 27.8 μm. Isth. 8 μm.

DISTRIBUTION: Minnesota.

PLATE LXXIV, fig. 10.

32j. Euastrum dubium var. **tritum** West & West 1907, Ann. Roy. Bot. Gard. Calcutta 6(2): 198. Pl. 14, Fig. 6.

A variety smaller than the typical, the cells broadly oval in outline; basal lobes narrowly rounded, the margins retuse to slightly produced upper, lateral swellings, converging to form a short, relatively narrow polar lobe, which is bilobed at the apex, with a relatively narrow and deep median notch; face of semicell smooth, without a median protrusion; lateral view narrowly oval, the poles rounded, the lateral margins smooth. L. 22–26 μm. W. 15.5 μm. Isth. 3.8–4.4 μm. Ap. 8–9 μm. Th. 8.5 μm.

This variety is an example of the great variation in the interpretations of *E. dubium* Näg. by different workers. *E. dubium* var. *tritum* and several other varieties very well deserve to be transferred to some other position. We are leaving the epithets, as they are pending a complete and critical review of the species.

DISTRIBUTION: Massachusetts, Oregon. Burma.

PLATE LXXIII, figs. 19, 19a; PLATE LXXIV, figs. 11, 11a.

33. Euastrum elegans (Bréb.) Ralfs 1848, Brit. Desm., p. 89. Pl. XIV, Figs. 7a–7c var. **elegans** f. **elegans**.

Cosmarium elegans Brébisson, in Meneghini 1840, Synop. Desmid., p. 222.
Euastrum elegans var. *poljanae* Pevalek 1925, Nuova Notarisia 36: 289. Fig. 1.
Euastrum elegans var. *bidentatum* (Nägeli) Jacobsen 1875, Bot. Tidsskr. 8: 191.

A small *Euastrum*, broadly oval in outline, 1.5 times longer than broad; basal lobes subrectangular and bilobed at the margins, which are deeply retuse to a subapical spine, the apical margin of the polar lobe highly elevated to a deep median incision; sinus narrow and closed; lateral view elliptic, the poles rounded and slightly produced above lateral, tubercular swellings, the lateral margins inflated in the midregion, with granular tubercles; vertical view elliptic, the poles narrowly rounded, the lateral margins slightly swollen in the midregion, bearing 2 tubercles. L. 26.6–39 μm. W. 14–30 μm. Isth. 4–8 μm. Ap. 12–18 μm. Th. 10–14.5 μm. Zygospore globular, the wall bearing numerous, sharp spines, 20–29 μm in diameter without spines; spines 7.5–9.5 μm. long.

DISTRIBUTION: Widely distributed throughout the United States and Canada. Cosmopolitan.

PLATE LXXVI, figs. 3–3b.

33a. Euastrum elegans var. **elegans** f. **prescottii** Förster 1972, Nova Hedwigia 25(2/3): 536. Pl. 8, Fig. 3.

Euastrum elegans var. *compactum* f. Prescott & Scott 1945, Amer. Mid. Nat. 34: 238. Pl. 2, Fig. 9.

A variety smaller than the typical and with reduced spinulation and granulation; basal lobes somewhat angular, with tubercles, the lateral margins retuse to the spines at the angles of the polar lobe, without upper, lateral spines on the margin as in the typical (and other varieties); facial tumor with 2 vertically arranged granular tubercles; lateral view elliptic in outline, the poles rounded and smooth, with a bilobed tumor near the base of the semicell. L. 30–32.5 μm. W. 21–23 μm. Isth. 5–6.5 μm. Th. 13.5 μm.

DISTRIBUTION: Florida.

PLATE LXXVI, figs. 4, 4a.

33b. Euastrum elegans var. **compactum** (Wolle) Krieger 1937, Rabenhorst's Kryptogamen-Flora 13(1:3): 593. Pl. 82, Figs. 1, 2.

Euastrum compactum Wolle 1884, Bull. Torr. Bot. Club 11: 15; Wolle 1884a, Desm. U.S., p. 107. Pl. 27, Figs. 28, 29.

A subrectangular variety in which the basal lobes are bilobed at the margin, the upper lobules extending further than the basal angles, the margins retuse to form a scarcely discernible, broad, polar lobe, the angles sharply rounded and furnished with a mucro, the apical margin elevated and rounded in the midregion, with a short, narrow polar incision; face of semicell with a low protrusion bearing a diamond-shaped pattern of 4 granules; wall with a granule within the margin of the upper lobule of the basal lobes. L. 28–50 μm. W. 20–33 μm. Isth. 5.8–9 μm. Th. 10–16 μm.

DISTRIBUTION: Colorado, Massachusetts, Montana, Pennsylvania. Québec. Australia, Africa.

PLATE LXXVI, figs. 5, 6, 6a (Zygospore).

33c. Euastrum elegans var. **granulosum** Prescott var. nov.

Varietas 1.5 plo longior quam lata; lobi basales late rotundati, marginibus retusis ad lobum polarem ubi parva protusio cum tuberculo intramarginali videtur; deinde margo apicalis ad 2 lobulos formandos elevatus, incisione media profunda angustaque; inflatio superficialis semicellulae 5 ordinibus verticalibus granulorum (2 vel 3 in omni ordine) praedita; membrana granulum manifestum intra utrumque lobulum polarem, atque granulum parvum intra marginem loborum basalium praebens. Long. 36.5 μm. Lat: 22.8 μm. Ist. 6.8 μm.

ORIGO: E stagno superiaciente super cinerem volvocanicum in loco Mt. Jefferson Park, Cascade Mts., Oregon dicto.

HOLOTYPUS: GWP Coll. OI-66(7).

ICONOTYPUS: Plate LXXVI, fig. 7.

A variety 1.5 times longer than broad; basal angles broadly rounded, the margins retuse to the polar lobe, where there is a slight protrusion, with an intramarginal tubercle, and then elevated to form 2 lobes, the apical incision deep and narrow; facial swelling of the semicell with 5 vertical rows of granules (2 or 3 in each row); wall with a prominent granule in each polar lobule, and a small granule within the margin of the basal lobes. L. 36.5 μm. W. 22.8 μm. Isth. 6.8 μm.

This variety might well be assigned to a place in *Euastrum fissum* West & West. It should be compared with var. *brasiliense* Krieger of that species. The apical lobules are remindful also of *E. fissum* var. *americanum* Cushman.

DISTRIBUTION: Oregon.

PLATE LXXVI, fig. 7.

33d. Euastrum elegans var. **novae-semliae** Wille 1879, Öfv. Kongl. Vet.-Akad. Förhandl. 1879(5): 32. Pl. 12, Fig. 8.

A variety slightly larger than the typical; sinus weakly open outwardly where the basal angles have a short spine; margins of the semicells converging to an upper, lateral spine, then retuse to the subapical spines; apical margin elevated to a narrow, apical notch; face of semicell with a slight protrusion bearing 3 prominent granules; wall with granules lateral to the apical notch, a pair of intramarginal granules at the upper, lateral spines and one within the basal angles. L. 28–53 μm. W. 18–34 μm. Isth. 3.8–9 μm. Ap. 20–22.5 μm. Th. 14–24 μm.

DISTRIBUTION: Alaska, Florida, Georgia, Utah. British Columbia, Ontario. Great Britain, Europe, Arctic.

PLATE LXXVI, fig. 8.

33e. **Euastrum elegans** var. **obtusum** Grönblad 1956. Soc. Sci. Fennica Comm. Biol. XV(12): 25. Figs. 33–35.

A variety smaller than the typical; the basal lobes of the semicells broadly rounded and somewhat angular, bearing a tubercle at the apex, the lateral margins smooth to broadly rounded lobules of the polar lobe, which is destitute of spines; face of semicell with a low protrusion that bears 3 prominent granules; lateral view narrowly oval, the poles broadly rounded, the lateral margins evenly inflated, bearing 2 granules in the midregion. L. 20 μm. W. 15 μm.

DISTRIBUTION: Massachusetts.

PLATE LXXVI, figs. 11, 11a.

33f. **Euastrum elegans** var. **ornatum** West 1892a, Jour. Roy. Microsc. Soc. 1892: 723. Pl. 9, Fig. 9.

A variety larger than the typical; basal angles somewhat sharply rounded and furnished with an aculeum, slightly biundulate to a prominent upper, lateral spine, the margins then retuse to the spines at the angles of the polar lobe, the apical margin elevated as in the typical; face of semicell with a low protuberance bearing 3 vertical pairs of granules; wall of cell with granulation similar to the typical; vertical view broadly oval, the poles broadly rounded, with a low median swelling bearing 3 granulations. L. 36–47 μm. W. 22–30 μm. Isth. 4.5–10 μm.

DISTRIBUTION: Louisiana, Massachusetts, New Hampshire, North Carolina. Newfoundland, Ontario. Great Britain, Europe.

PLATE LXXVI, fig. 10.

33g. **Euastrum elegans** var. **pseudelegans** (Turner) West & West 1905a, Monogr. II: 49. Pl. 38, Figs. 22, 23 f. **pseudelegans.**

Euastrum pseudelegans Turner 1885, Jour. Roy. Microsc. Soc., II, 5(6): 935. Pl. 15, Fig. 8.
Euastrum incurvatum Turner 1892, Kongl. Svenska Vet.-Akad. Handl. 25(5): 85. Pl. 11, Fig. 1.

A variety with urn-shaped semicells; basal angles broadly rounded, retuse and smooth margins to the spines at the angles of the polar lobe; wall granulations and median tumor as in the typical. L. 27–42 μm. W. 15–28 μm. Isth. 6–8 μm. Th. 10–13 μm.

DISTRIBUTION: Florida. British Columbia. Great Britain, Europe, Asia, Australia.

PLATE LXXVI, figs. 9, 13–13b.

33h. **Euastrum elegans** var. **pseudelegans** f. **quebecense** (Irénée-Marie) Prescott stat. nov.

Euastrum elegans var. *quebecense* Irénée-Marie 1956, Rev. Algol. 2(1/2): 115. Fig. 7.

A form similar to the typical variety, but with the apical margin less elevated, the apical lobules low; facial protrusion of the semicell with a transverse row of 3 granules. L. 40.2–45.1 μm. W. 26.6–28 μm. Isth. 6.4–6.5 μm.

DISTRIBUTION: Québec.
PLATE LXXVI, fig. 12.

33i. **Euastrum elegans** var. **spinosum** Ralfs 1848, Brit. Desm., p. 89. Pl. 14, Fig. 7f.

Euastrum spinosum Ralfs 1844, Ann. & Mag. Nat. Hist. 14: 193. Pl. 7, Fig. 6.

A variety with basal lobes subangular and bilobed, the lobules bearing a spiny tuberculation at the margins, retuse to the upper, lateral spines, which are longer and sharper than in the typical, retuse to the spines at the angles of the polar lobe; apical margin elevated as in the typical but bearing a sharp tuberculation on either side of the median notch; facial protrusion of the semicell with 4 granules; wall with 2 prominent tubercles on either side of the apex of the median notch, and with a pair of supraisthmial granules on the basal lobes; lateral view broadly elliptic, the poles truncate, the lateral margins biundulate to a basal protrusion which bears 2 swellings; vertical view elliptic, the poles narrow and bearing a short, sharp spine, lateral margins with undulations, the midregion with a bilobed tumor. L. 36–38.4 μm. W. 22–23 μm. Isth. 6 6.5 μm. Th. 13.5 μm.
DISTRIBUTION: New York. Great Britain, Europe.
PLATE LXXVI, figs. 18–18b.

34. **Euastrum elobatum** (Lund.) Roy & Bisset 1893, Ann. Scottish Nat. Hist. 1893(7): 176 var. **elobatum**.

Euastrum binale var. *subelobatum* West 1892, Linn. Soc. Jour. Bot. 29: 140. Pl. 20, Fig. 15.
Euastrum binale var. *elobatum* Lundell 1871, Nova Acta Reg. Soc. Sci. Upsal., III, 8(2): 22. Pl. 2, Fig. 7.

A small *Euastrum*, truncate-oval in outline; 1 1/3 times longer than broad; semicells truncate-pyramidal, the basal lobes narrowly rounded at the angles, sometimes slightly bilobed, the lateral margins converging, and with a low undulation, to the relatively broad and only slightly produced polar lobe, the angles sharply rounded, the apical margin nearly flat, with a slight invagination in the midregion; face of semicell with a simple median swelling; wall smooth; lateral view broadly elliptic, the poles narrow and rounded, slightly produced in the form of a knob, the lateral margins convex with a prominent median swelling bearing a simple knob; vertical view elliptic, the poles narrowly rounded, lateral margins with a slight undulation between the poles and the median tubercular protrusion. L. 21–28 μm. W. 14–20.5 μm. Isth. 4–5 μm. Ap. 10–14.5 μm. Th. 11–12.5 μm. Zygospore unknown.
DISTRIBUTION: Maine, North Carolina, Oregon. British Columbia, Québec. Great Britain, Europe, Asia, Africa, South America, Arctic.
PLATE LXXI, figs. 19–19b.

34a. **Euastrum elobatum** var. **simplex** Krieger 1932, Arch. f. Hydrobiol. Suppl. 11: 211. Pl. 20, Fig. 7.

A variety slightly smaller than the typical; angles of the basal lobes broadly rounded, the angles of the polar lobe smooth and convex; face of the semicell without a median swelling; lateral view narrowly oval, the poles rounded, the lateral margins smooth and convex; vertical view elliptic, the poles rounded and slightly produced, lateral margins smooth, with a slight median swelling. L. 20 μm. W. 14 μm. Isth. 5 μm. Th. 8 μm.
DISTRIBUTION: Wisconsin. Java.
PLATE LXXI, figs. 10–10b, 12.

35. **Euastrum erosum** Lundell 1871, Nova Acta Reg. Soc. Sci. Upsal., III, 8(2): 2. Pl. 2, Fig. 6 var. **erosum.**

A small *Euastrum* subrectangular in outline, up to 2 times longer than broad; semicells subquadrate, the basal lobes broadly rounded at the angles, the lateral margins retuse to an upper, lateral lobule, which extends as far as the basal lobes, then retuse to the relatively broad and truncate polar lobe, the angles rounded, the apical margin concave; face of the semicell without a median swelling; the wall smooth; lateral view narrowly oval, the poles truncate but 3-undulate, the margins smoothly convex. L. 32–39 μm. W. 20–24 μm. Isth. 7.5–8.5 μm. Ap. 14 μm. Th. 11–16 μm. Zygospore unknown.

DISTRIBUTION: United States. British Columbia. Great Britain, Asia, Europe, Africa, Arctic.

PLATE LXXIV, figs. 20, 20b.

35a. **Euastrum erosum** var. **granulosum** Cedercreutz 1932, Mem. Soc. Fauna Flora Fennica 7(3): 242. Figs. 5, 6.

A variety similar in size to the typical but with the lateral margins above the basal lobes more deeply retuse and the upper, lateral lobules more sharply rounded; apical notch more sharply V-shaped; wall with granules within the margin of the basal angles and the upper, lateral lobules, and with a prominent granule on either side of the median notch; lateral view subrectangular, the poles truncate and 3-undulate; vertical view rectangular, the poles broadly rounded and with 5 prominent granular tubercles, the lateral margins subparallel. L. 34–46 μm. W. 21–25 μm. Isth. 6–8 μm. Ap. 11–16 μm. Th. 16 μm.

DISTRIBUTION: Alaska. Europe, Asia, Arctic.

PLATE LXXIV, figs. 21–22a.

36. **Euastrum everettense** Wolle 1884a, Desm. U.S., p. 102. Pl. 28, Figs. 5, 6, 7.

A large *Euastrum*, broadly truncate-oval in outline, 2 times longer than broad; basal lobes broadly rounded at the angles, forming a widely open sinus, the lateral margins retuse to an upper, shoulderlike protrusion, then sharply and deeply retuse to the polar lobe, which is very short and inflated, the angles narrowly rounded, the apical margin flat, with a relatively deep, narrow V-shaped notch; face of semicell with 2 horizontally placed protrusions in the basal lobes, 2 smaller swellings in the upper, lateral inflations, and a minute one on either side of the apical notch; wall scrobiculate; sinus narrow but open within, enlarging outwardly; lateral view elliptic in outline, the poles tapered and narrowly rounded, smooth, the lateral margins with a median, sharply rounded protrusion; vertical view elliptic, the poles rounded and smooth, the lateral margins inflated in the midregion and showing 2 prominent protrusions. L. 87–106 μm. W. 48.5–66 μm. Isth. 15 μm. Ap. 25–27 μm.

Krieger (1937, p. 575) assigns Wolle's species to synonymy with *Euastrum ampullaceum* Ralfs. Comparison should be made with that species, but the shape of the semicell, while somewhat smaller, does not conform to the former; the basal lobes are of a different shape and it does not have the characteristic facial structures. The 2 plants are quite different in vertical and lateral views. Cf. *Euastrum didelta* var. *everettensiforme* (Wolle) Ducellier, which Krieger (1937, p.524) places under *E. humerosum.*

DISTRIBUTION: Massachusetts, New Hampshire, Utah, Vermont. Ontario. PLATE LXI, fig. 16.

37. **Euastrum evolutum** (Nordstedt) West & West 1896a, Trans. Linn. Soc. London, Bot., II, 5: 245. Pl. 14, Fig. 22 var. **evolutum**.

Euastrum coronatum Turner 1885, Jour. Roy. Microsc. Soc., II, 5(6): 935. Pl. 15, Fig. 9.
Euastrum abruptum var. *evolutum* Nordstedt 1877(1878), Öfv. Kongl. Vet.-Akad. Förhandl. 1877(3): 21. Pl. 2, Fig. 7.
Euastrum nordstedtianum Wolle 1884a, Desm. U.S., p. 105. Pl. 26, Figs. 7, 9–13; Pl. 52, Figs. 13–15.

Cells medium-sized, oval to subquadrate in outline, 1.5 times longer than broad; the semicells nearly semicircular-trapezoidal in outline; basal lobes narrowly rounded at the angles, the lateral margins deeply retuse to a prominent upper, lateral lobule, with 3 spines showing at the apex of the lobe; with a deep sinus between the lateral lobules and the polar lobe, which is anvil-shaped, with a subapical lobule, the apical margin flat, somewhat elevated at the margins of a deep median notch, and extended at the angles by a prominent, slightly upwardly directed spine immediately above the subapical lobule; face of semicell with a prominent median protrusion bearing 3 granules, with a horizontal pair of mucilage pores immediately above and to either side of the median protrusion; wall with an intramarginal granule on each side of the median notch, and a supraisthmial spine on the basal lobes; lateral view broadly oval, the poles truncate, with a prominent spine in the midregion of the apex and a pair of granular tubercles at each angle, the lateral margins concave to a very broad, bimammillate protrusion near the base of the semicell; vertical view subrectangular, the poles concave, the angles bearing a bilobed tubercle, the lateral margins slightly convex, with a bigranular protrusion in the midregion. L. 44–75 µm. W. 29–58 µm. Isth. 10–15 µm. Ap. 30–35 µm. Th. 30–38.5 µm. Zygospore unknown.

DISTRIBUTION: Alaska, Connecticut, Florida, Georgia, Illinois, Massachusetts, Michigan, Minnesota, Mississippi, New Hampshire, New Jersey, New York, Pennsylvania, South Carolina, Vermont, Washington. British Columbia, Ontario, Québec. Africa, South America, Panama Canal.
PLATE LXXIX, figs. 2–3.

37a. **Euastrum evolutum** var. **glaziovii** (Börges.) West & West 1898, Linn. Soc. Jour. Bot. 33: 292.

Euastrum glaziovii Börgesen 1890, Desmidiaceae, in Warming's Symbolae ad floram Brasiliae centralis cognoscendam. Vid Medd. Naturh. Foren. 1890: 941. Pl. 3, Fig. 23.

A variety usually somewhat smaller than the typical, rectangular in outline, the semicells quadrate, the basal lobes narrow and sharply pointed at the angles, the lateral margins diverging slightly and concave to the upper, lateral, bispinate lobules, a broad sinus between the upper, lateral lobules and the polar lobe, which is bispinate at the angles, the upper one being prominent and extending as much as or more than the basal lobes, the apical margin elevated to the median notch, which is bordered by hornlike thickenings of the wall; face of semicell with a prominent protrusion and with wall markings as in the typical; lateral view broadly elliptic, the poles narrowly rounded and somewhat produced, the lateral margins showing a knoblike protrusion below the apex, and a prominent, bimammillate protrusion near the base of the semicell; vertical view subrectangular, the poles truncate and spiny, the lateral margins convex with a slight protrusion in the midregion. L. 58–68.5 μm. W. 37–42.8 μm. Isth. 21 μm. Ap. 34–40 μm. Th. 6.5–11 μm.

DISTRIBUTION: Alaska, Florida, Georgia, Kentucky, Louisiana, Michigan, Mississippi, Vermont, Washington, Wisconsin. Québec. Europe, Africa, South America, Panama Canal.

PLATE LXXIX, figs. 4, 4a.

37b. **Euastrum evolutum** var. **guianense** (Raciborski) West & West 1898, Linn. Soc. Jour. Bot. 33: 292; Krieger 1937, Rabenhorst's Kryptogamen-Flora 13(1:3): 615. Pl. 87, Fig. 21.

Euastrum evolutum f. *minor* West & West 1898, Linn. Soc. Jour. Bot. 33: 292. Fig. 1c.
Euastrum nordstedtiana Wolle 1884a, Desm. U.S., p. 105. Pl. 26, Fig. 8.

A variety similar in shape and proportions to the typical but smaller throughout and with lobules and spines distinctly reduced in prominence (length). L. 42–47 μm. W. 25–30 μm. Isth. 5–8 μm.

DISTRIBUTION: Florida, Maine, Minnesota, Montana, Pennsylvania. British Columbia, Québec. Africa, South America.

PLATE LXXIX, figs. 6, 7A.

37c. **Euastrum evolutum** var. **incudiforme** (Börges.) West & West 1898, Linn. Soc. Jour. Bot. 33: 292. No. fig.

Euastrum incudiforme Börgesen 1890, Vid. Medd. Naturh. Foren. Kjöbenhavn 1890: 940. Pl. 3, Fig. 22.

A variety averaging smaller than the typical, similar in shape but lacking the subapical lobules; the margins of the polar lobules with 2 minute teeth on either side of the median notch; intramarginal granules of the polar lobules lacking. L. 50–60 μm. W. 36–39 μm. Isth. 7.5–10 μm. Ap. 35 μm.

DISTRIBUTION: United States. (West & West, *l.c.*, discuss this variety in their report on desmids of the United States, but they do not indicate a specific locality. Reference is made to Brazil, hence this may not be a North American record.) South America.

PLATE LXXIX, figs. 5–5b.

37d. **Euastrum evolutum** var. **integrius** West & West 1896a, Trans. Linn. Soc. London, Bot. 5: 244. Pl. 14, Figs. 23–25 f. **integrius.**

Euastrum spinosum Ralfs, in Wolle 1884a, Desm. U.S., p. 106. Pl. 27, Figs. 4, 5.

A variety with configurations simpler than in the typical; basal lobes broadly rounded, with a pair of short spines at the lower angle and a pair immediately above on the upper angle; upper, lateral lobules lacking, the margin with a deep sinus forming the polar lobe, which has a simple, subapical spine rather than a lobule as in the typical. L. (27)46–61(64.4) μm. W. 31–41 μm. Isth. 7–10 μm. Ap. 24–31 μm. Th. 22–27 μm.

Wolle (*l.c.*) records *Euastrum spinosum* Ralfs and illustrates it with 4 figures (Pl. 27, Figs. 4–7). His figures 4, 5 appear to be incomplete illustrations of var. *integrius*, whereas his figure 6 is assignable to typical *E. evolutum*.

DISTRIBUTION: Connecticut, Florida, Georgia, Louisiana, Michigan, Minnesota, Mississippi, New Hampshire, New York, North America, Pennsylvania, Virginia. Ontario. Asia, Africa, India.

PLATE LXXIX, figs. 8–8b; LXXXV, fig. 8.

37e. **Euastrum evolutum** var. **integrius** f. **turgidum** Scott & Prescott 1952, Hydrobiologia 4(4): 381. Pl. 3, Figs. 4, 5.

A variety relatively broader than the typical, the cells subrectangular, the basal lobes briefly rounded at their angles, furnished with a granular tubercle, the margins of the semicells converging and somewhat concave above the basal angles to an upper, lateral, bispinate swelling which forms a pronounced shoulder between the basal lobes and the polar lobe, which is somewhat anvil-shaped, the apical margin elevated to the median notch, the angles bearing a short, stout, spine, a minute subapical spine on the margins of the polar lobe in lieu of the bispinate lobule characteristic of the typical species; lateral view broadly elliptic, the poles narrowly rounded, a subapical tubercle on either side, the margins symmetrically convex with a bimammillate protrusion. L. 50-60 μm. W. 36-39 μm. Isth. 8-12 μm. Th. 20-22 μm.

DISTRIBUTION: Alaska, Florida, Louisiana, North Carolina.

PLATE LXXIX, fig. 9, 9a.

37f. **Euastrum evolutum** var. **monticulosum** (Taylor) Krieger 1937, Rabenhorst's Kryptogamen-Flora 13(1:3): 616. Pl. 88, Figs. 7, 8 f. **monticulosum**.

Euastrum monticulosum Taylor 1935a, Amer. Jour. Bot. 22: 774. Pl. 3, Fig. 2.

A variety relatively broader than the typical, the basal lobes longer and extending beyond the upper, lateral lobules; mucilage pores in the face of the semicell apparently lacking. L. 60-75 μm. W. 52-56 μm. Isth. 10-13 μm. Th. 34 μm.

DISTRIBUTION: Mississippi. South America.

PLATE LXXXV, fig. 7, 7a.

37g. **Euastrum evolutum** var. **monticulosum** f. **poriferum** Scott & Prescott 1952, Hydrobiologia 4(4): 381. Pl. 3, Fig. 2.

Euastrum turneri var. *poriferum* (Presc. & Scott) Jackson 1971, Dissert., p. 174. Pl. 37, Figs. 7, 8.

A form differing from the typical by having 2 mucilage pores, one on either side and above the facial protuberance, and in having the wall finely scrobiculate in a circular area surrounding the facial protuberance, the wall distinctly punctate throughout. L. 71-81 μm. W. 48-51 μm. Isth. 12 μm. Th. 28-34 μm.

DISTRIBUTION: Florida, Montana. Newfoundland.

PLATE LXXIX, fig. 10; PLATE LXXX, figs. 1, 1a.

37h. **Euastrum evolutum** var. **reductum** Scott & Prescott 1952, Hydrobiologia 4(4): 382. Pl. 3, Fig. 3. f. **reductum**.

A variety differing from the typical by having the lobes, lobules, and spines decidedly shorter and less produced. L. 58-60 μm. W. 39-40 μm. Isth. 12 μm.

It is questionable whether this form of *Euastrum evolutum* deserves a variety status. Additional collections may reveal that this is a habitat variable. In any event it seems appropriate to have this expression to be indicated as a forma taxon rather than as a variety. The following form illustrates a further reduction in the wall markings.

DISTRIBUTION: Florida.

PLATE LXXX, figs. 2, 2a.

37i. Euastrum evolutum var. **reductum** f. **simplicissimum** Prescott f. nov.

Forma paululo minor quam forma typica, granulis marginalibus atque verrucis ad angulous reductis; lobuli loborum basalium leviter rotundati. Long. 50 μm. Lat. 33–35 μm. Ist. 10 μm.

ORIGO: E loco Wood Pond, Cape Cod, Massachusetts dicto.
HOLOTYPUS: GWP Coll. 1-Ma-135.
ICONOTYPUS: Plate LXXX, fig. 3.
A form smaller than the typical variety, with a reduction of marginal granules and verrucae at the angles; the lobules of the basal lobes smoothly rounded. L. 50 μm. W. 33–35 μm. Isth. 10 μm.
DISTRIBUTION: Massachusetts.
PLATE LXXX, fig. 3.

38. Euastrum fissum West & West 1902a, Trans. Linn. Soc. London, Bot., II, 6: 154. Pl. 20, Fig. 17 var. **fissum.**

A small *Euastrum*, broadly oval in outline, 2 times longer than broad; semicell rhomboid in outline, the basal lobes broadly rounded, the lateral margins above retuse to a prominent, hornlike process which extends as far as the basal lobes, the polar lobe not defined, the margins convex to a broadly truncate, nearly flat apex, the angles rounded, the median incision deep and narrow; face of the semicell with a prominent protrusion, with a circle of tubercles; cell wall with a tubercle in the apical lobules opposite the marginal, upper, lateral tubercles; sinus narrow, opening slightly outward; lateral view narrowly oval, the poles rounded and smooth, lateral margins with a subapical tubercle and with a prominent protrusion at the base of the semicell. L. 34.5–41 μm. W. 19–22 μm. Isth. 3–4 μm.
DISTRIBUTION: Massachusetts, New Hampshire. Québec. Asia, Ceylon, Africa.
PLATE LXXVI, figs. 17, 17a.

38a. Euastrum fissum var. **americanum** Cushman 1905, Rhodora 7: 253. Pl. 64, Fig. 2.

A variety relatively narrower than the typical; semicell margins with a low undulation or prominence between the basal lobes and the upper, lateral tubercle; face of semicell with a semicircle of granules surrounding the median protrusion; cell wall with an intramarginal granule at the base of each polar lobule. L. 44 μm. W. 25 μm. Isth. 3.5–6 μm.
DISTRIBUTION: New Hampshire. Newfoundland.
PLATE LXXVI, fig. 16.

38b. Euastrum fissum var. **angustum** Scott & Prescott 1952, Hydrobiologia 4(4): 382. Pl. 3, Fig. 7.

Cells relatively narrower than the typical; basal lobes rounded at the angles but bearing a tubercle, with a granule on the swelling half the distance to the subapical spine; face of semicell with a protrusion bearing 3 granules and a tubercle, with 3 granules on either side of the polar incision; wall with intramarginal granules at all lobes and swellings; lateral view narrowly elliptic, the poles sharply rounded and granular; vertical view oval, the poles broadly rounded and granular, lateral margins with a median protrusion. L. 38–41 μm. W. 19–20 μm. Isth. 5 μm. Th. 15 μm.

DISTRIBUTION: Florida.
PLATE LXXVI, fig. 14; PLATE LXXIX, figs. 7–7b.

38c. **Euastrum fissum** var. **decoratum** Scott & Prescott 1952, Hydrobiologia 4(4): 383. Pl. 1, Figs. 1, 2.

A variety larger than the typical, 2 times longer than broad but narrower at the poles, the apical margin much elevated, forming lobules on either side of a deep median notch; basal lobes briefly rounded at the angles and furnished with a sharp granule, margins of the semicells concave and smooth to the polar area; facial protrusion of the semicell bearing 3 vertical rows of granules; wall with intra-marginal granules at the basal angles, and with a prominent tubercle or verruca on either side of the apical notch, and usually a granule on either side of the apex of the notch; lateral view elliptic, the poles narrowly rounded, with a subapical granule on the margin and an upper, lateral verruca, the margin concave to the granular, basal swelling of the semicell and bearing a granule immediately beneath the verrucae. L. 40–59 μm. W. 24–30 μm. Isth. 7–9 μm. Th. 18 μm.

This variety should be compared with *Euastrum elegans* (Bréb.) Ralfs and some of its variations. It is distinguished by the pincers type of polar lobe, with a deep apical notch.

DISTRIBUTION: Florida. Ontario.
PLATE LXXVI, fig. 15.

38d. **Euastrum fissum** var. **hians** Scott & Grönblad 1957, Acta Soc. Fennica, II, B, II(8): 12. Pl. 1, Fig. 4 f. **hians.**

A variety differing from the typical by the inflated, pincers type of polar lobe, the apical notch deep but much wider than in the typical, and by the open sinus, acute at the apex, also by the margin of the semicells being undulate below the polar lobe; lateral view narrowly oval, the poles rounded and granular, and with granules along the margin, sharply inflated in the midregion, which shows 3 crenations. L. 41–42 μm. W. 22–24 μm. Isth. 6–7 μm. Th. 12 μm.

DISTRIBUTION: Florida.
PLATE LXXVI, figs. 25, 25a.

38e. **Euastrum fissum** var. **hians** f. **minus** Prescott f. nov.

Forma a forma typicali differens quod minor et quod in superficie semi-cellulae protrusionem mediam 2 ordines verticales granulorum ferentem habet; membrana pari granulorum magnorum intra marginem lobulorum polarium prae-dita. Long. 34.6 μm. Lat. 18.5 μm. Ist. 5 μm.

ORIGO: E loco Sam Pond, Cape Cod, Massachusetts dicto.
HOLOTYPUS: GWP Coll. Ma-7005.
ICONOTYPUS: Plage LXXVI, figs. 23, 23a.

A form differing from the typical by its smaller size and by having a median protrusion on the face of the semicell which has 2 vertical rows of granules, the wall with a pair of large granules within the margin of the polar lobules. L. 34.6μm. W. 18.5μm. Isth. 5 μm.

DISTRIBUTION: Massachusetts.
PLATE LXXVI, figs. 23, 23a.

38f. **Euastrum fissum** var. **pseudoelegans** Scott & Prescott 1952, Hydrobiologia 4(4): 384. Pl. 1, Fig. 3.

A variety larger than the typical, 2 times longer than broad; semicells wider at the apex than at the base; basal lobes sharply rounded at the angles, the lateral margins converging throughout two-thirds of the distance to the apex and then diverging to the upper, lateral angles, which bear a straight, horizontally directed spine; the apical margin of the polar lobes sloping gently upward to the deep median notch, the apex broadly rounded; basal lobes with a short, blunt tooth at the angles and with 3 intramarginal granules in a triangular pattern; a very slight protuberance on the margin of the semicell about half-way to the apex; face of semicell with a median protrusion bearing 3 vertical rows of granules; wall with a protuberance bearing a circular pattern of granules in the midregion of the polar lobules, and with a bidentate, intramarginal granule on either side of the apical notch; sinus linear and closed throughout; lateral view of semicell oval in outline, the poles sharply rounded, the lateral margins convex at the base and extending into prominent lateral protuberances furnished with 3 crenations, the upper, lateral margins deeply retuse between the basal protuberances and a bidentate, subapical protuberance on either side, 2 granules showing on each side between the upper, lateral protuberances and the apex; vertical view oval with a tricrenate swelling in the midregion on either side, the poles broadly rounded and showing a pair of subpolar granules on each side. L. 58 μm. W. 31 μm. Isth. 10 μm. Ap. 33 μm. Th. 23 μm.

DISTRIBUTION: Florida.

PLATE LXXVI, figs. 26–26b.

39. **Euastrum floridense** Scott & Prescott 1952, Hydrobiologia 4(4): 385. Pl. 1, Fig. 11.

A large *Euastrum*, 2 times longer than broad, subrectangular in outline, with the polar lobe as wide as the basal lobes of the semicells; basal lobes truncately rounded at the angles, the margins of the semicells sharply incised to an upper, lateral lobule which extends as far as the basal lobes or further, a narrow, sharp sinus separating the lateral lobules from a much inflated polar lobe with broadly rounded and somewhat angular margins to the truncate apical margin, which is flat, the median incision narrow and closed; face of semicell without a median protrusion, but with 3 basal protrusions; the wall minutely scrobiculate; lateral view oval, the poles broadly rounded, the lateral margins convex and smooth. L. 104 μm. W. 48 μm. Base of semicell 46 μm. Isth. 16 μm. Th. 32 μm. Zygospore unknown.

This species should be compared with *Euastrum allenii* Cushman, from which it differs especially in the form of the inflated polar lobes and the narrow, deep sinus between the basal lobes and the polar lobe.

DISTRIBUTION: Florida.

PLATE LXIII, figs. 5, 5a.

40. **Euastrum formosum** Wolle 1880, Bull. Torr. Bot. Club 7: 45. Pl. 5-B; 1892, Desm. U.S., p. 103. Pl. 26, Fig. 16. (Non *Euastrum formosum* Gay 1884a, Monogr. Locale des Conjuguées, p. 55. Pl. 1, Fig. 55).

A medium-sized *Euastrum*, broadly oval in outline, 1.5 times longer than broad; semicells semicircular in outline, almost equally 6-lobed, the basal lobes bifid, the other lobes trifid; basal lobes horizontally directed, the margins incised to form upper, lateral lobes which extend as much as the basal lobes; polar "lobe" consisting of 2 diverging lobules; sinus narrow but open; facial protuberance (?); lateral view (?). L. 62 μm. W. 40 μm.

DISTRIBUTION: New Jersey.

PLATE LXXIII, fig. 6; Plate LXXX, fig. 15.

41. **Euastrum gayanum** De Toni 1889, Sylloge algarum, p. 1075; Krieger 1937, Rabenhorst's Kryptogamen-Flora 13(1:3): 565. Pl. 77, Figs. 9-11 f. **gayanum**.

Euastrum formosum Gay 1884a, Bull. Soc. Bot. France 31: 55. Pl. 1, Fig. 9.

A small *Euastrum*, subquadrangular in outline, slightly longer than broad; semicells transversely rectangular in outline, the basal lobes broadly rounded, with minute teeth on the margin of the angles, lateral margins of semicells retuse to the broad polar lobe, which is nearly as wide as the basal lobes, the apical margin angularly bilobed, the median notch a broad V, with a short spine at the lateral angles of the polar lobe, and 2 marginal granules on each lobule; face of the semicell with a simple protrusion; wall with small granules within the polar lobules and a supraisthmial granule in the basal lobes; sinus narrow and closed; lateral view narrowly oval, the poles sharply rounded, with an apical spine and a small subapical spine on either side, lateral margins convex with a truncate protrusion in the midregion; vertical view narrowly oval, the poles truncate, with a sharp granule at either angle and a similar subapical granule on either side, lateral margins with a truncate protrusion in the midregion. L. 10.6-18.8 μm. W. 10.6-13.5 μm. Isth. 2-4.6 μm. Ap. 10 μm. Th. 6-6.6 μm. Zygospore unknown.

DISTRIBUTION: Massachusetts, Montana. Great Britain, Europe, Asia, Africa, South America, Azores.

PLATE LXXII, figs. 20-20b.

41a. **Euastrum gayanum** f. **fennicum** (Grönblad) Krieger 1937, Rabenhorst's Kryptogamen-Flora 13(1:3): 565. Pl. 77, Figs. 12, 13.

Euastrum sibiricum Boldt f. *fennica* Grönblad 1921, Acta Soc. Fauna Flora Fennica 49(7): 13. Pl. 13, Figs. 17, 18.

A form slightly larger than the typical with reduced wall markings; basal lobes broadly rounded, margins of semicell slightly retuse to form a broad, flat polar lobe, the apical margin slightly retuse and without a vertical incision; vertical view oval, the poles rounded and smooth, the lateral, median swelling scarcely evident. L. 19.8 μm. W. 15 μm. Isth. 4.6 μm. Th. 9.2 μm.

DISTRIBUTION: Minnesota. Europe, Arctic.

PLATE LXXII, fig. 21.

42. **Euastrum gemmatoides** Irénée-Marie 1959, Hydrobiologia 13(4): 352. Pl. 5, Fig. 6.

A small *Euastrum*, broadly oval in outline, nearly twice as long as broad; semicells subrectangular in outline, the basal lobes sharply rounded and narrow at the angles, the lateral margins concave but diverging to upper, lateral lobules which extend further than the basal lobes; the polar lobe with 2 upwardly directed lobules with a deep median notch; face of semicell swollen with a slight protrusion or thickening in the midregion; wall smooth (?); sinus narrow and closed; lateral view truncate-oval in outline, the poles broadly truncate, the lateral margins concave and sloping to the basal inflations, giving the semicell a vase-shaped appearance. L. 34.5-40 μm. W. 25-30 μm. Isth. 8-10 μm. Ap. 12-18 μm.

Irénée-Marie (*l.c.*) compares the general shape of this species to *Euastrum gemmatum* Brébisson.

DISTRIBUTION: Québec.

PLATE LXXXI, figs. 11, 11a.

43. **Euastrum gemmatum** (Bréb.) Ralfs 1848, Brit. Desm. p. 87. Pl. XIV, Figs. 4a-4e var. **gemmatum**.

Cosmarium gemmatum Brébisson, in: Meneghini 1840, Synop. Desm., p. 221.

A medium-sized *Euastrum*, subquadrate in outline, 1.5 to 1 2/3 times longer than broad; semicells horizontally subrectangular in outline, 6-lobed; basal lobes broadly rounded with granular margins; upper, lateral lobules prominent and extended as far as the basal lobes, the margins deeply retuse to the bilobed, short polar lobe, with angles broadly rounded, the median notch merely an invagination between the 2 polar lobules; face of the semicell with 3 protrusions, the median one more prominent, with large granules, the 2 lateral protrusions less prominent and with smaller granules; wall scrobiculate and granular on the margins; sinus narrow and closed; lateral view broadly oval, the poles deeply bilobed, the margins strongly concave to the much swollen basal lobes; vertical view subquadrate in outline, the poles narrowly rounded and produced beyond lateral lobes, the median region with a prominent protrusion on either side. L. 43–70 μm. W. 35–52(69) μm. Isth. 8–15 μm. Ap. 15–22 μm. Th. 20–30 μm. Zygospore unknown.

This species shows several variations. Pl. LXXXII, fig. 2 illustrates a form with widely open sinus which possibly merits a taxon epithet. It should be compared with *Euastrum germanicum* (Schmidle) Krieger, which is similar in appearance in face view, but is quite different in vertical and lateral views.

DISTRIBUTION: Widely distributed in the United States and Canada. Cosmopolitan.

PLATE LXXXII, figs. 1, 2.

43a. Euastrum gemmatum var. monocyclum Nordstedt 1880, Acta Univ. Lund 16: 8. Pl. 1, Fig. 13.

Euastrum spinulosum subsp. *inermius* var. *laticeps* Borge 1903, Ark. f. Bot. 1: 113. Pl. 5, Fig. 4.

A variety differing from the typical by having a shorter polar lobe, the polar lobules not extended but with the apical margin slightly retuse; the basal portion of the cell stout, with the upper, lateral lobules not so long as in the typical; wall with granules on the lobes only; vertical view subrectangular, the poles slightly retuse, the median swelling of the sides not so prominent as in the typical. L. 47–68(76) μm. W. 44–57 μm. Isth. 13–16 μm. Ap. 18–26 μm. Th. 21–29 μm.

DISTRIBUTION: Alaska, Oregon. Europe, Asia, Africa, South America.

PLATE LXXXII, figs. 4, 4a.

43b. Euastrum gemmatum var. taftii Prescott nom. nov.

Euastrum gemmatum f. *latior* Grönblad 1945, Acta Soc. Sci. Fennicae, II, B, II(6): 13. Pl. 3, Fig. 57.

Euastrum gemmatum var. *alatum* Taft 1949, Trans. Amer. Microsc. Soc. 68(3): 214. Pl. 2, Fig. 6 (Non *Euastrum gemmatum* var. *alatum* Kossinskaja 1949, Not. Syst. e Sect. Crypt. Inst. Bot. nom. V. L. Komarovii Acad. Sci. URSS 6(1/6): 43. Pl. 1, Fig. 4.).

A variety differing from the typical by having an open sinus, the basal lobes slightly produced and down-turned at the angles, basal and upper, lateral lobes narrower and proportionately longer than in the typical. L. 47–56 μm. W. 36–42 μm. Ap. 20–25 μm. Isth. 10–11 μm.

Grönblad's form (*l.c.*) is somewhat shorter than the plant described by Taft (*l.c.*), but both have similar characteristics and a description could not separate the two.

DISTRIBUTION: California, Oklahoma. South America.

PLATE LXXVI, fig. 22; Pl. LXXXII, figs. 3, 5.

43c. **Euastrum gemmatum** var. **tenuius** Krieger 1937, Rabenhorst's Kryptogamen-Flora 13(1:3): 641. Pl. 92, Figs. 12, 13.

Euastrum gemmatum Brébisson f. Borge 1918, Ark. f. Bot. 15(13): 61. Pl. 5, Fig. 8.

A variety smaller than the typical, and more slender, the invagination between the basal angles and the upper, lateral lobules much less, and the upper, lateral sinus between the base and the polar lobe shorter, so that the polar lobe is relatively broader in appearance. L. (36)38–49 μm. W. 24–31 μm. Isth. 7–11 μm.
DISTRIBUTION: Mississippi. South America.
PLATE LXXXII, figs. 7, 7a.

44. **Euastrum giganteum** (Wood) Nordstedt, in DeToni 1889, Sylloge algarum, p. 1106; Krieger 1937, Rabenhorst's Kryptogamen-Flora 13(1:3): 482. Pl. 57, Figs. 4, 5.

Tetmemorus giganteus Wood 1869, Proc. Philadelphia Acad. Sci. 1869: 19; 1872, Smithson. Contrib. to Knowledge 241: 117. Pl. 12, Fig. 7.
Euastrum giganteum var. *latum* Taylor 1935, Pap. Michigan Acad. Sci., Arts & Lettr. 20: 203. Pl. 41, Figs. 8, 8a.

A large *Euastrum*, elongate-rectangular in outline, up to 3 times longer than broad; semicells truncate-trapezoid, the basal lobes sharply rounded at the angles, the lateral margins convex, undulate, and converging toward the apex, the polar lobe wanting; apical margin flat, slightly elevated at the margins of a deep, linear apical notch; face of semicell with 3 prominent protrusions at the base; cell wall coarsely punctate; lateral view broadly oval, the poles truncate and smooth, the margins diverging at first and then subparallel to the base of the semicell. L. (202)220–251 μm. W. 60–80 μm. Isth. 28–35 μm. Ap. 15–45 μm. Th. (59) 60–73 μm. Zygospore unknown.

It seems probable that Taylor's var. *latum* reported from Newfoundland is the same plant as described by Wood as *Tetmemorus giganteus*. Wood's figure and analysis are incomplete, but he mentions the tendency for the lateral margins to converge slightly to the apex, although not depicted in his illustration. Taylor properly described var. *latum* with some reservation and gives an excellent illustration. Krieger (1937) places it in synonymy.
DISTRIBUTION: Pennsylvania. Newfoundland.
PLATE LVIII, figs. 13, 13a.

45. **Euastrum humerosum** Ralfs 1848, Brit. Desm., p. 82. Pl. 13, Fig. 2 var. **humerosum**.

Euastrum humerosum f. *scrobiculata* Nordstedt 1873, Acta Univ. Lund 9: 8.

A large *Euastrum*, broadly oval in outline, 2 times longer than broad; semicells subpyramidal in outline, the basal lobes bilobed at the margin, the lateral margins of the semicell retuse and converging to prominent upper, lateral lobules which are somewhat upwardly directed, a deep sinus between the lateral lobules and the polar lobe, which is inflated, with diverging margins and rounded angles, the apical margin flat, the median notch narrow and closed; face of semicell with a median, mucilage pore (usually present), bordered on either side by a swelling, and 3 prominent basal swellings, the two lateral protuberances bilobed; wall coarsely punctate throughout; sinus narrow and closed; lateral view broadly oval, the poles truncate and slightly bilobed, the apical margins retuse, lateral margins subparallel below the apex for a short distance and then convex, with a marginal mucilage pore showing in the midregion; vertical view elliptic, the poles narrowly rounded and produced, the lateral margins triundulate. L. 102–160 μm. W. 51–82 μm. Isth. 14–24(29) μm. Ap. 26.4–42 μm. Th. 40–47 μm. Zygospore unknown.

DISTRIBUTION: Florida, Maine, Massachusetts, Michigan, Montana, New Jersey, New York, North Carolina, Pennsylvania, Utah, Washington. British Columbia, Labrador, Québec. Great Britain, Europe, Asia, South America, Arctic, Azores, Faeroes.

PLATE LXIII, figs. 7, 7a.

45a. Euastrum humerosum var. **evolutum** Krieger 1937, Rabenhorst's Kryptogamen-Flora 13(1:3): 525. Pl. 70, Figs. 1, 2.

Euastrum humerosum Ralfs f. Prescott 1935, Rhodora 37: 24. Pl. 325, Figs. 11, 12.

A variety with the basal lobes subrectangular, bilobed at the margins, sharply delimited from the upper, lateral lobules by a deeply retuse invagination of the margin, sinus between the lateral lobules and the polar lobe narrow; polar lobe anvil-shaped with the lobules broadly extended, much more than in the typical; wall coarsely punctate, especially in the areas of the lobes. L. 108–114(118) μm. W. 60–68(71) μm. Isth. 16–20.5 μm.

DISTRIBUTION: Louisiana, Massachusetts, Mississippi. Québec.

PLATE LXIII, figs. 10, 10a, 14, 14a.

45b. Euastrum humerosum var. **mammosum** Schmidle 1893a, Ber. d. Naturf. Ges. Freiburg i Br. 7(1): 106. Pl. 6, Figs. 9, 10.

A variety with cells more oval in outline and relatively narrower than the typical; semicells subpyramidal, the basal lobes rounded and bilobed, the lateral margins converging and retuse to very short, upper lateral lobules; the polar lobe exserted with slightly diverging lateral margins to a relatively narrow apex, the apical margin flat, the angles broadly rounded; face of semicell with the supraisthmial protuberances prominent, and with a protuberance immediately below the apex of the median notch of the polar lobe, and with a protuberance on either side of the face of the upper, lateral lobules. L. 100 μm. W. 60 μm.

Krieger (1937, p. 524) places this variety in synonymy with the typical, but the plant described by Schmidle (*l.c.*) is quite different in shape and proportions. We prefer to set this variety apart.

DISTRIBUTION: Florida. Europe.

PLATE LXIII, fig. 8.

45c. Euastrum humerosum var. **parallelum** Krieger 1937, Rabenhorst's Kryptogamen-Flora 13(1:3): 526. Pl. 69, Fig. 9.

A variety with basal lobes angular, the margins subparallel but slightly bilobed; the upper, lateral lobules forming short shoulders; the polar lobe relatively narrower and not inflated, the lateral margins parallel, the angles smoothly rounded; face of semicell without a central protuberance and without a mucilage pore; other swellings on the face much reduced.

DISTRIBUTION: Colorado. New Brunswick. Europe.

PLATE LXIII, fig. 13.

46. Euastrum hypochondroides West & West 1895, Trans. Linn. Soc. London, Bot., II, 5: 49. Pl. 6, Fig. 8.

A medium-sized *Euastrum*, cells nearly round in outline, about 1.25 times longer than broad; the semicells distinctly 3-lobed, the basal lobes sharply rounded, the lateral margins of the semicell sloping slightly upward to a narrow, much produced, collarlike polar lobe with parallel sides, the angles of the apex broadly rounded, the apical margin flat and without a median notch; face of the semicell with a low median protrusion bearing a pattern of granules enclosed by a circle of granules, all lobes of the semicell bearing granules at the margins and intramarginally; sinus narrow and closed; lateral view narrowly oval, the poles rounded and granular, the lateral margins slightly concave to the swollen, granular basal portion of the semicell; vertical view very narrowly oval, the poles rounded and granular, the margins bearing a granular swelling in the midregion. L. 49–52 μm. W. 37–41 μm. Isth. 10 μm. Ap. 13–14 μm. Th. 16 μm. Zygospore unknown.

DISTRIBUTION: Montana. Africa.

PLATE LXXX, figs. 16, 16a.

47. **Euastrum hypochondrum** Nordstedt 1880, Acta Univ. Lund 16: 8. Pl. 1, Fig. 11 var. **hypochondrum** f. **hypochondrum**.

A small *Euastrum*, nearly round in outline, but broader than long; the basal lobes broadly rounded and much extended, bearing on their upper margins a low, bispinate protrusion, margins forming a broad sinus between it and the polar lobe, which has subparallel margins, slightly diverging to the angles of the apex, which are rounded and granular, the apical margin slightly retuse, forming 2 apical lobules; face of semicell with a median, supraisthmial tubercle below a circular pattern of granules; wall with vertical rows of granules near the apex of the basal lobes and intramarginal granules at the apical lobules; lateral view truncate-oval, the poles broad and flat, granular, the margins subparallel for half the distance and then abruptly inflated to form a granular protrusion; vertical view elliptic, the poles narrowly rounded and granular, margins with a median granular swelling. L. 50–62 μm. W. 50–62 μm. Isth. 14.5 μm. Ap. 20–23 μm. Th. 26 μm. Zygospore unknown.

DISTRIBUTION: Florida. Africa, South America. Panama.

PLATE LXXXI, fig. 1.

47a. **Euastrum hypochondrum** var. **hypochondrum** f. **decoratum** Scott & Prescott 1952, Hydrobiologia 4(4): 386. Pl. 3, Fig. 10.

A form larger throughout than the typical, with the median facial protuberance more prominent and the wall decorated with large, round or conical granules; vertical view with a relatively deep invagination of the wall on either side of the median protrusion. L. 60–72 μm. W. 57–60 μm. Isth. 12–15 μm. Th. 35–39 μm.

DISTRIBUTION: Florida, Michigan.

PLATE LXXXI, figs. 4–4b.

47b. **Euastrum hypochondrum** var. **hypochondrum** f. **prominens** Scott & Prescott 1952, Hydrobiologia 4(4): 387. Pl. 3, Fig. 13.

A form similar in size to the typical but with the median protuberance narrow and more prominent as seen in lateral and vertical views; the basal lobes encircled by 3 rows of granules. L. 54–56 μm. W. 49–55 μm. Isth. 11–12 μm. Th. 25–28 μm.

DISTRIBUTION: Florida.

PLATE LXXXI, figs. 2–2b.

47c. **Euastrum hypochondrum** var. **acephalum** Scott & Grönblad 1957, Acta Soc. Fennicae, II, B, II(8): 13. Pl. 3, Figs. 6, 7.

A variety averaging somewhat larger than the typical, nearly circular in outline, differing from the typical principally by the reduction in the polar lobe, which is only slightly produced from the broadly rounded basal lobes, resulting in a *Cosmarium*-like aspect in facial view; median protrusion of the semicell with a circular pattern of granules, with 1 or 2 tubercles below the protrusion; sinus narrow and closed; lateral view broadly oval, the semicells nearly circular in outline, with a very slight swelling in the midregion; vertical view elliptic, the poles rounded and granular, the margins with a truncate protrusion and with larger granules in the midregion. L. 61-65 μm. W. 57-60 μm. Isth. 15-16 μm. Th. 35-37 μm.

DISTRIBUTION: Florida.

PLATE LXXXI, figs. 3-3b.

47d. **Euastrum hypochondrum** var. **ohioense** (Taft) Prescott comb. nov.

Euastrum ohioense Taft 1945, Ohio Jour. Sci. 45: 190. Pl. 2, Fig. 17.

A variety differing from the typical by having a more subpyramidal semicell with thicker (vertically) basal lobes, the angles more narrowly rounded, with the secondary protuberances on the upper margin less produced, and in having the polar lobes shorter; face of semicell with protuberances and granulations similar to the typical. L. 55 μm. W. 37 μm. Isth. 6 μm. Ap. 23 μm.

This variety should be compared with the much larger *Euastrum verrucosum* var. *subplanctonicum* Grönblad, with which it has some similarities in face view. The vertical view of the latter is quite different, being subrectangular in outline, whereas *E. hypochondrum* is narrowly elliptic.

DISTRIBUTION: Ohio, Oklahoma.

PLATE LXXXI, figs. 5, 5a.

48. **Euastrum incavatum** Josh. & Nordstedt, in Wittrock & Nordstedt 1886, Bot. Notiser 1866, Alg. Exsicc. No. 637 var. **incavatum.**

Cells small, truncate-oval in outline, 1.5 to 1 1/3 times longer than broad; semicells urn-shaped, the basal lobes broadly rounded at the angles, the lateral margins retuse and converging symmetrically to the slightly inflated polar lobe, the angles of the polar lobe sharp and spinelike, the apical margin truncate and retuse in the midregion to form a U-shaped invagination; face of semicell without a median protrusion; sinus narrow and closed; lateral view oval, the poles broadly rounded, the semicells symmetrically oval with convex margins; vertical view oval, the sides convex and smooth. L. 26-43 μm. W. 16-24 μm. Isth. 4.5-6 μm. Ap. 12 μm. Th. 11-16 μm. Zygospore unknown.

DISTRIBUTION: Florida. Great Britain, South America, Jamaica.

PLATE LXXII, figs. 5-5b, 7-7b.

48a. **Euastrum incavatum** var. **floridense** Scott & Grönblad 1957, Acta Soc. Sci. Fennicae, II, B, II(8): 13. Pl. 3, Fig. 16 f. **floridense.**

A variety smaller than the typical; basal lobes with the lateral margins angular and subparallel, the margins of the semicell sharply retuse to the smooth, truncate polar lobes, the median notch slight and broadly V-shaped; lateral view more broadly oval than the typical. L. 26-28 μm. W. 15-16 μm. Isth. 5-6 μm. Th. 11 μm.

This variety should be compared with *Euastrum crassicolle* Lund. as seen in front view, a species which is quite different in lateral view.

DISTRIBUTION: Florida, Louisiana.

PLATE LXXII, figs. 6-6a.

48b. Euastrum incavatum var. **floridense** f. **pyramidatum** Scott & Grönblad 1957, Acta Soc. Sci. Fennicae, II, B, II(8): 13. Pl. 3, Figs. 17, 18.

A form somewhat larger than the typical variety, the semicells pyramidal; basal lobes angular, the lateral margins with an undulation immediately above the basal lobes, then sharply retuse to form the polar lobe, which is similar in form to the typical but with the angles bearing a sharp tubercle; lateral view oval, the poles broadly rounded, a round tubercle in the midregion of the lateral margins. L. 36-37 μm. W. 22-24 μm. Isth. 6-9 μm. Th. 13-15 μm.

DISTRIBUTION: Florida, Louisiana, Mississippi.

PLATE LXXII, figs. 8-8b, 12, 12a.

48c. Euastrum incavatum var. **wollei** Raciborski 1889, Pam. Wydz. Mat.-Przyr. Akad. Umiej. 17: 103. Pl. 2, Fig. 4.

A variety with the angles of the basal lobes slightly divergent and angular, symmetrically retuse to the narrow polar lobe, which is truncate, with a broad V-shaped invagination; vertical view oval, the poles rounded, with a subapical tubercle on either side. L. 39-43 μm. W. 23-25 μm. Isth. 7-8 μm. Ap. 13 μm. Th. 14-15 μm.

DISTRIBUTION: Pennsylvania. Europe.

PLATE LXXII, figs. 9, 9a.

49. Euastrum incertum Fritsch & Rich 1924, Trans. Roy. Soc. South Africa 11: 334. Fig. 11.

Cells medium-sized, broadly truncate-oval in outline, 1.5 times longer than broad; semicells truncate-pyramidal, the basal lobes convexly rounded at the angles, bearing 2 or 3 sharp spines, with a bispinate process on the upper margins of the basal lobes, the margins of the semicell sharply incised to the polar lobe, which has subparallel margins, bearing a sharp spine midway to the apex, which is truncate and furnished with 1 or 2 spines at the angles, the apical margin truncate and slightly undulate, with a deep, linear median incision; face of semicell profusely decorated with granules, a circle of large granules on a prominent median protrusion, a circular pattern of granules on either side of the median notch, a circular pattern of granules supraisthmial on the basal lobes and a scattering of granules within the margin of the basal lobes; sinus narrow and closed; lateral view elliptic, the poles truncate but furnished with a short, granular protrusion, bordered by a lateral tubercle, a promiment, trigranulate, subapical protrusion, with the lateral margins retuse to a prominent, truncate and granular, collarlike protrusion in the midregion. L. 60 μm. W. 44-46 μm. Isth, 9-10 μm. Th. 20 μm. Zygospore unknown.

DISTRIBUTION: Maine. South Africa.

PLATE LXXV, fig. 22.

50. Euastrum incrassatum Nordstedt 1887, Bot. Notiser 1887: 156; 1888a, Kongl. Svenska Vet.-Akad. Handl. 22(8): 35. Pl. 3, Fig. 12.

A small *Euastrum* broadly oval in outline, 1.5 to 2 times longer than broad; semicells nearly semicircular, the lateral margins undulate and convex from a narrow, closed sinus to the apex, which is bilobed, the median notch a relatively deep V; the polar lobe scarcely evident; face of semicell with a low median swelling; the wall bearing 2 intra-marginal rows of crenations (series of low swellings), with a supraisthmial tubercle at the base of the semicell; lateral view narrowly oval, the poles rounded-truncate, the margins retuse to a median swelling in the midregion; vertical view oval, the poles broadly rounded and smooth, the lateral margins with a rounded protrusion. L. 29–49 μm. W. 19–26 μm. Isth. 6–9 μm. Th. 10μm. Zygospore unknown.

DISTRIBUTION: Michigan. Europe, Australia, New Zealand, South Africa.
PLATE LXXIII, figs. 9–9b.

51. **Euastrum indicum** Krieger 1932, Arch. f. Hydrobiol. Suppl. 11: 213. Pl. 22, Fig. 5.

A medium-sized *Euastrum*, oval in outline, 2 times longer than broad; semicells vase-shaped, the basal lobes with broadly rounded to angular margins, then retuse to the long polar lobe with parallel margins, the apex truncate with rounded angles, a short, closed median notch; face of semicell prominently swollen in the midregion near the base and with a central mucilage pore, bordered by a low facial swelling; sinus short, narrow, and closed; wall lightly punctate; lateral view elliptic in outline, the poles rounded, the sides smooth and converging slightly to a basal, smooth swelling; vertical view oval, the poles rounded, with 2 lateral convolutions on either side, the aspect presenting a 6-undulate figure. L. 63–76 μm. W. 31–36 μm. Isth. 10.5–11 μm. Th. 19 μm.

DISTRIBUTION: Mississippi. Sumatra.
PLATE LXVII, figs. 4–4b.

52. **Euastrum inerme** (Ralfs) Lundell 1871, Nova Acta Reg. Soc. Sci. Upsal., III, 8(2): 20. Pl. 2, Fig. 3 var. **inerme**.
Euastrum elegans var. *inerme* Ralfs 1848, Brit. Desm., p. 89. Pl. 17, Fig. 7e.

A small *Euastrum*, broadly oval in outline, 1.5 times longer than broad; semicells truncate-pyramidal, the basal lobes broadly rounded, the lateral margins converging, with a conspicuous undulation halfway to the apex; the polar lobe only slightly produced, the apex truncate with broadly rounded angles, the median incision short, narrow, and closed; face of semicell with a low median protrusion in which there is a mucilage pore, a supraisthmial swelling median at the base of the semicell; wall smooth; sinus narrow and closed; lateral view oval, the poles broadly rounded, the lateral margins symmetrically convex to the base of the semicell, where there is a slight thickening of the wall, a mucilage pore evident in the midregion of the lateral margins; vertical view oval, the poles symmetrically rounded and smooth, the lateral margins slightly triundulate. L. 42–66 μm. W. 28–42 μm. Isth. 7.2–10 μm. Ap. 14–19.2 μm. Th. 20–23.5 μm. Zygospore unknown.

DISTRIBUTION: Louisiana, Massachusetts, Mississippi, Nebraska, New Hampshire, New Jersey, North Carolina, Utah, Washington. British Columbia, Labrador, Newfoundland, Québec. Great Britain, Europe, Asia, Africa, South America, Faeroes.
PLATE LXI, figs. 5, 5a.

52a. **Euastrum inerme** var. **depressum** Wolle 1883, Bull. Torr. Bot. Club 10: 18; 1892, Desm. U.S., p. 115. Pl. 35, Fig. 8; Krieger 1937, Rabenhorst's Krypto-gamen-Flora 13(1:3): 507. Pl. 63, Fig. 26.

A variety relatively broader than the typical; semicells broadly pyramidal with a narrower apex. L. 50–55 μm. W. 36–40 μm.

DISTRIBUTION: New Jersey.

PLATE LX, fig. 16.

53. **Euastrum infernum** Irénée-Marie 1956, Rev. Algol. 2(1/2): 116. Fig. 9.

A large *Euastrum*, broadly truncate-oval in outline, 1.75 times longer than broad; the semicells pentagonal in outline, the basal lobes narrowly rounded and somewhat downwardly directed, the lateral margins abruptly diverging to form an invaginated upper, lateral lobule, a narrow but open sinus between the lateral lobules and the polar lobe, which is anvil-shaped and inflated, the apical margin flat, with a very short, closed median incision; face of semicell with a supraisthmial protrusion in the midregion, and with a low swelling within the margins of the apical lobules; wall punctate throughout; sinus open within but closed outwardly. (Lateral view not described). L. 165–170 μm. W. 96–97 μm. Isth. 25.5–26.6 μm. Ap. 64–65 μm.

Although this species is incompletely described, it presents a very distinct aspect as seen in face view and should be sought so as to establish it as an acceptable taxon.

DISTRIBUTION: Québec.

PLATE LXV, fig. 3.

54. **Euastrum informe** Borge, Emend. Prescott 1935, Rhodora 37: 25. Pl. 325, Figs. 15–17 f. **informe**.

Euastrum informe Borge 1925, Ark. f. Bot. 19(17): 23. Pl. 1, Figs. 9, 10.

A small *Euastrum*, narrowly rectangular in outline, 2 times longer than broad; the semicells semiquadrate, the basal lobes relatively wide (vertically), the lower and upper angles furnished with a tubercle, an abrupt incision above the basal lobes forming a sinus between the polar lobe, which has the apical margin elevated to a deep median, V-shaped notch, bordered by hornlike thickenings of the wall, giving the polar lobe a pincerlike aspect, a tubercle on the outer margin of the polar lobe; face of semicell with 3 protrusions across the base of the lower lobes, a prominent protrusion in the midregion of the polar lobules, and a tuberculation (granule) within the apical margin; lateral view narrowly elliptic, the poles apiculate, with subapical tubercles, a truncate protrusion near the base of the polar lobe and a larger swelling on either side of the basal lobes; vertical view oval, the poles apiculate, the margins with 3 swellings in the midregion. L. 37.5–42.7 μm. W. 18.5–21 μm. Isth. 4.5–7 μm. Ap. 14.5–15.5 μm. Zygospore unknown.

DISTRIBUTION: Massachusetts. Newfoundland, Québec. South America.

PLATE LXXVIII, figs. 21–21b.

54a. **Euastrum informe** f. **oculatum** Scott & Prescott 1952, Hydrobiologia 4(4): 387. Pl. 3, Fig. 8.

A variety differing from the typical by the possession of a large mucilage pore medially located immediately above the central facial protuberance of the semicell. L. 39–44 μm. W. 18–19 μm. Isth. 5 μm. Th. 14 μm.

DISTRIBUTION: Florida, Mississippi.

PLATE LXXVIII, figs. 22, 22a.

55. **Euastrum insigne** Hassall ex Ralfs 1848, Brit. Desm., p. 83. Pl. XIII, Figs. 6a–6e var. **insigne** f. **insigne.**

Euastrum insigne Hassall 1845, British Freshwater Algae II, p. 211. Pl. 91, Fig. 2.
Euastrum mammillosum Wolle 1883, Bull. Torr. Bot. Club 10: 13. Pl. 27, Fig. 21.

A relatively large *Euastrum*, truncate-oval in outline, 2 times longer than broad; the semicells vase-shaped, the basal lobes narrowly rounded and much extended, the lateral margins converging and retuse to form a prominent polar lobe which is inflated and expanded at the apex to form 2 narrow lobules that are horizontally directed, the apical margin flat, with a short, median, closed incision, with a prominent swelling within the margin of the apical lobules; face of semicell with 2 prominent downwardly directed lobes nearly meeting the facial lobes of the other semicell; wall coarsely granular throughout; sinus closed within but opening wide outwardly; lateral view truncate-oval, the poles flat, truncate, and extended laterally to form lobules, the margins deeply retuse to the much inflated basal lobes, with shoulderlike upper angles and subparallel margins to the basal angles; vertical view subrectangular, the poles extended to form a prominent lobe, with a subpolar protrusion, the sides retuse in the midregion. L. 80–124 μm. W. (30)43–76 μm. Isth. 11–20 μm. Ap. 22–42 μm. Th. 28.8–39 μm. Zygospore unknown.

DISTRIBUTION: Florida, Georgia, Massachusetts, Michigan, Minnesota, New Jersey, Pennsylvania, Rhode Island. Labrador, Québec. Great Britain, Europe, Arctic, Faeroes.

PLATE LXV, fig. 11.

55a. **Euastrum insigne** var. **insigne** f. **poriferum** Graffius 1963, Dissert., p. 125. Pl. 7, Fig. 6.

A form differing from the typical by having reduced apical lobules and abbreviated facial protuberances, the median sinus of the cell widely open and shorter than in the typical, and with less extended basal lobes; a mucilage pore between 2 facial protuberances. L. 94 μm. W. 55 μm. Isth. 13 μm. Th. 35 μm.

DISTRIBUTION: Michigan.

PLATE LXV, fig. 12.

55b. **Euastrum insigne** var. **elevatum** Prescott var. nov.

Varietas a varietate typica differens ut minor et ut lobos basales breves, super quos lobi secondarii laterales videntur, habet; lobus polaris multo brevior quam in specie typica; superficies semicellulae 3 protuberationes trans basim atque par protuberationum horizontaliter positorum media in parte; semicellulae a latere visae 2 protrusiones in parte basali praebentes. Long. 82 μm. Lat. 41 μm. Ist. 12 μm. Crass. 24 μm.

ORIGO: E palude distante 7 milia passuum (11.3 km.) ad meridiem a loco Sumatra, Florida dicto.

HOLOTYPUS: Scott Coll. Fla-62.

ICONOTYPUS: Plate LXV, figs. 14, 14a.

A variety differing from the typical by being smaller and by having shorter basal lobes, above which are lateral, secondary lobes, the polar lobe much shorter than the typical; face of semicell with 3 protuberances across the base of the semicell and a pair of horizontally placed protuberances in the midregion of the semicell; lateral view with 2 protrusions in the basal part of the semicell. L. 82 μm. W. 41 μm. Isth. 12 μm. Th. 24 μm.

DISTRIBUTION: Florida.
PLATE LXV, fig. 14, 14a.

55c. Euastrum insigne var. **lobulatum** Prescott & Scott 1945, Amer. Mid. Nat. 34(1): 240. Pl. 2, Fig. 16 f. **lobulatum.**

Euastrum insigne Hassall, in: Prescott 1935, Rhodora 37: 25. Pl. 325, Figs. 9, 10.
Euastrum insigne f. Grönblad 1921, Acta Soc. Fauna Flora Fennica 49(7): 35. Pl. 3, Fig. 35.

A variety differing from the typical by having narrow and more produced basal lobes, the upper margins of the basal lobes with a protrusion; the polar lobe with 4 lobules as in the typical; wall coarsely scrobiculate. L. 96.6–115 μm. W. 54–66 μm. Isth. 13–15 μm. Ap. 28–34 μm.
DISTRIBUTION: Florida, Louisiana, North Carolina. Québec. Europe.
PLATE LXVI, fig. 1.

55d. Euastrum insigne .var. **lobulatum** f. **acutilobum** Scott & Prescott 1952, Hydrobiologia 4(4): 388. Pl. 2, Fig. 4.

A form differing from the typical variety by having more extended and narrower basal lobes, the sinus more widely open; protuberances of the face of the semicell longer and bilobed, the lower lobule somewhat produced and in contact with the lobules of the other semicell. L. 111 μm. W. 78 μm. Isth. 15 μm.
DISTRIBUTION: Florida, North Carolina.
PLATE LXVI, fig. 2.

55e. Euastrum insigne var. **lobulatum** f. **taylorii** Prescott and Scott 1945, Amer. Mid. Nat. 34(1): 241. Pl. 7, Fig. 6.

Euastrum insigne f. Taylor 1935, Pap. Michigan Acad. Sci., Arts & Lettr. 20: 205. Pl. 45, Fig. 8.

A form differing from the typical variety by being smaller and stouter, and in having shorter, less extended basal lobes; the facial protuberances shoulderlike, somewhat bilobed on the margin. L. 89–108 μm. W. 53–53.9 μm. Isth. 13–17.7 μm. Ap. 17 μm below the apical lobules. Ap. 29–32.2 μm.
DISTRIBUTION: Louisiana. Newfoundland, Québec.
PLATE LXVI, fig. 3.

55f. Euastrum insigne var. **pulchrum** Krieger 1937, Rabenhorst's Kryptogamen-Flora 13(1:3): 531. Pl. 7, Figs. 4, 5.

A variety differing from the typical by having subquadrate, bilobed basal lobes and by having 3 swellings between the 2 principal facial protuberances, 1 supraisthmial and 2 in the midregion of the semicell. L. 120 μm. W. 60 μm. Isth. 13 μm. Ap. 33 μm. Th. 40 μm.
DISTRIBUTION: Minnesota. Germany.
PLATE LXVI, fig. 4.

56. Euastrum insulare (Wittr.) Roy 1877, Scott. Nat. 1877: 68. Pl. 40, Figs. 11–13 var. **insulare.**

Euastrum binale var. *insulare* Wittrock 1872, Bih. Kongl. Vet.-Akad. Handl. 1: 49. Pl. 4, Fig. 7.

A small *Euastrum,* quadrate in outline, 1.5 times longer than broad; semicells quadrangular, the basal lobes quadrangular and weakly bilobed with the margins subparallel; polar lobe short and broad, with parallel margins, the apical angles furnished with a blunt, spinelike thickening, the apical margin flat but retuse in the midregion; face of the semicell with a smooth median protrusion; sinus narrow and closed throughout; lateral view oval, the poles broadly rounded and smooth, the margins convex with a median protrusion; vertical view oval, the poles rounded, the margins convex with a median protrusion. L. 17–34 μm. W. 11.5–24 μm. Isth. 3.3–6 μm. Ap. 8–10(12) μm. Th. 7.5–12 μm. Zygospore globose, the wall beset with long, sharp spines, 20 μm in diameter without spines.

DISTRIBUTION: Colorado, Connecticut, Florida, Iowa, Kentucky, Massachusetts, Michigan, Minnesota, Montana, New Hampshire, New York, North Carolina, Oklahoma, South Carolina, Utah, Virginia, Washington. British Columbia, Newfoundland, Québec. Great Britain, Europe, Asia, Africa, South America, Arctic.

PLATE LXXI, figs. 8–9b, 11–11b, 13–13b.

56a. Euastrum insulare var. **basichondrum** Messikommer 1938, Hedwigia 78: 168. Pl. 2, Figs. 14, 15.

A variety differing from the typical by having semicells somewhat pyramidal in outline, the basal lobes not quadrate; with a supraisthmial tubercle in the midregion. L. 30–31 μm.

DISTRIBUTION: Michigan. Europe.

PLATE LXXI, figs. 6, 6a.

56b. Euastrum insulare var. **lacustre** (Messik). Krieger 1937, Rabenhorst's Kryptogamen-Flora 13(1:3): 556. Pl. 75, Figs. 15–18 f. **lacustre**.

Euastrum binale var. *lacustre* Messikommer 1927, Mitt. Bot. Mus. Univ. Zürich 122: 98. Pl. 1, Fig. 16.

A variety similar in shape to the typical but with the wall bearing numerous irregularly distributed pores, and with pores forming from 4 to 6 vertical rows in the polar lobe. L. 23–32 μm. W. 19–22 μm. Isth. 4.5–6 μm. Ap. 9.5–14 μm. Th. 12–14.5 μm.

This variety has been found as a fossil in peat and lake chalk in Switzerland.

DISTRIBUTION: Louisiana, Mississippi. Europe.

PLATE LXXI, figs. 5–5b.

56c. Euastrum insulare var. **lacustre** f. **rectum** Prescott f. nov.

Forma minor quam forma typica, angulis lobi polaris late rotundatis, margine apicali profunde invaginato; anguli loborum basalium subquadrati, marginibus subparallelis, tuberculo parvo ad apices viso; protuberatio superficialis nulla, tuberculum, autem, admodum infra apicem incisurae polaris mediae visum; semicellulae a latere visae ovatae, polis subrotundatis atque paululum productis, tuberculum subapicale in margines praebentibus; lateribus convexis. Long. 20 μm. Lat. 13 μm. Ist. 4 μm.

ORIGO: E palude Sphagni pleno distante 2 milia passuum (3.2 km.) ad orientem a loco Pearl River, Louisiana dicto.

HOLOTYPUS: Scott Coll. La-18.

ICONOTYPUS: Plate LXXI, figs. 4, 4a.

A form smaller than the typical variety, the angles of the polar lobe broadly rounded, the apical margin deeply invaginated; angles of the basal lobes subquadrate with subparallel margins, and with a small tubercle at the apices; facial protuberance lacking but with a tubercle immediately below the apex of the median apical notch; lateral view oval, the poles rounded and slightly produced, with a subapical tubercle on the margin of the convex sides. L. 20 μm. W. 13 μm. Isth. 4 μm.

DISTRIBUTION: Louisiana.

PLATE LXXI, figs. 4, 4a.

56d. **Euastrum insulare** var. **silesiacum** (Grönblad) Krieger 1937, Rabenhorst's Kryptogaman-Flora 13(1:3): 557. Pl. 76, Figs. 19–21 f. **silesiacum.**

Euastrum insulare f. *silesiaca* Grönblad 1926, Soc. Sci. Fennica. Commen. Biol. 2(5): 13. Pl. 1, Figs. 26, 27.

A variety shorter and relatively broader than the typical; basal angles with slightly diverging margins; lateral view oval, the poles somewhat produced, broadly expanded in the midregion of the semicell; in vertical view similar to the typical. L. 16–22 μm. W. 14–17 μm. Isth. 4.5–5 μm. Th. 12 μm.

DISTRIBUTION: Alaska, Colorado, Ohio, Oregon. Labrador, Europe, Asia, Africa, South America, Panama Canal, West Indies.

PLATE LXXI, figs. 1–1b.

56e. **Euastrum insulare** var. **silesiacum** f. **minus** Prescott & Scott 1945, Amer. Mid. Nat. 34(1): 241. Pl. 2, Fig. 2.

A form similar to the typical variety but decidedly smaller throughout. L. 19 μm. W. 14.6 μm. Isth. 3.5 μm.

DISTRIBUTION: Louisiana. South America.

PLATE LXXI, figs. 2–2b, 3.

57. **Euastrum integrum** (Wolle) Wolle 1892, Desm. U.S., p. 117. Pl. 31, Figs. 18-22.

Euastrum simplex Wolle 1884a, Desm. U.S., p. 106. Pl. 27, Figs. 18–22.
(Non *Euastrum simplex* Gay 1884a, Essai d'une monographie locale des Conjuguées, p. 62. Pl. 2, Fig. 2).
Euastrum binale var. *hians* West, (p.p.), Conn & Webster 1908, Connecticut Geol. & Nat. Hist. Surv. 10: 62. Fig. 105; Krieger 1937, Rabenhorst's Kryptogamen-Flora 13(1:3): 552. Pl. 75, Fig. 16.

Cells small, truncate-oval in outline, 1.5 times longer than broad; semicells urn-shaped, the basal lobes broadly rounded with smooth margins, retuse to the short polar lobe which has diverging margins, the apical margin flat, sharply pointed at the angles, the median notch a relatively deep, narrow V; lateral view (?); vertical view (?). W. 16–36 μm. Zygospore unknown.

This is an incompletely described species and at present an invalid taxon. It is included here for the sake of the record because as described and illustrated by Wolle (*l.c.*) it appears to be a distinct species. Krieger (*l.c.*) has placed the species as interpreted by Conn & Webster (1908, p. 62, Fig. 105) in synonymy with *Euastrum binale* var. *hians* W. West. The polar lobe is especially so distinctive that we cannot agree that *E. integrum* is related to *E. binale*, even though the illustration used by Conn & Webster is not entirely clear as to its identity. Kreiger (*l.c.*) has indicated the plant as described by Wolle as of uncertain position.

DISTRIBUTION: Connecticut, Florida, Massachusetts, Minnesota, Vermont.

PLATE LX, fig. 6.

58. **Euastrum intermedium** Cleve 1864, Öfv. Kongl. Vet.-Akad. Förhandl. 20(10): 484. Pl. 4, Fig. 1 var. **intermedium.**

Euastrum purum Wolle 1887, Freshwater Alg. U.S., p. 37, Pl. 58, Figs. 9-11.

A medium-sized *Euastrum*, truncate-oval in outline, 1.75 times longer than broad; the semicells vase-shaped, the basal lobes narrowly rounded and laterally extended, the margins at the angles smooth, deeply retuse above to the polar lobe, which has diverging margins and an inflated apex, the apical margin broadly convex, the angles rounded, the median notch short and narrow but not quite closed; face of semicell with 2 median swellings near the base; wall smooth; sinus acute within, opening widely outward; lateral view narrowly oval, the poles inflated with the apex broadly rounded, the lateral margins retuse and then symmetrically swollen in the basal region; vertical view broadly elliptic, the poles narrowly rounded and somewhat produced, margins with 2 undulated swellings. L. 51-85.5 µm. W. 28.5-48 µm, Isth. 8-13 µm. Ap. 19.2-26.4 µm. Th. 20-28 µm.

DISTRIBUTION: Alaska, Florida, Georgia, Louisiana, Massachusetts, New Hampshire, New Jersey, North Carolina, Pennsylvania, South Carolina, Washington. British Columbia. Great Britain, Europe, Arctic, Azores.

PLATE LXVI, figs. 6-6b.

58a. Euastrum intermedium var. **longicolle** Borge 1925, Ark. f. Bot. 19(17): 21. Pl. 2, Figs. 30-32.

Euastrum intermedium (Cleve) West & West 1896a, Trans. Linn. Soc. London, Bot., II, 5: 242. Pl. 14, Figs. 18, 19.

A variety with the basal lobes narrower, more sharply rounded at the angles, and extended, the lateral margins of the semicell more deeply retuse to form a more slender polar lobe, the apex narrower than in the typical. L. 60-86 µm. W. 38.5-50 µm. Isth. 7-10 µm. Ap. 17-20 µm. Th. 16.5-18.5 µm.

DISTRIBUTION: Florida, Georgia, Louisiana, Michigan. Ontario, Québec. South America.

PLATE LXVI, figs. 8, 8a (zygospore).

58b. Euastrum intermedium var. **orientale** (Turner) Krieger 1937, Rabenhorst's Kryptogamen-Flora 13(1:3): 533. Pl. 71, Figs. 11, 12.

Euastrum orientale Turner 1892, Kongl. Svenska Vet.-Akad. Handl. 25(5): 79. Pl. 10, Fig. 34.

A variety shorter and stouter than the typical, the basal lobes rounded at the angles, the lateral margins not so deeply retuse, forming a polar lobe with parallel margins, the apical margin flat and with a small V-shaped notch. L. 64-70.2 µm. W. 42-46.8 µm. Isth. 12 µm. Th. 28 µm.

DISTRIBUTION: Florida, Wisconsin. Asia (India).

PLATE LXVI, Figs. 11, 11a; Pl. LXVII, Fig. 1.

58c. Euastrum intermedium var. **scrobiculatum** (Schmidle) Krieger 1937, Rabenhorst's Kryptogamen-Flora 13(1:3): 533. Pl. 71, Fig. 13.

Euastrum intermedium f. *scrobiculata* Schmidle 1898, Bih. Kongl. Svenska Vet.-Akad. Handl. 24, Afd. III(8): 46. Pl. 2, Fig. 27.

A variety differing from the typical by having the polar lobe more widely expanded at the apex, the angles sharply rounded and somewhat produced, the apical margin flat, with an intramarginal protuberance (lobule) at the apex; face of semicell with a pair of mucilage pores in vertical arrangement in the midregion between 2 protrusions. L. 64-96 µm. W. 32-54 µm. Isth. 6.5-13 µm. Ap. 20-27.5 µm. Th. 26.2-27.5 µm.

DISTRIBUTION: Michigan, Mississippi, North Carolina. Newfoundland, Ontario. Carinthia, Arctic.

PLATE LXVII, fig. 2.

58d. Euastrum intermedium var. **validum** West & West 1898, Linn. Soc. Jour. Bot. 33: 288. Pl. 16, Fig. 8.

A variety differing from the typical by being stouter, the basal lobes shorter, with somewhat truncate angles which are slightly upturned; apex with intra-marginal lobules as in var. *scrobiculatum*. L. 70–88 μm. W. 40–54 μm. Isth. 6.5–8 μm. Th. 23.5–40 μm.

DISTRIBUTION: Alaska, Florida. Québec.

PLATE LXVII, fig. 3.

59. Euastrum invaginatum Scott & Prescott 1952, Hydrobiologia 4(4): 388. Pl. 1, Fig. 8.

Cells medium-sized, narrowly oval in outline, about 2 times longer than broad; semicells semioval with a very deep polar notch that is broadly open internally to form a prominent invagination that extends almost to the midregion of the semicell, the resulting lobules of the semicell pincer-shaped; margin of the semicell showing 9 crenulations between the sharp basal angles and the apex; face of semicell with an almost imperceptible swelling in the midregion; the wall thin and coarsely punctate; median incision deep, the sinus narrow and closed through-out; in lateral view the cell fusiform in outline with a V-shaped incision in the midregion, the apices truncate, the base of the semicell only slightly swollen, the margins undulate; vertical view quadrate-oval, the lateral margins nearly parallel, with a slight protrusion on each side, the poles broadly convex and crenate, with three or four series of crenations within the margin. L. 42–45 μm. W. 20–25 μm. Isth. 7–8 μm. Th. 12–13 μm.

DISTRIBUTION: Florida.

PLATE LXXIII, fig. 13–13b.

60. Euastrum jenneri Archer, in Pritchard 1861, A History of Infusoria, p. 730; Krieger 1937, Rabenhorst's Kryptogamen-Flora 13(1:3): 505. Pl. 63, Fig. 8.

Euastrum circulare var. *ralfsii* Brébisson 1856, Mém. Soc. Impér. Sci. Cherbourg 4: 122.

A medium-sized *Euastrum*, quadrate in outline, 1.5 times longer than broad; semicells rectangular in outline, the basal lobes rounded at the angles, the margins retuse to upper, lateral lobules which extend as far as the basal lobes, establishing a shoulderlike incision to the polar lobe, which is short and collarlike, with subparallel margins, the apex slightly inflated, with rounded angles, the apical margin flat, the median incision narrow and closed; face of semicell with 3 transverse rows of protuberances with mucilage pores, a row of 4 in the area of the upper, lateral lobules, 3 in a median row, and 4 in the region of the basal lobes; wall punctate. L. 72–76 μm. W. 45–48 μm. Isth. 15 μm. Zygospore unknown.

DISTRIBUTION: Michigan, New Hampshire. Great Britain, Europe.

PLATE LXI, fig. 4.

61. Euastrum johnsonii West & West 1898, Linn. Soc. Jour. Bot. 33: 288. Pl. 16, Fig. 9 var. **johnsonii** f. **johnsonii**.

A medium-sized *Euastrum*, broadly oval in outline, 1.5 times longer than broad; semicells bell-shaped, the basal lobes rounded, with a slight elevation on the upper margins, lateral margins symmetrically retuse to form a broad, truncate polar lobe with broadly rounded angles, apical margin with a subapical circle of granules; face of semicell with a broad, low protrusion; basal part of the semicell with scattered patterns of granules across the face of the basal lobes, especially near the angles; lateral view broadly oval, the poles broadly rounded and showing a subapical circle of granules, the margins retuse to a broad, basal swelling that is bigranulate at the margin. L. 60 μm. W. 40 μm. Isth. 11 μm. Ap. 17 μm. Th. 20 μm. Zygospore unknown.

DISTRIBUTION: Connecticut, Florida, Kentucky.

PLATE LXIX, figs. 5, 5a.

61a. Euastrum johnsonii var. johnsonii f. pyriforme Prescott f. nov.

Forma a forma typica differens quod margo apicalis distincte retusus, non convexus, et quod semicellula 2 protrusiones media in parte et protrusionem supraisthmialem ad basim, necnon protrusionem intramarginalem parvam vicinitate inflationum lateralium superiorum habet, membrana sine granulis nisi intra marginem loborum basalium ad apices; semicellulae a latere visae late ovatae, polis late rotundatis, marginibus lateralibus retusis et ad inflationem prominentem media in regione semicellulae divergentibus, deinde convergentibus, inflationem parvam admodum super sinum formantibus. Long. 57 μm. Lat. 38 μm. Crass. 25 μm. Lat. ad isthmum 11 μm.

ORIGO: E palude distante 8 milia passuum (12.9 km.) ad septentriones et orientem a loco Deland, Florida dicto.

HOLOTYPUS: Scott Coll. Fla-99.

ICONOTYPUS: Plate LXIX, figs. 8-8b.

A form differing from the typical by having the apical margin distinctly retuse rather than convex, and in having 2 protuberances in the midregion of the semicell, and with a supraisthmial protrusion at the base of the semicell; a small intramarginal protrusion in the region of the upper, lateral swellings; cell wall with granules only within the margin of the basal lobe angles; lateral view broadly oval, the poles broadly rounded, the margins retuse and diverging to a prominent swelling in the midregion of the semicell, and then converging, forming a low swelling immediately above the sinus. L. 57 μm. W. 38 μm. Isth. 11 μm. Th. 25 μm.

DISTRIBUTION: Florida.

PLATE LXIX, figs 8-8b.

61b. Euastrum johnsonii var. nudum Prescott 1942, Trans. Amer. Microsc. Soc. 61(2): 117. Figs. 11, 12.

A variety differing from the typical by being broader, the basal lobes undulate at the margin, beyond which is a slight swelling; the apical margin truncate but slightly retuse in the midregion; face of semicell with a transverse row of 3 protuberances which are smooth, and with an intramarginal protrusion in the region of the upper, lateral swellings; lateral view broadly oval, the poles truncate and broadly convex, the margin with 2 swellings in the basal region. L. 66-68 μm. W. 45.6-46 μm. Isth. 10.5 μm. Th. 26.1-27.5 μm.

This variety should be compared with *Euastrum pectinatum* var. *planum* Krieger.

DISTRIBUTION: Louisiana.

PLATE LXIX, figs. 7, 7a.

61c. **Euastrum johnsonii** var. **porrectum** (Borge) Irénée-Marie 1947, Nat. Canadien 74: 118. Pl. 2, Fig. 2.

Euastrum porrectum Borge 1903, Ark. f. Bot. 1: 115. Pl. 5, Fig. 8.

A variety differing from the typical by having a smooth cell wall; basal lobes broadly rounded at the angles, the margins symmetrically retuse to a relatively narrow polar lobe, apical margin truncate but slightly retuse in the midregion; face of semicell with 2 protuberances in the midregion; lateral view broadly truncate-oval, the poles truncate, the margins retuse and diverging to form a smooth, basal swelling; vertical view oval, the poles rounded and slightly produced, with a subapical swelling, and with a prominent protrusion in the midregion which is biundulate. L. 58–67.5 μm. W. 38.6–48 μm. Isth. 13–16.2 μm. Ap. 15.5–17 μm. Th. 28–30 μm.

Krieger (1937, p. 657) assigns *Euastrum porrectum* to *Cosmarium*. This is not understandable inasmuch as the lateral and vertical views especially show *Euastrum* characteristics.

DISTRIBUTION: Québec. South America.

PLATE LXIX, figs. 9–9b.

62. **Euastrum kriegeri** Prescott nom. nov.

Euastrum crameri var. *tropicum* Krieger 1937, Rabenhorst's Kryptogamen-Flora 13(1:3): 620. Pl. 90, Figs. 4–6.
Euastrum crameri f. Krieger 1932, Arch. f. Hydrobiol. Suppl. 11: 211. Pl. 20, Fig. 12.

A medium-sized *Euastrum*, trapezoidal or truncate-oval in outline, 1.25 times longer than broad; semicells truncate-pyramidal, the basal lobes extended and biundulate, with a crenate protrusion between the basal lobes and the polar lobe; polar lobe with subparallel margins and with a slight, protruding undulation immediately below the apex, which is flat, with a deep median incision and with the angles extended into slightly diverging, hornlike processes; face of semicell with a very slight median swelling, above which is a horizontally arranged pair of mucilage pores; lobules of the polar lobe with a conspicuous granule, and with intramarginal granules on either side of the polar incision as well as within the margins of the basal lobes; sinus narrow and closed; lateral view narrowly oval, the poles truncate with a protruding process and 2 lateral processes immediately below the apex; base of semicell broadly rounded and smooth; vertical view narrowly oval, the poles rounded and furnished with 3 granulations, slightly convex in the midregion of the lateral margins. L. 43–52 μm. W. 30–34 μm. Isth. 7–8 μm. Th. 17–18 μm.

Although this form has been assigned to *Euastrum crameri* by Krieger (*l.c.*), we believe that the shape of the semicell, especially the polar lobe, and the lack of facial protrusions and tubercles, in addition to the difference in size and in cell proportions require a separation. The lateral view of *E. crameri* and *E. kriegeri* emphasizes the difference. The only feature which the 2 plants have in common is the pair of mucilage pores, which are, however, median in the former species, supramedian in the latter.

DISTRIBUTION: Florida. Sumatra.

PLATE LXXIV, figs. 24–24b; Pl. LXXX, figs. 10, 10a.

63. **Euastrum lapponicum** Schmidle 1898, Bih. Kongl. Svenska Vet.-Akad. Handl. 24: 47. Pl. 2, Fig. 29 var. **lapponicum** f. **lapponicum.**

A small *Euastrum*, rectangular in outline, 1.5 to 1 1/3 times longer than broad; semicells rectangular, the angles of the basal lobes rounded and bearing 2 granules, one above the other, the lateral margins of the semicell retuse above the basal angles to an upper, lateral lobule which bears a granule at the apex, the lobules extending as far as the basal lobes; polar lobe short and broad with parallel margins, the apical margin flat, the angles furnished with a tubercle, the apical notch a short V in the midregion; face of semicell with a prominent protuberance bearing a pair of large granules; wall with a conspicuous granule and 2 smaller granules within the angles of the basal lobe, a tubercular granule within the margin of the upper, lateral lobules, 3 small granules within the apical margin on either side of the median notch and a larger granule in the midregion of the apical lobules; sinus narrow and closed throughout; vertical view oval, the poles knoblike, with a subapical swelling on either side, the lateral margins convex and with a bimammillate swelling in the midregion; lateral view narrowly oval or broadly elliptic, the poles rounded and bearing a hornlike protrusion, with subapical tubercles on either side, the midregion of the semicell much inflated and showing a bimammillate protuberance. L. 35–44 μm. W. 27–36 μm. (with spines). Isth. 6–9 μm. Ap. 22–23 μm. Th. 15–20 μm. Zygospore unknown.

DISTRIBUTION: Alaska, Idaho, Michigan, Mississippi. Newfoundland, Ontario, Québec. Europe, Arctic.

PLATE LXXVII, figs. 12, 12a, 15.

63a. Euastrum lapponicum var. lapponicum f. laeve Prescott f. nov.

Forma minor quam forma typica, granulationes in marginibus lobulorum necnon in superficie semicellularum reductae; par horizontale pororum mucosum admodum super tumorem medium et ad utrumque latus; incisura media profundior, extrinsecus aperta et ad extremitatem apertam per incrassationem membranae praetexa. Long. 32 μm. Lat. 18 μm. Lat. ad isthmum 6 μm.

ORIGO: E stagno prope Via 26 dicta, distante 3 milia passuum (4.8 km.) ad occidentem a flumine Pascagoula dicto, Mississippi.

HOLOTYPUS: Scott Coll. Miss-74.

ICONOTYPUS: Plate LXXVII, fig. 13.

A form smaller than the typical, and with reduced granulations on the margins of the lobules, as well as on the face of the semicell; median protrusion with a horizontal pair of mucilage pores immediately above and to either side; median notch of the polar lobe deeper, open outwardly, and bordered by a wall thickening at the open end. L. 32 μm. W. 18 μm. Isth. 6 μm.

DISTRIBUTION: Mississippi.

PLATE LXXVII, fig. 13.

63b. Euastrum lapponicum var. mauritianum Prescott comb. nov.

Euastrum dubium Nageli f. *mauritiana* Irénée-Marie 1958, Nat. Canadien 85(5): 124. Fig. 15.

A form with the sinus of the cell shorter and opening outwardly into broadly rounded basal lobes, with upper, lateral lobules immediately above which are equal to the basal lobes in prominence; polar lobe with subparallel margins; face view of semicells with a simple, low protrusion in the midregion, and with a large granule immediately below the apical notch; lateral view subrectangular, the poles quadrately truncate and furnished with a short spine, the lateral margins undulate. L. 32.3–35 μm. W. 19.4–21 μm. Isth. 6.4 μm.

This variety is reminiscent of var. *ornatum* Wolosz. in its contour, but the facial markings are more simple, and the apical notch is even deeper. That feature especially excludes the form from *E. dubium*, to which Irénée-Marie (*l.c.*) assigned it.

DISTRIBUTION: Québec.
PLATE LXXIV, figs. 2, 2a.

63c. Euastrum lapponicum var. **protuberans** Prescott var. nov. f. **protuberans**.

Euastrum lapponicum f. Prescott & Scott 1945, Amer. Mid. Nat. 34(1): 242. Pl. 3, Fig. 10.

Varietas a varietate typica differens ut relative longior, 2 plo longior quam lata, differens necnon praesentia pororum mucosorum admodum super tumorem medium, et ad utrumque latus; tumor medius (a latere visus) maxime prominens; semicellulae a latere visae conico-pyramidales, protrusionem prominentem utroque in latere paululum ultra isthmium praebentes; margines superiores 3 granula utroque in latere ferentes; lobus polaris angustior quam in varietate typica, incisuram verticalem profundiorum quam in varietate typica et spinam brevem in angulis exterioribus habens. Long. 37 μm. W. 19 μm. Lat. ad isthmum 7 μm. Crass. 16 μm.

ORIGO: E stagno liliis favente distante 29 milia passuum (46.8 km.) ad meridiem et orientem a loco Panama City, Florida dicto.

HOLOTYPUS: Scott Coll. Fla-16.

ICONOTYPUS: Plate LXXVI, figs. 19, 19a.

A variety differing from the typical by its relatively greater length, 2 times longer than broad, by the presence of mucilage pores immediately above and to either side of the median tumor, which is very prominent (in lateral view); in lateral view the semicellls conical-pyramidal, with a promiment protrusion on either side a short distance beyond the isthmus, the upper margins bearing granules on either side; vertical incision of the polar lobe deeper than in the typical, and with a short spine on the outer angles of the polar lobe, which is relatively narrower than the typical. L. 37 μm. W. 19 μm. Isth. 7 μm. Th. 16 μm.

DISTRIBUTION: Florida.
PLATE LXXVI, figs. 19, 19a.

63d. Euastrum lapponicum var. **protuberans** f. **obliquum** Prescott f. nov.

Forma minor quam forma typica, granulatione superficiali aliquantum reducta; semicellula a vertice vise ovata, protrusionem prominentem utroque in latere in regionibus mediis praebens, hac protrusione, autem, oblique ordinata in rationem alterius semicellulae; membrana levis; sine protrusione inter inflationem mediam et polos. Long. 32 μm. Lat. 20 μm. Lat. ad isthmum 4.5 μm. Crass. 16 μm.

ORIGO: E stagno distante 3 milia passuum (4.8 km.) ad accidentem a loco Sorrento, Florida dicto.

HOLOTYPUS: Scott Coll. Fla-140.

ICONOTYPUS: Plate LXXVI, figs. 20–20b.

A form smaller than the typical variety, with the facial granulation somewhat reduced; in vertical view oval with a prominent protrusion on either side in the midregion but obliquely arranged in reference to one another, the wall smooth, without protrusion between the median swelling and the poles. L. 32 μm. W. 20 μm. Isth. 4.5 μm. Th. 16 μm.

DISTRIBUTION: Florida.
PLATE LXXVI, figs. 20–20b.

63e. Euastrum lapponicum var. **quebecense** Irénée-Marie 1958, Nat. Canadien 85(5): 131. Fig. 16.

A variety differing from the typical by having the basal lobes distinctly wider (vertically) than the typical, and with the upper, lateral lobules more prominent; apical margin undulate, and with a deeper, closed median incision; cell wall with fewer and larger tubercles (granules). L. 39.4–40.5 µm. W. 33.5 µm. Isth. 8.5–9.3 µm.

DISTRIBUTION: Québec.
PLATE LXXVI, fig. 24.

64. Euastrum latipes Nordstedt 1869, Vid. Medd. Naturh. Foren. Kjöbenhavn 1869, III, 1(14/15): 216. Pl. 2, Fig. 9 var. **latipes.**

A medium-sized *Euastrum*, 1.75 times longer than broad; the semicells flask- or vase-shaped, with a broad base, the basal angles sharply rounded, the margins converging into the polar lobe, with a low marginal protrusion about 1/3 the distance to the apex; apex decidedly inflated, the angles bluntly pointed and horizontally directed, the apical margin flat, with a slight median notch; face of semicell with a supraisthmial protuberance in the midregion and 2 horizontally disposed protuberances opposite the upper, marginal inflations; wall punctate; sinus closed for most of its length but open outwardly; lateral view narrowly truncate-oval, the poles slightly inflated and 3-lobed at the apical margin, the lateral margins diverging to the base and with a slight protuberance above the basal angles; vertical view elliptic, the poles narrowly rounded and produced, the lateral margins with 5 undulations. L. 113–139 µm. W. 65–82 µm. Isth. 15–23 µm. Th. 35–39 µm. Zygospore unknown.

DISTRIBUTION: Typical form not found in North America. South America.
PLATE LXXVIII, figs. 1–1b.

64a. Euastrum latipes var. **evoluta** (Grönblad) Taft 1949, Trans. Amer. Microsc. Soc. 68(3): 214. Pl. 1, Fig. 1.

Euastrum latipes f. *evoluta* Grönblad 1945, Acta Soc. Sci. Fennica, II, B, II(6): 13. Fig. 58.

A variety differing from the typical by being relatively shorter and stouter; basal lobes more rounded at the angles and bearing a prominent toothlike projection; the median notch of the polar lobe deeper and closed, bordered outwardly by a prominent thickening of the wall; midregion of the semicell with a mucilage pore. L. 129–132 µm. W. 82–86 µm. Isth. 16–17 µm. Ap. 36–39 µm.

The narrow and relatively deep apical incision, the presence of a facial mucilage pore, and the prominent protrusions of the apical margin on either side of the polar lobe incision give this form an appearance that suggests it does not belong with *E. latipes*. We have left the epithet as is until specimens can be seen and further studied.

DISTRIBUTION: New Jersey. Brazil.
PLATE LX, fig. 1.

65. Euastrum longicolle Nordstedt 1887, Bot. Notiser 1887: 156; 1888, Kongl. Svenska Vet.-Akad. Handl. 22: 33. Pl. 3, Fig. 5; Krieger 1937, Rabenhorst's Kryptogamen-Flora 13(1:3): 493. Pl. 60, Figs. 7–9 var. **longicolle.**

A large *Euastrum*, elongate-oval in outline, 2.5 times longer than broad; the semicells vase-shaped, distinctly 3-lobed, the basal lobes high and somewhat angular on the margins, which are subparallel, sharply retuse to a long, columnar polar lobe with slightly diverging margins, the apex inflated, the angles rounded, the apical margin flat but elevated in the region of the median notch, which is short and closed; face of semicell with 2 transverse rows of protuberances, the median one of the upper row containing a mucilage pore; wall punctate; sinus narrow and closed; lateral view subrectangular, truncately rounded at the apex, the margins sloping outwardly to the basal swelling, a mucilage pore showing in the midregion. L. 87 μm. W. 38 μm. Isth. 10 μm. Ap. 23–26 μm.

DISTRIBUTION: Virginia. New Zealand, Asia.

PLATE LX, fig. 8.

65a. **Euastrum longicolle** var. **capitatum** f. **extensum** (Scott & Grönblad) Prescott f. nov.

Euastrum longicolle var. *capitatum* f. Scott & Grönblad 1957, Acta Soc. Sci. Fennicae, II, B, II(8): 13. Pl. 3, Fig. 1.

Forma a forma typica differens ut lobus polaris multo elongatior, apex minus inflatus, incisura media paululo profundior et inaperta, incrassatione membranae apicalis praetexa. Long. 93–109 μm. Lat. 45–50 μm. Lat. lobi polaris 23 μm. Lat. ad isthmum 12–15 μm. Crass. 31 μm.

ORIGO: E stagno liliis favente, distante 6 milia passuum (9.7 km.) a loco Kinder, Louisiana dicto.

HOLOTYPUS: Scott Coll. La-118.

ICONOTYPUS: Plate LX, figs. 4, 4a, 9, 9a.

A form differing from the typical variety by having the polar lobe much more elongate and the apex less inflated, the median notch slightly deeper and closed, bordered by a thickening of the apical wall. L. 93–109 μm. W. 45–50 μm. W. pol. lob. 23 μm. Isth. 12–15 μm. Th. 31 μm.

The illustration, Pl. LX, fig. 4 has features more nearly like the typical variety. The latter has basal lobe margins which are diverging; a much shorter polar lobe; the lateral margins above the basal lobes deeply concave to distinctly swollen apices.

DISTRIBUTION: Louisiana.

PLATE LX, figs. 4, 4a, 9, 9a.

66. **Euastrum luetkemuelleri** Ducellier 1918a, Bull. Soc. Bot. Genève, II, 10: 134. Textfig. 123a var. **luetkemuelleri**.

A small *Euastrum*, cells truncate-oval in outline, 1.5 times longer than broad; semicells truncate-pyramidal, the basal lobes narrowly rounded at the angles, the margins retuse and converging to the apex; polar lobe scarcely evident; angles of the polar lobe bluntly pointed, the apical margin flat but retuse in the midregion, with a slight notch in evidence; face of semicell with a broad, low protuberance with a mucilage pore in the center; sinus narrow and closed throughout; lateral view narrowly oval, the poles rounded, the lateral margins slightly retuse and then convex in the region of the lower basal swelling; vertical view oval, the poles rounded, the midregion of the sides with a low swelling in which there is a mucilage pore. L. 20.5–27 μm. W. 12.7–19 μm. Isth. 4.5–7 μm. Ap. 12 μm. Th. 9.7–13 μm. Zygospore unknown.

DISTRIBUTION: Massachusetts, Mississippi, Montana, North Carolina, Ohio, Pennsylvania, Virginia. Québec. Great Britain, Europe, Asia, Africa, South America, Arctic.

PLATE LXXII, figs. 2, 2a.

66a. **Euastrum luetkemuelleri** var. **carniolicum** (Lütk.) Krieger 1937, Rabenhorst's Kryptogamen-Flora 13(1:3): 561. Pl. 80, Figs. 5–7.

Euastrum crassangulatum var. *carniolicum* Lütkemüller 1900, Ver. k.k. Zool.-Bot. Ges. Wien 50:
73. Pl. 1, Figs. 20-22.
Euastrum scaphaephorum Skuja (1932)1934, Acta Horti Bot. Latvia 7(1932): 95. Fig. 95.

A variety similar in size to the typical but with the polar lobe more in
evidence, short with subparallel margins, the apex slightly inflated, margins of the
semicell above the basal angles slightly undulate; wall with an internal nodular
thickening below the apical notch; median pore characteristic of the typical
lacking. L. 25-39 µm. W. 17.5-25 µm. Isth. 5-9 µm. Ap. 13.5 µm. Th. 12.5-17
µm.

DISTRIBUTION: Massachusetts, Mississippi, Montana. Europe, Asia,
Australia, Africa, South America.

PLATE LXXII, figs. 3-3b.

66b. **Euastrum luetkemuelleri** var. **floridanum** Scott & Grönblad 1957, Acta Soc.
Sci. Fennicae, II, B, II(8): 13. Pl. 1, Fig. 5.

A variety differing from the typical by being relatively stouter, the apical
lobe scarcely evident, the apical margin broadly convex without a median notch;
facial protuberance of the semicell more elevated, the mucilage pore showing
toward the apical margin; lateral view broadly oval, the semicells nearly round,
without a median protrusion. L. 36-39 µm. W. 24-25 µm. Isth. 9-10 µm. Th.
18-20 µm.

DISTRIBUTION: Florida.

PLATE LXXII, figs. 1-1b.

66c. **Euastrum luetkemuelleri** var. **sublaticolle** Scott & Grönblad 1957, Acta Soc.
Sci. Fennicae, II, B, II(8): 13. Pl. 3, Fig. 11.

A variety similar in size to the typical but with a more angular cell; basal
lobes of the semicell with diverging margins, bluntly angled, the lateral margins
deeply retuse to a short, truncate polar lobe, the apex flat and slightly concave,
with a subapical tubercle on the inner face of the wall as in var. *carniolicum*. L.
33 µm. W. 27 µm. Isth. 9 µm. Th. 16 µm.

DISTRIBUTION: Florida.

PLATE LXXII, figs. 4, 4a.

67. **Euastrum marianopoliense** Irénée-Marie 1938, Flore Desmid. Montréal, p.
140. Pl. 69, Fig. 4.

A medium-sized *Euastrum*, broadly truncate-oval in outline, about 1.5 times
longer than broad; semicells truncate-pyramidal, the basal lobes angular at the
margins which are subparallel, the lateral margins of the semicell retuse to an
upper, lateral lobule which is upwardly directed, a narrow, deep sinus between the
lobules and the cuneate polar lobe; lateral angles of the polar lobe narrowly
rounded, the apical margin elevated and then flat or slightly retuse in the
midregion; face of semicell with a smooth median protuberance; sinus narrow and
closed; lateral view narrowly oval in outline, the poles broadly rounded, the sides
convex to a median protrusion; vertical view narrowly oval, the poles narrowly
rounded, with subapical protrusions, concave to the bimammillate, median protrusion. L. 60 µm. W. 44 µm. Isth. 10 µm. Th. 26 µm. Zygospore unknown.

DISTRIBUTION: Québec. Venezuela.

PLATE LXXIX, figs. 1-1b.

68. **Euastrum montanum** West & West 1905, Trans. Proc. Bot. Soc. Edinburgh 23: 14. Pl. 1, Figs. 11, 12 f. **montanum**.

A small *Euastrum*, quadrate in outline, 1.5 times longer than broad; basal lobes with diverging margins, the lateral margins of the semicell undulate to a brief polar lobe which has rounded lateral angles, truncate but retuse in the midregion of the apical margin; face of semicell with a low protuberance in the midregion of the apical margin; wall smooth, sinus narrow and closed; lateral view oval, the poles broadly rounded, the semicells rhomboid with a prominent median swelling; vertical view oval, the poles rounded and smooth, with a slight protrusion laterally in the midregion. L. 17–28.6 μm. W. 14.5–26 μm. Isth. 3.5–6.5 μm. Ap. 9–15 μm. Th. 9–14.5μm. Zygospore globose, with blunt spines, 29 μm in diameter without spines; 44 μm in diameter with spines.

DISTRIBUTION: Colorado, Idaho, New Hampshire, Oklahoma. British Columbia, Labrador, Newfoundland. Great Britain, Europe, Asia, South America, Arctic.

PLATE LXX, figs. 20–20b.

68a. **Euastrum montanum** f. **reductum** Croasdale, in Croasdale & Grönblad 1964, Trans. Amer. Microsc. Soc. 83(2): 168. Pl. 7, Figs. 33, 34.

A variety differing from the typical by having the upper, lateral lobules on the margin of the semicell more developed, extending further than the basal lobes; apical margin flat with a slight retuseness in the midregion; facial protuberance less pronounced; in lateral view the semicells smoothly oval. L. 23–25 μm. W. 17–18 μm. Isth. 5–6 μm. Th. 11–12 μm.

DISTRIBUTION: Labrador.

PLATE LXX; figs. 24, 24a.

69. **Euastrum notatum** Taft 1947, Ohio Jour. Sci. 47: 86. Fig. 2.

A small *Euastrum*, truncate-oval in outline, 1.5 times longer than broad; semicells truncate-pyramidal, the basal lobes broadly rounnded from a widely open and very short sinus, the lateral margins slightly retuse and converging to a broad, truncate polar lobe, the angles rounded, the apical margin retuse; face of semicell with a broad, low protuberance; wall smooth; isthmus very wide, the sinus a broad V in the cell; lateral view truncate-oval, the poles truncate, slightly retuse, the lateral margins retuse to the smooth, median swelling, with a slight undulation at the base of the semicell; vertical view quadrate, the poles truncate and bilobed, the lateral margins with a smooth protuberance in the midregion. L. 37–43 μm. W. 27–29 μm. Isth. 17–19 μm. Ap. 16–17 μm. Zygospore unknown.

An unusual feature of this species is the very wide isthmus, giving a *Cosmarium*-like aspect when seen in facial view.

DISTRIBUTION: New Mexico.

PLATE LXVIII, figs. 6–6b.

70. **Euastrum obesum** Joshua 1886, Linn. Soc. Jour. Bot. 21: 638. Pl. 23, Figs. 19, 20 var. **obesum**.

Euastrum ralfsii Rabenhorst (*p.p.*), Wood 1872, Smithson. Contrib. to Knowledge 19: 139. Pl. 13, Fig. 1.

A small *Euastrum*, broadly truncate-oval in outline, up to 2 times longer than broad; semicells pyramidal, the basal lobes broadly rounded and wide (vertically), the upper margins of the semicell retuse to form a short, truncate polar lobe, the angles rounded, the apex relatively narrow with 2 lobules formed by a short, narrow, median notch; face of semicell with a broad, low protrusion; wall smooth; sinus short, closed internally but opening widely outward; lateral view broadly oval, the poles truncate and broadly rounded, the lateral margins retuse below the apex and then broadly inflated in the basal part of the semicell. L. 40-111 μm. W. 23-59 μm. Isth. 9-21 μm. Ap. 16-28 μm. Th. 18-24 μm. Zygospore unknown.

DISTRIBUTION: Alaska, California, Kentucky, Montana, Utah, Virginia. British Columbia, Labrador, Newfoundland, Québec. Great Britain, Europe, Asia, East Indies, Africa, Arctic.

PLATE LX, fig. 7.

70a. Euastrum obesum var. **subangulare** West & West 1895, Trans. Linn. Soc. London, Bot. II, 5: 50. Pl. 6, Fig. 15.

Euastrum obesum f. Irénée-Marie 1938, Flore Desm. Montréal, p. 125. Pl. 8, Fig. 9.

A variety differing from the typical by having more sharply rounded angles of the basal lobes, the lobes not as wide (vertically); facial protuberance low and broad; vertical view narrower than the typical, the poles more broadly rounded and the lateral, median swelling less. L. 53-77 μm. W. 32-44 μm. Isth. 9-13.5 μm. Ap. 16-22 μm. Th. 18-29.7 μm.

DISTRIBUTION: Alaska, California. Québec. Europe, Africa.

PLATE LX, figs. 5, 5a, 10.

70b. Euastrum obesum var. **trapezicum** (Börges.) Krieger Rabenhorst's Kryptogamen-Flora 13(1:3): 497. Pl. 59, Figs. 15-17.

A variety more slender than the typical, more than 2 times longer than broad; angles of basal lobes more narrowly rounded, the apex relatively broader; semicells in lateral view symmetrically oval. L. 76-94 μm. W. 29-44 μm. Th. 31 μm.

DISTRIBUTION: Alaska. South America.

PLATE LX, figs. 11, 11a.

71. Euastrum oblongum (Grev.) Ralfs 1848, Brit. Desm., p. 80. Pl. XII var. **oblongum**.

Echinella oblonga Greville in Hooker 1830, English Flora, II: 398.
Euastrum oblongum f. *scrobiculata* Nordstedt 1873, Acta Univ. Lund 9: 7.
Euastrum pecten Ehrenberg 1838, Infusionsthierchen, p. 162. Pl. 12, Fig. IV.

A large *Euastrum*, broadly oval in outline, 2 times longer than broad; semicells broadly oval in outline from a broad, truncate base; basal lobes subrectangular at the margin, bilobed and shoulderlike, a deep incision between the basal lobes and a bilobed upper, lateral lobule and a narrow, deep incision between the upper, lateral lobules and the polar lobe, which is anvil-shaped, the angles narrowly rounded, the apical margin convex, with a deep, narrow median incision; face of semicell with a supraisthmial protuberance near the midregion of the semicell, with a central mucilage pore; wall punctate throughout; sinus narrow and closed; lateral view narrowly truncate-oval in outline, the poles truncate and slightly produced, with a median, apical swelling and a swelling at the angles, lateral margins of the semicell convex and 3-undulate, with a mucilage pore on the median protrusion; vertical view broadly elliptic, the poles narrowly rounded and slightly produced, margin with 5-undulations and a mucilage pore showing in the midregion. L. 107-205 μm. W. 48-107 μm. Isth. 13-26.5 μm. Ap. 31-52 μm. Th. 32-65 μm. Zygospore globose with blunt spines, 92-123 μm in diameter.

DISTRIBUTION: Widely distributed throughout the United States and Canada. Great Britain, Europe, Asia, Africa, South America, Arctic.
PLATE LXV, fig. 4.

71a. **Euastrum oblongum** var. **angustum** Prescott var. nov.

Varietas a varietate typica differens quod angustior et quod sinus inter lobos laterales superiores et lobum polarem late apertus; differens necnon quod protrusiones media in regione semicellulae utroque in latere pori mucosi centralis multo reductae vel carentes (?). Long. 120 μm. Lat. 58 μm. Lat. ad isthmum 18 μm.

ORIGO: E stagno distante 8 milia passuum (12.9 km.) a loco Brookhaven, Mississippi dicto.

HOLOTYPUS: Scott Coll. Miss-40.

ICONOTYPUS: Plate LXV, fig. 6.

A variety differing from the typical by being narrower, and in having the sinus between the upper, lateral lobules and the polar lobe widely open, the protrusions in the median part of the semicell on either side of the mucilage pore much reduced or lacking (?). L. 120 μm. W. 58 μm. Isth. 18 μm.

DISTRIBUTION: Mississippi.

PLATE LXV, fig. 6.

71b. **Euastrum oblongum** var. **cephalophorum** West, in West & West 1894, Jour. Roy. Microsc. Soc. 1894: 4; West & West 1905a, Monogr. II: 14. Pl. 35, Fig. 1.

A variety similar to the typical but differing by having the invaginations of the lateral margins between the basal lobes and the upper, lateral lobules wider; the angles of the polar lobe relatively more produced and usually closing the sinus between the polar lobe and the upper, lateral lobules; face of semicell without a mucilage pore. L. 151–155 μm. W. 73–90 μm. Isth. 27 μm.

DISTRIBUTION: Kentucky, Massachusetts, Michigan, Oklahoma, Washington. British Columbia, Labrador, Newfoundland, Québec. Great Britain, Europe.

PLATE LXV, figs. 7, 10.

71c. **Euastrum oblongum** var. **depauperatum** West & West 1905a, Monogr. II: 15. Pl. 34, Fig. 10.

A variety differing from the typical by having slightly retuse margins between the basal lobes and the upper, lateral lobules, which themselves are decidedly emarginate or bilobed, the upper protrusion of the lateral lobules less extended than in the typical; the sinus between the upper, lateral lobules and the polar lobe narrow and often closed; face of semicell without a mucilage pore. L. 140–185 μm. W. 63–85 μm. Isth. 17 μm. Ap. 50 μm.

DISTRIBUTION: Massachusetts. Québec. Great Britain.

PLATE LXV, fig. 9.

71d. **Euastrum oblogum** var. **ellipticum** (Irénée-Marie) Irénée-Marie 1958, Nat. Canadien 85(5): 133.

Euastrum oblongum f. *elliptica* Irénée-Marie 1947, Nat. Canadien 74(3/4): 120. Pl. II, Fig. 5.

A variety somewhat larger than the typical, the cells broadly oval in outline; basal lobes quadrate and slightly bilobed, a deep, narrow U-shaped sinus between the basal lobes and the upper, lateral lobules, which are emarginate; a similar sinus between the upper, lateral lobules and the polar lobe, which is broadly cuneate, the apical margin broadly convex, the median incision deep and closed; face of semicell with a row of 3 protuberances across the base of the semicell, 2 in the midregion, and 1 within each of the upper, lateral lobules. L. 139–158 μm. W. 76.5–84 μm. Isth. 16.5–28 μm. Ap. 38.6–45 μm.

DISTRIBUTION: Québec.

PLATE LXV, fig. 5.

72. **Euastrum octogibbosum** Krieger 1937, Rabenhorst's Kryptogamen-Flora 13(1:3): 586. Pl. 80, Figs. 27-29.

Euastrum coralloides Joshua var. *subintegrum* West & West 1907, Ann. Roy. Bot. Garden Calcutta VI(2): 198. Pl. 14, Fig. 8.

A small *Euastrum*, rectangular in outline, slightly longer than broad; semicells subquadrangular-trapezoidal, the apex nearly as wide as the basal lobes; basal angles quadrately rounded, margin invaginated by a narrow incision immediately above, forming upper, lateral lobules, with a deep incision between the lobules and the short, somewhat capitate polar lobe; the apical margin flat, the angles bearing a toothlike spine, the median notch short and nearly closed; facial protuberance of the semicell prominent, furnished with a circle of about 4 large granules; wall with a pair of intramarginal granules at the apical lobules, a low protrusion at the base of each apical lobule, and a similar swelling in the midregion of each basal lobe; sinus narrow and closed throughout; lateral view oval in outline, the poles convexly rounded and furnished with a toothlike spine, the lateral margins diverging to form prominent, granular protrusions in the basal part of the semicell; vertical view oval, the poles rounded and smooth, with a subapical tubercular enlargement on either side, and a promiment granular swelling in the midregion. L. 24.5 μm. W. 21 μm. Isth. 5.6 μm. Ap. 15 μm. Th. 13 μm.

DISTRIBUTION: Montana. Asia (Burma).

PLATE LXXV, figs. 13A, 13Aa.

73. **Euastrum oculatum** Börgesen 1890, Vid. Medd. Naturh. Foren. Kjöbenhavn 1890: 941. Pl. 3, Fig. 2 var. **oculatum**.

A medium-sized *Euastrum*, rectangular in outline, 2 times longer than broad; semicells rectangular in outline, the basal lobes bispinate at the angles, the margins retuse to an upper, lateral, bigranular protrusion, then abruptly invaginated to the polar lobe, with subparallel margins, the angles bearing a long, stout, somewhat upwardly directed spine, the apical margin flat but with a swelling of the wall in the region of the median incision, which is short and closed outwardly; face of semicell with a low protuberance in the midregion, with a pair of mucilage pores immediately above; wall smooth but for minute granules within the angles of the basal lobes; sinus narrow and closed; lateral view elliptic, the poles narrowly rounded and furnished with an apiculation, the lateral margins diverging and undulate to the inflation at the base of the semicell; vertical view oval, the poles with an apiculation, the lateral margins convex with a low median swelling. L. 56 μm. W. 32 μm. Isth. 8 μm. Th. 22 μm. Zygospore unknown.

DISTRIBUTION: Connecticut, Florida, Kansas, Massachusetts. Africa, South America.

PLATE LXXX, figs. 4–4b.

73a. **Euastrum oculatum** var. **prescottii** Förster 1972, Nova Hedwigia 23(2/3): 539. Pl. 8, Figs. 1, 2.

Euastrum oculatum Borge var. *tonsum* West & West, f. Prescott & Scott 1942, Trans. Amer. Microsc. Soc. 61(1): 9. Pl. 1, Fig. 27.

A variety shorter than the typical, truncate-oval in outline, about 1.5 times longer than broad; semicells truncate-pyramidal, the basal lobes with sharply rounded angles, bearing a pair of sharp granules, the upper, lateral swellings on the margin less developed than in the typical, the polar lobe shorter; lateral view with a prominent conical protrusion at the base of the semicells. L. 38–56 μm. W. 24–34 μm. Isth. 5–8 μm. Ap. 15–17 μm. Th. 16–20 μm.

DISTRIBUTION: Florida, Michigan, Mississippi. Québec.

PLATE LXXX, figs. 5, 5a.

73b. Euastrum oculatum var. **tonsum** West & West 1896a, Trans. Linn. Soc. London, Bot. 5: 244. Pl. 14, Fig. 26 f. **tonsum**.

A variety in which the basal lobes and the upper, lateral lobules are reduced and with short spines or mucros, the polar lobe shorter, spines at the angles less developed and with a mere notch or slight invagination in the midregion of the apical margin rather than a closed, narrow incision. L. 38.5–48 μm. W. 24–30 μm. Isth. 5.6–8 μm. Th. 17–19 μm.

DISTRIBUTION: Florida, Louisiana, Michigan, Mississippi, North America (West & West). Québec.

PLATE LXXX, fig. 6.

73c. Euastrum oculatum var. **tonsum** f. **mucronatum** Irénée-Marie 1952, Hydrobiologia 4(1/2): 172.

A form differing from the typical variety in its possession of a mucro or tooth at the angles of the basal lobes. L. 45–58 μm. W. 29–34 μm. Isth. 7.2–10 μm. Ap. 19.3–23.3 μm. Th. 20–22 μm.

Population studies likely would show this to be a growth variable.

DISTRIBUTION: Alaska. Québec.

74. Euastrum ovale Prescott sp. nov.

Euastrum magnitudine minus quam mediocre, fere circulare, aliquantum longius quam latum; semicellulae semicirculares, marginibus in tres lobulos bispinatos subaeque divisis; lobus polaris vix evidens, brevissimus, sinu inter lobulos laterales superiores et lobum polarem paulo acutiore profundioreque quam incisuare quae lobulos inferiores formant; margo apicalis rectus incisuram mediam forma litterae V habens, angulis lobi polaris pari spinarum brevium acutarum praeditis; superficies semicellulae protuberationem parvam, atque granulum magnum conspicuum utroque in latere praebens; membrana semicellulae granulis binis intra omnes lobulos praedita; sinus angustus et prorsus inapertus; semicellulae a latere visae anguste subrectangulares, polis truncatis, angulis rotundatis, incisura marginum infra apicem capitulationem efficiente, marginibus 2- vel 3-undulatis. Long. 32.8 μm. Lat. 24.4 μm. Lat. ad isthmum 11.5 μm. Lat. ad apicem 14.4 μm. Crass. 16.5 μm.

ORIGO: E fonte prope matrem per castores constructam in loco Vilas County, Wisconsin dicto.

HOLOTYPUS: GWP Coll. 2W-263.

ICONOTYPUS: Plate LXXX, figs. 8, 8a.

A less than average-sized *Euastrum*, nearly circular in outline, slightly longer than broad; semicells semicircular in outline, the lateral margins about equally, divided into 3 bispinate lobules; the polar lobe scarcely evident, very short, the sinus between the upper, lateral lobules and the polar lobe somewhat sharper and deeper than the incisions forming the lower lobules; apical margin straight with a

V-shaped median incision, the angles of the polar lobe with a pair of sharp, short spines; face of semicell with a low protuberance, with a conspicuous, large granule on either side; wall of semicell with granules in pairs within all lobules; sinus narrow and closed throughout; lateral view narrowly subrectangular, the poles truncate, the angles rounded, an incision of the margins below the apex resulting in a capitulation, the lateral margins 2- or 3-undulate. L. 32.8 μm. W. 24.4 μm. Isth. 11.5 μm. Ap. 14.4 μm. Th. 16.5 μm.

DISTRIBUTION: Wisconsin.

PLATE LXXX, figs. 8, 8a.

75. Euastrum pectinatum Brébisson ex Ralfs 1848, Brit. Desm., p. 86. Pl. XIV, Figs. 5a-5f var. **pectinatum** f. **pectinatum.**

Cosmarium pectinatum Brébisson, in: Meneghini 1840, Synop. Desm., p. 222.

A medium-sized *Euastrum*, subrectangular in outline, 1.5 to 1 2/3 times longer than broad; semicells quadrate in outline, the basal lobes narrowly rounded, the margins retuse to an upper, lateral lobule which extends as far as the basal lobes, the incision deep above the upper, lateral lobules and the margin nearly horizontal; polar lobe short, with diverging margins to bluntly pointed angles, the apical margin flat, slightly retuse in the midregion; face of semicell with a prominent, smooth protrusion, with a similar protuberance on either side, and with 2 smaller protuberances within the apical margin, one on either side of the median invagination; wall smooth; sinus narrow and closed except for the exterior, where it opens slightly; lateral view subrectangular, the poles truncate and slightly inflated, the lateral margins retuse below the apex and then expanding to form a prominent protrusion; vertical view subquadrangular, the poles truncate but bilobed, the margins with 3 prominent undulations. L. 48-84 μm. W. 40-49 μm. Isth. 8-16 μm. Ap. 30-36 μm. Th. 21-40 μm. Zygospore globose to elliptic, with simple, papillalike granules or blunt spines, 50-58 μm in diameter, or 46-54 μm x 57-61 μm.

DISTRIBUTION: Colorado, Michigan, Minnesota, New Jersey, Oklahoma, Utah, Wyoming. British Columbia, Labrador. Great Britain, Europe, Asia, Africa, South America, Arctic.

PLATE LXVII, figs. 7, 7a.

75a. Euastrum pectinatum var. **pectinatum** f. **elongata** Irénée-Marie 1956, Rev. Algol. 2(1/2): 117. Fig. 10.

A form differing from the typical by having a somewhat more exserted polar lobe which is anvil-shaped, the angles narrowly rounded. L. 60 μm. W. 38.6-39 μm. Isth. 9.7 μm. Ap. 13 μm.

DISTRIBUTION: Québec.

PLATE LXVII, fig. 6.

75b. Euastrum pectinatum var. **brachylobum** Wittrock 1872, Bih. Kongl. Vet.-Akad. Handl. 1: 48. Pl. 4, Fig. 5 f. **brachylobum.**

A variety differing from the typical mostly by the reduction in the angles of the polar lobe, which is truncate with rounded angles, the lateral margins subparallel; in vertical view showing only 1 swelling in the midregion of the lateral margins. L. 55-85 μm. W. 37-56 μm. Isth. 9-15 μm. Ap. 21-32 μm. Th. 28-48 μm.

DISTRIBUTION: Alaska, Louisiana, Michigan, Minnesota, Mississippi. Newfoundland, Québec. Great Britain, Europe, Arctic, Faeroes.

PLATE LXVII, figs. 5-5b.

75c. **Euastrum pectinatum** var. **brachylobum** f. **majus** Taylor 1935, Pap. Michigan Acad. Sci., Arts & Lettr. 20: 205. Pl. 40, Fig. 16.

A form differing from the typical by its larger size, and in having the angles of the polar lobe slightly extended. L. 77-84 μm. W. 48-56 μm. Isth. 10 μm. Th. 36-40 μm.

This form is intermediate between *Euastrum pectinatum* and its variety *brachylobum* in respect to the morphology of the polar lobe and its angles.

DISTRIBUTION: Alaska. Newfoundland.

PLATE LXVII, figs. 8, 8a.

75d. **Euastrum pectinatum** var. **brachylobum** f. **triquetra** (Kaiser) Krieger 1937, Rabenhorst's Kryptogamen-Flora 13(1:3): 538. Pl. 72, Figs. 10, 11.

A form with much inflated semicells, with a broad median protrusion so that the semicells are triangular in vertical view; polar lobe short and broad, the apical margin showing a broad, low protrusion; sinus narrow and closed. L. 60-78 μm. W. 30-48 μm. Isth. 14-16 μm. Ap. 26-30 μm.

DISTRIBUTION: Labrador. Europe, Arctic.

PLATE LXVII, fig. 10.

75e. **Euastrum pectinatum** var. **fennicum** Grönblad 1942, Acta Soc. Sci. Fennicae, II, B, II(5): 36. Pl. 1, Fig. 28.

A variety with much inflated cells, triangular in vertical view; the semicells rhomboid in outline, the margins of the basal lobes diverging to upper, lateral protuberances, the margins converging to a broad, truncate, slightly produced polar lobe which is broad and flat; sinus closed within but open outwardly. L. 64-70 μm. W. 42-47 μm. Isth. 14 μm.

DISTRIBUTION: Labrador. Europe, Arctic.

PLATE LXVII, figs. 9, 9a.

75f. **Euastrum pectinatum** var. **inevolutum** West & West 1905, Trans. & Proc. Bot. Soc. Edinburgh 23: 17. Pl. 1, Figs. 13, 14.

Euastrum pectinatum var., Wolle 1887, Freshwater Alg. U.S., p. 36. Pl. 54, Figs. 10-12; 1892, Desm. U.S., p. 109. Pl. 12, Figs. 10-12.

A variety averaging smaller than the typical, differing mostly in the form of the polar lobe, the angles only slightly produced, the apical margin elevated and then slightly retuse in the midregion; sinus narrow and closed; vertical view similar in configuration to the typical. L. 42-67.8 μm. W. 30.8-48.8 μm. Isth. 10-11.5 μm. Ap. 19.2-30 μm. Th. 20.4-29 μm.

DISTRIBUTION: Alaska, Michigan, Washington. British Columbia, Newfoundland, Québec. Great Britain, Europe, Arctic.

PLATE LXVII, figs. 12, 12a.

75g. **Euastrum pectinatum** var. **lagenale** Boldt 1888, Bih. Kongl. Svenska Vet.-Akad. Handl. 13, Afd. II (5): 6. Pl. 1, Fig. 4.

A variety with very much reduced lobes, the basal lobes with slightly biundulate margins, less invaginated above the basal lobes so that the polar lobe is very broad, smoothly rounded at the angles, and slightly convex at the apex;

lateral view broadly oval, the poles truncately rounded, the margins smooth but slightly inflated in the midregion, the semicells nearly circular in outline; vertical view broadly oval, the poles broadly rounded, the margins slightly triundulate. L. 62.4–68 μm. W. 40–42 μm. Isth. 9.6–13 μm. Th. 31.2–34 μm.

Krieger (1937, p. 637) places this variety in synonymy with the typical but we believe the distinct configuration of the cells, especially of the polar lobe, justifies the retention of var. *lagenale* as a taxon.

DISTRIBUTION: Labrador. Greenland.

PLATE LXVII, fig. 11.

75h. Euastrum pectinatum var. **lobuliferum** Scott & Prescott 1952, Hydrobiologia 4(4): 389. Pl. 1, Fig. 12.

Differing from the typical by having thicker lateral lobules in the polar lobe, their apices more broadly rounded, and in having thicker, bilobate, lower lateral lobes; the sinus narrow and closed throughout; also differing by having a supra-centric mucilage pore in the face of the semicell surrounded by 3 low swellings, and with an intramarginal mucilage pore on each side of the polar incision which is characteristically very shallow; wall finely scrobiculate; in lateral view semicell thicker than in the typical, with marginal protuberances just below the poles rather than an invagination, also with secondary lobules at the base of the semicell where the median incision is narrow and deep; in vertical view broadly elliptic in outline, with 2 prominent swellings in the midregion on each side, and with 2 lower swellings between the midregion and the poles (6 marginal swellings on each side). L. 70–74 μm. W. 52–54 μm. Isth. 11–12 μm. Th. 30 μm.

DISTRIBUTION: Florida.

PLATE LXVII, figs. 15, 15a.

75i. Euastrum pectinatum var. **minnesotense** Prescott var. nov.

Varietas a varietate typica differens possessione 3 lobulorum lateralium in parte basali semicellulae, marginibus divergentibus et retusis ab angulo basali inferiore ad lobulum prominentem qui ultra angulum basalen extendit, deinde retusis ad lobulum lateralem superiorem, qui aeque ac angulus basalis extendit; lobus polaris brevis, angulis rotundatis reductisque, margine apicali perspicue reducto, protrusionem intramarginalem media in regione habente; semicellulae a vertice visae quadrangulares, polis rectangularibus et ad marginem retusis, et lobulum exsertum intermarginalem atque protrusionem prominentem media in regione marginum lateralium habentibus. Long. 63 μm. Lat. 46.5 μm. Lat. ad isthmum 8.7 μm.

ORIGO: E palude Sphagni pleno prope locum Lower Sand Lake, Minneosta dictum.

HOLOTYPUS: GWP Coll. Minn-31.

ICONOTYPUS: Plate LXVII, figs. 13, 13a.

A variety differing from the typical by having 3 lateral lobules in the basal part of the semicell, the margins diverging and retuse from the lower basal angle to a prominent lobule extending further than the basal angle, then retuse to the upper, lateral lobule, which extends as far as the basal angle; polar lobe short with reduced, rounded angles, conspicuously retuse in the apical margin, with an intramarginal protrusion in the midregion; vertical view quadrangular, the poles rectangular and retuse at the margin, and with an intramarginal, exserted lobule and a prominent protrusion in the midregion of the lateral margins. L. 63 μm. W. 46.5 μm. Isth. 8.7 μm.

DISTRIBUTION: Minnesota.

PLATE LXVII, figs. 13, 13a.

75j. Euastrum pectinatum var. **planum** Krieger 1937, Rabenhorst's Kryptogamen-Flora 13(1:3): 539. Pl. 72. Fig. 14.

A variety similar in shape to var. *brachylobum* but with a smoothly rounded semicell face, without a median protuberance; polar lobe with sharply rounded angles. L. 63.7–76 μm. W. 42–48 μm. Ap. 23–30 μm. Th. 29–34 μm.

This variety possibly should be assigned to a form of var. *brachylobum*. Larsen (1904, p. 92, Fig. 2) reports it as a form of this variety without an epithet from Greenland. Although Larsen does not describe or illustrate a median facial protuberance, one is illustrated by Croasdale & Grönblad (1964, Pl. 8, Fig. 4). They also show a form with a widely open sinus, whereas Larsen's plant has a sinus closed for most of its length. Examination of Larsen's specimens is necessary to determine whether this is a justifiable taxon differentiation, and whether Croasdale and Grönblad's form is identical.

DISTRIBUTION: Labrador. Greenland.

PLATE LXVII, figs. 16–16b.

75k. Euastrum pectinatum var. **reductum** Taylor 1935, Pap. Michigan Acad. Sci., Arts & Lettr. 20: 206. Pl. 40, Fig. 14 f. **reductum**.

A variety with much reduced configurations; the semicells truncate-pyramidal; basal lobes with a mucro at the angles, the lateral margins vertical and subparallel, retuse to a broad polar lobe which is truncate with broadly rounded angles, very slightly retuse in the midregion; wall strongly punctate; sinus narrow and closed; face of semicell with a broad, flat protrusion (or none); lateral view broadly oval in outline, the poles truncately rounded, the margins diverging symmetrically to a median protrusion. L. 50–70 μm. W. 35–44 μm. Isth. 8.5–11.5 μm. Th. 22–28 μm.

DISTRIBUTION: Florida, Louisiana. Newfoundland. Europe, Greenland,

PLATE LXVII, figs. 14–14b; Plate LXVIII, fig. 1.

75l. Euastrum pectinatum var. **reductum** f. **simplicissimum** Prescott f. nov.

Forma magnitudine formae typicae similis, lobi basales, autem, late rotundati, et margo inter angulos inferiores superioresque aliquantum retusus; semicellulae truncato-pyramidales; lobus polaris late rotundatus, margine apicali convexo; membrana minute scrobiculata. Long. 54 μm. Lat. 36 μm. Lat. ad isthmum 10 μm. Crass. 27 μm.

ORIGO: E loco Honey Island swamp near Mitchell's Hammock, Louisiana dicto.

HOLOTYPUS: Scott Coll. La-78.

ICONOTYPUS: Plate LXVIII, fig. 2.

A form similar to the typical in size but with the basal lobes broadly rounded, with a slight retuseness between the lower and upper angles; semicells truncate-pyramidal in outline; polar lobe broadly rounded, the apical margin convex; wall minutely scrobiculate. L. 54 μm. W. 36 μm. Isth. 10 μm. Th. 27 μm.

DISTRIBUTION: Mississippi.

PLATE LXVIII, fig. 2.

75m. **Euastrum pectinatum** var. **rostratum** (Taylor) Krieger 1937, Rabenhorst's Kryptogamen-Flora 13(1:3): 539. Pl. 73, Figs. 4, 5.

Euastrum pectinatum var. *brachylobum* f. *rostrata* Taylor 1935, Pap. Michigan Acad. Sci., Arts & Lettr. 20: 206. Pl. 41, Fig. 4.

A variety differing from the typical by having inflated semicells; the sinus opening widely outward, and with the angles of the basal lobes extended to form hornlike processes, distinctly longer than the upper, lateral lobules; polar lobe short and with bilobed angles with processes somewhat longer than in other varieties. L. 65–71 μm. W. 44.4–52 μm. Isth. 11–13 μm. Ap. 22 μm. Th. 30–32 μm.

DISTRIBUTION: Alaska, Michigan. Newfoundland.
PLATE LXVIII, figs. 3–5a.

75n. **Euastrum pectinatum** var. **scrobiculatum** Prescott 1940, Pap. Michigan Acad. Sci., Arts & Lettr. 25: 98. Pl. 1, Fig. 22.

A variety with the polar lobe short, the apex convex; basal lobes quadrate with deep emarginations in the lateral walls; wall coarsely scrobiculate; 2 large tuberculations within the margin of the basal lobules; lateral view subquadrangular, a large median tuberculation on the lateral margins and a smaller one immediately below it on the basal angles. L. 73–76.4 μm. W. 47–49.4 μm. Isth. 15.2 μm.

DISTRIBUTION: Michigan.
PLATE LXVIII, fig. 8.

76. **Euastrum personatum** West & West 1895, Trans. Linn. Soc. London, Bot., II, 5: 52. Pl. 6, Fig. 19.

A somewhat less than medium-sized *Euastrum*, truncate-oval in outline, 1.5 times longer than broad; semicells truncate-pyramidal; angles of the basal lobes narrowly rounded, the margins retuse and converging to an upper, lateral lobule, then sharply incised to form a short polar lobe, broad and truncate, with the angles extended to form a slightly upturned spine, the apical margin flat, with a short, narrow median incision, a slight protrusion or thickening of the wall immediately below the lateral spines at the polar angles; face of semicell with a smooth median protrusion below a pair of mucilage pores; wall granular within the margins of all lobes, larger granules intramarginal at the base of the apical spines; sinus narrow and closed; lateral view elliptic, the poles narrow and extended into a vertical spine, with a sharp projection on either side at the polar angles, margin of the semicell smoothly convex with a median mucilage pore; vertical view narrowly oval, the poles rounded and smooth, with a low swelling in the midregion on either side. L. 36–48 μm. W. 16–18 μm. Isth. 6–9 μm. Th. 16–18 μm. Zygospore unknown.

DISTRIBUTION: Massachusetts. Africa, Madagascar, Ceylon.
PLATE LXXVIII, fig. 18.

77. **Euastrum pictum** Börgesen 1890, Vid. Medd. Naturh. Foren. Kjöbenhavn 1890: 939. Fig. 19 var. **pictum**.

A less than medium-sized *Euastrum*, rectangular in outline, 1.5 times longer than broad; semicells quadrate, the angles of the basal lobes bi- or trispinate, the margins invaginated to an upper, lateral, bispinate lobule, with a deep, sharp incision to the polar lobe, which is as broad as the basal lobes, the angles extended into horizontally directed spines, the apical margin elevated to the short, broad median incision; face of semicell with a median protuberance bearing a pattern of granules; wall with 2 circular patterns of granules within the apical lobules, and similar patterns of granules within the lower lobules; sinus narrow and closed; vertical view broadly elliptic, the poles extended into a spine, with a subapical spine on either side, the margins with 3 granular swellings, the median one being more prominent. L. 56–72 μm. W. 38–44 μm. Isth. 8–12 μm. Th. 27–28 μm. Zygospore unknown.

DISTRIBUTION: Connecticut, New York. British Columbia, Newfoundland. Great Britain, Europe, South America.

PLATE LXXVIII, fig. 6A.

77a. **Euastrum pictum** var. **subrectangular** West & West f. Taylor 1935, Pap. Michigan Acad. Sci., Arts & Lettr. 20: 207. Pl. 40, Fig. 12.

A form differing from the typical in the shape of the polar lobe and in the reduction of spines on the margins of the lateral lobules; apical margin elevated at the median incision and then sloping sharply downward to a subapical, prominent spine; subapical sinus of the polar lobe pronounced; lobules of the basal lobes with smooth margins; facial protuberance with 3 vertical, knoblike tubercles. L. 70 μm. W. 52 μm. Isth. 8.5 μm.

Krieger (1937, p. 601) places var. *subrectangulare* in synonymy with *Euastrum bidentatum* Nägeli but recognizes *E. pictum* Börgesen. Undoubtedly these 2 species (*E. bidentatum* and *E. pictum*) are closely related, as indicated by details of their morphology. For the time being we recognize the desirability of keeping var. *subrectangulare* under *E. pictum*.

DISTRIBUTION: New Jersey. Newfoundland.

PLATE LXXVIII, fig. 2.

78. **Euastrum pingue** Elfving 1881, Acta Soc. Fauna Flora Fennica 2(2): 7. Pl. 1, Fig. 3.

A species under medium size, broadly oval in outline, 2 times longer than broad; semicells truncate-pyramidal in outline, the basal lobes broadly rounded, the margins symmetrically retuse to form a short, little-defined polar lobe which is truncate, the angles rounded, the apical margin slightly convex, median incision short and open; face of semicell with a large median protrusion with a circular pattern of granules and with a conspicuous mucilage pore excentrically located above and to one side of the facial protuberance, a tubercular granule supra-isthmial below the protrusion; wall coarsely scrobiculate; sinus narrow and closed; lateral view broadly oval in outline, the poles rounded, slightly produced, the lateral margins convex, inflated in the midregion and roughened by wall scro-biculations, with a tubercle showing at the base of the semicell; vertical view oval, the poles broadly rounded, the wall roughened by scrobiculations, lateral margins inflated in the midregion by a low, broad protrusion. L. 50–60 μm. W. 35–45 μm. Isth. 9.6–12 μm. Ap. 18–21 μm. Th. 26–28 μm. Zygospore unknown.

DISTRIBUTION: Florida, Maine, Massachusetts, Michigan, New Jersey, New York. Newfoundland. Great Britain, Europe, Asia, Arctic.

PLATE LX, figs. 12–12b.

79. **Euastrum pinnatum** Ralfs 1848, Brit. Desm., p. 81, Pl. XIII, Fig. 1 var. **pinnatum**.

Euastrum multilobatum Wood 1870, Proc. Acad. Nat. Sci. Philadelphia 1869: 16.

A large *Euastrum*, cells broadly truncate-oval in outline, 2 times longer than broad; semicells rectangular in outline, the basal lobes quadrate, bilobed at the margins, the lateral margins sharply and deeply invaginated and then diverging to form prominent, horizontally directed upper, lateral lobules, then deeply incised with a broad sinus to form a polar lobe which is anvil-shaped, the angles narrowly rounded, the apical margin flat, with a deep, narrow median incision; face of semicell with 2 median protuberances and a mucilage pore between, a protuberance within each basal lobe and a supraisthmial protrusion in the midregion below the mucilage pore; wall coarsely punctate-scrobiculate; sinus narrow and closed; lateral view truncate-oval, the poles truncate and slightly inflated, flat at the apex, the lateral margins diverging to a prominent median protrusion, below which is a secondary swelling at the base of the semicell; vertical view broadly quadrate-oval in outline, the poles extended to form a prominent lobelike protrusion, the angles of the poles extended to form lobelike protrusions, the lateral margins triundulate. L. 110–158 μm. W. 57.6–80 μm. Isth. 15–30 μm. Ap. 26–43 μm. Th. 40–50 μm. Zygospore unknown.

DISTRIBUTION: Widely distributed in the United States and Canada. Great Britain, Europe, Asia, Africa, South America, Arctic.

PLATE LXV, fig. 1.

79a. **Euastrum pinnatum** var. **capitatum** Krieger 1937, Rabenhorst's Kryptogamen-Flora 13(1:3): 523. Pl. 68, Figs. 8, 9.

A variety smaller than the typical, the lobules of the polar lobe more rounded and not extended as much; the upper lateral lobules occurring higher in the semicell than in the typical; mucilage pore lacking (?). L. 91–94 μm. W. 43–47 μm. Isth. 16 μm. Th. 31–36 μm.

Euastrum magnificum Wolle (1887, Freshwater Alg. U.S., p. 37. Pl. 58, Figs. 6–8) possibly is related to *Euastrum pinnatum*. The former is an incompletely described species which has similarities with var. *capitatum*, reported from New Jersey.

DISTRIBUTION: Louisiana. Asia.

PLATE LXIII, figs. 11–11b.

79b. **Euastrum pinnatum** var. **pres-scottii** Irénée-Marie 1958, Nat. Canadien 85(5): 137. Figs. 18, 19.

Euastrum pinnatum f. Prescott & Scott 1945, Amer. Mid. Nat. 34(1): 243. Pl. 7, Figs. 7–10.

A variety narrower and longer than the typical, the margins less incised, the lobules shorter, especially those of the polar lobe, which are not horizontally extended as in the typical; wall markings similar to those of the typical. L. 113–138 μm. W. 52.3–62 μm. Isth. 16–17.7 μm.

DISTRIBUTION: Georgia, Kentucky, Michigan, North Carolina. Québec.

PLATE LXIV, fig. 13.

79c. **Euastrum pinnatum** var. **scottii** Prescott var. nov.

Varietas potius rectangularis, lobuli laterales superiores atque lobuli lobi polaris fere aeque longe ac lobi basales extendentes, incisuris marginalibus relative profundioribus atque lobulis tenuioribus; sinus extrinsecus inapertus, intus, autem, apertus; membrana grosse scrobiculata. Long. 131 μm. Lat. 68 μm. Lat. ad isthmum 29 μm.

ORIGO: E fossa distante 8.2 milia passuum (13.2 km.) ad orientem a loco Pearlington, Mississippi dicto.

HOLOTYPUS: Scott Coll. Miss-47.

ICONOTYPUS: Plate LXV, figs. 2, 2a.

A variety more nearly rectangular in outline, the upper, lateral lobules and the lobules of the polar lobe extended nearly as far as the basal lobes, the marginal incisions relatively deeper and the lobules more slender; the sinus closed outwardly but open within; wall coarsely scrobiculate. L. 131 μm. W. 68 μm. Isth. 29 μm.

DISTRIBUTION: Mississippi.

PLATE LXV, figs. 2, 2a.

80. **Euastrum platycerum** Reinsch 1875, Contrib. Algal. et Fungolog. I, p. 85. Pl. 12, Fig. 6 var. **platycerum**.

Cells less than medium-sized, 1.1 to 1.2 times longer than broad; sinus narrow and closed inwardly, opening outwardly with the extensions of the basal lobes broadly rounded and horizontally extended, the lateral margins with a shoulderlike incision between the basal lobes and the polar lobe, which has subparallel margins, the angles rounded and spiny, the apical margin flat or slightly retuse in the midregion; face of semicell with a low median protrusion bearing a circular pattern of granules; wall with a scattering of granules within the lobes, and a patch of granules on low protuberances on either side of the median swelling. L. 45–54 μm. W. 40–49 μm. Isth. 9–13 μm. Ap. 13–16 μm. Th. 23–24 μm. Zygospore unknown.

DISTRIBUTION: Typical form not reported from the United States. Europe, Asia, Indonesia, Africa.

PLATE LXXXI, figs. 6, 6a.

80a. **Euastrum platycerum** var. **acutilobum** Krieger 1937, Rabenhorst's Krypto-gamen-Flora 13 (1:3): 631. Pl. 91, Fig. 2 f. **acutilobum**.

A variety with basal lobes more slender and extended further than in the typical, the lobes with a rounded basal angle, the margins slightly retuse from a granular base, forming a slender lobule that extends further than the basal lobes. L. 60–70 μm. W. 52–60 μm. Isth. 12–15 μm. Ap. 21–22 μm.

A form collected in Florida is slightly different in that the lower angles of the basal lobes do not have a granular swelling as in the typical.

DISTRIBUTION: Florida. South America.

PLATE LXXXI, fig. 7.

80b. **Euastrum platycerum** var. **acutilobum** f. **dentiferum** Scott & Prescott 1952, Hydrobiologia 4(4): 391. Pl. 3, Fig. 12.

A form differing from the typical by having a more highly decorated wall in the form of granules both on the face of the semicell and around the polar lobules, and by having a tubercular tooth supraisthmial in the midregion of the semicell base; lateral lobules of the semicell more nearly horizontal than in the typical. L. 57–60 μm. W. 54–56 μm. Isth. 11–12 μm. Th. 27 μm.

DISTRIBUTION: Florida.

PLATE LXXXI, fig. 8.

80c. **Euastrum platycerum** var. **eximium** f. **clausum** Grönblad & Scott, in Grönblad, Prowse & Scott 1958, Acta Bot. Fennica 58: 14. Fig. 50.

A form in which the sinus is less open outwardly, and with the basal angles not bearing teeth. L. 101 μm. W. 94 μm. Isth. 22 μm.

This form should be compared with *Euastrum subhypochondrum* Fritsch & Rich, and with *E. platycerum* var. *acutilobum* Krieger.

DISTRIBUTION: Virginia. Africa.

PLATE LXXXI, fig. 9.

81. Euastrum prescottii Jackson 1971, Dissert., p. 178. Pl. 36, Fig. 4.

A small *Euastrum*, quadrangular in outline, 1 to 1.2 times longer than broad; semicells subquadrangular in outline, the basal lobes rounded and extended horizontally, the lateral margins deeply retuse to the short polar lobe, which is broad and truncate, with subparallel margins; the apex trilobulate; apical margin slightly retuse; face of semicell with a broad central protrusion in which there are minute pores (faintly discerned by staining); sinus narrow and closed to the base of the lobes; lateral view subrectangular, the poles broad and with 3 lobules, the upper margins subparallel, midregion swollen; vertical view broadly oval, the poles truncate and bilobed, the lateral margins undulate with a pronounced median swelling. L. 17-19 μm. W. 13.5-16 μm. Isth. 6.2-7 μm. Th. 10.5-11.5 μm. Zygospore unknown.

DISTRIBUTION: Montana.

PLATE LXII, figs. 5, 5a.

82. Euastrum prominens Scott & Grönblad 1957, Acta Soc. Sci. Fennicae, II, B, II(8): 14. Pl. 3, Fig. 15.

A small *Euastrum*, subrectangular in outline 1.25 times longer than broad; semicells trapeziform, the basal lobes narrowly rounded at the angles, the lateral margins converging and retuse to an upper, lateral protrusion, and then sharply incised to the polar lobe, which has diverging margins, the apical angles sharply rounded and bearing a mucro, the apical margin broad and truncate, slightly retuse in the midregion; face of semicell with a round tubercle in the midregion of the polar lobe and with a supraisthmial tubercle in the midregion of the semicell base; wall punctate; sinus narrow and closed; lateral view oval, the poles broadly rounded, with a subapical tubercle on either side, the lateral margins smoothly convex to the base, where there is a small tubercle; vertical view oval, the poles broadly rounded and smooth, with a median tubercle in the midregion on either side. L. 18-18.8 μm. W. 14.5-16 μm. Isth. 7.5-8.5 μm. Zygospore unknown.

DISTRIBUTION: Florida.

PLATE LXXI, figs. 17-17b.

83. Euastrum protuberans Scott & Grönblad 1957, Acta Soc. Sci. Fennicae, II, B, II(8): 14. Pl. 1, Fig. 6.

A small-sized *Euastrum*, truncate-oval in outline, about 1/4 times longer than broad; semicells subpyramidal; basal lobes with diverging margins and slightly bilobed, a tubercular spine at each angle, the upper angles extended to form the widest part of the semicell; lateral margins of semicell converging to an upper, lateral, tooth-bearing protuberance, then sharply incised to the polar lobe, which is short, with subparallel margins, the angles rounded and bearing a short spine, the apical margin slightly elevated to the deep, widely open median incision; face of semicell with a prominent central protuberance, above which is a horizontal pair of mucilage pores; wall with a granule on either side of the median incision near the apical margin, a granule on either side of the apex of the median incision, and intramarginal granules at each lobe of the semicell, wall smooth otherwise; sinus narrow and closed throughout; lateral view elliptic, the poles narrowly rounded with 2 small apical spines on either side, the margins of the semicell diverging greatly to the prominent median protrusion (the semicell pyramidal in outline). L. 34 μm. W. 19 μm. Isth. 4 μm.

DISTRIBUTION: Florida.

PLATE LXXVI, figs. 21, 21a.

84. **Euastrum pseudoberlinii** Prescott sp. nov.

Species mediocris, c. 1.5 plo longior quam lata, subrectangularis; sinus profundus inapertusque; margines laterales loborum basalium divergentes, lobulum basalem parvum ad isthmum habentes, lobis lateralibus rotundatis atque aliquantulum productis, marginibus semicellulae profunde concavis et ad lobum polarem quadratum formandum subparallelis factis; lobus polaris media in regione marginis apicalis retusus; superficies semicellularum unam inflationem prominentem media in regione, atque inflationem intramarginalem intra lobos laterales, atque inflationem intra duos lobulos apicales habens; semicellulae a vertice visae late ovatae, polis truncatis atque bilobatis atque tumores medias prominentes utroque in latere praebentes. Long. 37 μm. Lat. 22 μm. Lat. ad isthmum 9 μm.

ORIGO: E palude Sphagni pleno prope locum Whitesand Lake, Vilas Co., Wisconsin.

HOLOTYPUS: GWP Coll. W-44.

ICONOTYPUS: Plate LXII, fig. 2.

A species of medium size, about 1.5 times longer than broad, subrectangular in outline; sinus deep and closed; lateral margins of the basal lobes diverging, with a small basal lobule at the isthmus; the lateral angles rounded and somewhat produced; margins of the semicell deeply concave and becoming subparallel to form a quadrate polar lobe which is retuse in the midregion of the apical margin; face of semicells with 1 prominent swelling in the midregion, and a swelling intramarginal of the lateral lobules, and with a swelling within the 2 apical lobules; vertical view broadly oval, the poles truncate and bilobed, with prominent swellings on either side of the midregion. L. 37 μm. W. 22 μm. Isth. 9 μm.

DISTRIBUTION: Wisconsin.

PLATE LXII, fig. 2.

85. **Euastrum pseudoboldtii** Grönblad 1921, Acta Soc. Fauna Flora Fennica 49: 11. Pl. 3, Figs. 5–7.

A small *Euastrum*, quadrangular in outline, 1.25 times longer than broad; semicells quadrangular, the basal lobes slightly produced and equally bilobed, the semicell margins above the basal lobes retuse and then slightly diverging in the polar lobe, which is equal in height to the basal lobes, the apex truncate, the margin flat but slightly retuse in the midregion, the angles of the polar lobe mucronate; face of semicell with a small low protuberance in the midregion, subtended by 2 small supraisthmial granules; wall with a few granules within the margin of each lobule of the cell; sinus narrow and closed; lateral view rectangular, the poles truncate with rounded margins, equal in width to the base of the semicells, lateral margins slightly retuse but with a small tubercle in the midregion; vertical view broadly elliptic, the poles sharply rounded, the lateral margins smooth but with a low, nodular protrusion in the midregion. L. 20–22.4 μm. W. 16–16.5 μm. Isth. 3.9–5 μm. Th. 9.9–12 μm. Zygospore unknown.

Grönblad, Scott & Croasdale (1964, Pl. 2, Figs. 23, 24) illustrate a form of this species which probably should be given a varietal assignment. The basal lobes are slightly different in shape, and are not as wide (vertically). The apical margin is furnished with a pair of sharp granules on either side of the median invagination. In vertical view the cells are oval rather than elliptic as in the typical. It is illustrated Plate LXXII, figs. 19, 19a.

DISTRIBUTION: Colorado. Québec. Europe, Asia, Africa, Arctic.

PLATE LXXII, figs. 18–19a.

86. **Euastrum pseudocoralloides** Fritsch 1918, Ann. South African Mus. 9: 549. Textfig. 25.

A small *Euastrum*, rhomboid to nearly sphaeroidal in outline, 1 1/5 to 1 1/3 times longer than broad; semicells subquadrate or trapezoidal, the basal lobes rounded, the margins retuse to an upper, lateral lobule which is mucronate, and then sharply incised to form a shoulder between the lateral lobules and the polar lobe, which is very short, with undulate margins, the angles of the polar lobe mucronate, the apical margin flat, with a short, V-shaped median incision; face of semicell with a smooth protuberance; wall smooth; sinus narrow and closed; lateral view elliptic in outline, the poles rounded and smooth, the lateral margins diverging abruptly to the prominent, median protrusion; vertical view narrowly oval, the poles rounded, the lateral margins with a median swelling. L. 21.5–25 μm. W. 18–21 μm. Isth. 5–6 μm. Th. 9–10 μm. Zygospore unknown.

DISTRIBUTION: California, Massachusetts, Wisconsin. South Africa.

PLATE LXXIV, fig. 25.

87. **Euastrum pseudopectinatum** Schmidle 1898a, Engler's Bot. Jahrb. 26: 46. Pl. 2, Fig. 39.

A medium-sized *Euastrum*, subrectangular in outline, 1.5 times longer than broad; semicells rectangular in outline, the basal lobes bispinate at the margin, the lateral margins retuse to a shoulderlike upper, lateral lobule which is broadly rounded, extending nearly as far as the basal lobes, a sharp, open sinus between the upper, lateral lobules and the polar lobe, which is short and wedge-shaped, the lateral margins diverging, the angles with a sharp mucro, the apical margin truncate, flat but with a slight median invagination; face of semicell with 2 widely spaced protuberances near the base, a mucilage pore in the center of the polar lobe, and a small protrusion near the apical margin below the median invagination; wall smooth; sinus narrow and closed throughout; vertical view broadly oval in outline, the poles broadly rounded-truncate, the lateral margins triundulate. L. 42–46 μm. W. 28–35.5 μm. Isth. 8 μm. Ap. 20–21.5 μm.

DISTRIBUTION: Michigan. East Africa.

PLATE LXVIII, fig. 7.

88. **Euasturm pulchellum** Brébisson 1856, Mém. Soc. Sci. Impér. Nat. Cherbourg 4: 124. Pl. 1, Fig. 5 var. **pulchellum**.

A small *Euastrum*, rectangular in outline, 1 1/3 to 1.5 times longer than broad; semicells quadrate, the basal lobes broadly rounded and granular at the margin, with a mucro at the lower angle, the lateral margins retuse from the basal lobes to the polar lobe, which is approximately equal in height to the basal lobes, the angles extended to form horizontally directed spines, the apical margin elevated to the narrow median incision; face of semicell with a prominent protuberance bearing 3 large granules with a mucilage pore above and to either side; wall with granules within the margin of all lobes; sinus narrow and closed, but slightly open outwardly; lateral view narrowly oval, the poles rounded and furnished with an apiculate spine, the margins convex to the promiment, granular protuberance, with a mucilage pore in the wall immediately above; vertical view oval, the poles rounded and bearing 4 spines at the margin, lateral margins broadly convex with a median protrusion. L. 26–41 μm. W. 20–33 μm. Isth. 6.5–8 μm. Ap. 14–20 μm. Th. 16.5–20 μm. Zygospore globose with stout spines, 27–30 μm in diameter without spines.

DISTRIBUTION: Alaska, Connecticut, Florida, Kentucky, Illinois, Indiana, Massachusetts, Michigan, Mississippi, New Hampshire, New York, Oklahoma, Utah, Virginia, Washington, Wisconsin. British Columbia, Québec. Great Britain, Europe, Asia, South America, Central America, Arctic.

PLATE LXXV, figs. 14–14b.

88a. Euastrum pulchellum var. **protrusum** Grönblad & Scott, in Grönblad, Prowse & Scott 1958, Acta Bot. Fennica 58: 15. Figs. 29, 30.

A variety differing from the typical by the polar lobe being more extended and by the very prominent central protuberance of the semicell. L. 29–31 μm. W. 21 μm. Th. 16 μm. Isth. 6–7 μm.

DISTRIBUTION: Virginia (Rep. without illust.). Africa.

PLATE LXXV, figs 15, 15a.

88b. Euastrum pulchellum var. **retusum** West & West 1905a, Monogr. II: 47. Pl. 64, Fig. 17.

A variety differing from the typical by having the upper, lateral margins of the semicell retuse and converging symmetrically to the polar lobe, which is narrower than in the typical, the apical margin elevated to a deep, V-shaped median incision; with a large granule or tumor on either side of the median notch; without a subapical spine at the angles of the polar lobe as in the typical; wall with granulations arranged as in the typical. L. 32–40 μm. W. 25–28 μm. Isth. 4.5–7.5 μm. Th. 16.5 μm.

Krieger (1937, p. 588) places this variety in synonymy with the typical, but we believe that it is a distinct taxon, especially because of the difference in the shape of the polar lobe.

DISTRIBUTION: Kentucky, Massachusetts. Labrador, Newfoundland, Québec. Great Britain, Europe, Asia.

PLATE LXXV, fig. 25.

88c. Euastrum pulchellum var. **subabruptum** Grönblad 1921, Acta Soc. Fauna Flora Fennica 49(7): 12. Pl. 3, Figs. 27, 28.

Euastrum pulchellum var. *abruptum* Grönblad, Miscr. Prescott & Scott 1942, Trans. Amer. Microsc. Soc. 61(1): 9. Pl. 4, Fig. 3.

A variety differing from the typical by having the basal lobes more rounded and more extended, the upper, lateral margins deeply retuse and diverging to a broad apex, the apical angles furnished with a stout, upwardly directed spine; with 3 verrucae within the margins of the polar lobules and similar granules within the margin of the basal lobes; a pair of supraisthmial granules on either side of the isthmus.

DISTRIBUTION: Mississippi. Europe.

PLATE LXXV, fig. 24.

89. Euastrum pyramidatum West 1892, Linn. Soc. Jour. Bot. 29: 139. Pl. 20, Fig. 13.

This truncate-pyramidal species has been incompletely described and must be regarded as uncertain. West & West's (1888, Pl. 16, Fig. 14) is included for reference. It seems to be closely related to *Euastrum luetkemuelleri* Ducellier, but smaller. West & West (1905a, p. 72) refer it to *E. crispulum* (Nordst.) West & West.

Semicells truncate-pyramidal, the basal lobes narrowly rounded at the angles; lateral margins converging to a short polar lobe, with a slight undulation on the margins between the basal angles and the polar lobe; apical margin flat and straight, the wall thickened internally in the midregion. L. 25–28 μm. W. 16–20 μm. Isth. 3.5–7.5 μm. Ap. 9–11 μm. Th. 12.5–15 μm. (as *E. crispulum*).

DISTRIBUTION: Florida. Great Britain.

PLATE LXXII, fig. 13.

90. Euastrum quadrilobum Scott & Grönblad 1957, Acta Soc. Sci. Fennicae, II, B, II(8): 14. Pl. III, Figs. 8, 9.

A small *Euastrum*, subrectangular in outline 1 1/3 times longer than broad; semicells quadrate, the basal lobes sharply rounded at the angles and bearing a mucro, the lateral margins diverging to an upper protrusion of the basal lobes, then retuse to an upper, lateral swelling, then abruptly retuse to the short polar lobe with subparallel margins; apical angles sharply rounded, bearing a short, stout spine, the apical margin truncate but slightly elevated to a deep, open median incision; face of semicell with a pronounced median projection, with a mucilage pore above and to either side; wall with small granules on either side of the apex of the median notch, one within the margin of the apical lobules, and within the margin of the lower lobules, a mucro on the margin of all lobules, wall smooth otherwise; sinus closed and narrow; lateral view elliptic in outline, the semicells cone-shaped, the poles bluntly rounded and bearing 2 short, granular spines on either side, the lateral margins diverging and undulate to the prominent median protrusion. L. 23–36 μm. W. 19–36 μm. Isth. 6–8.5 μm. Th. 12 μm. Zygospore unknown.

Although the form described by Prescott & Scott (*l.c.*) from Mississippi is similar to the plant described from Florida and Louisiana, the former averages larger throughout (L. 36 μm. W. 35–36 μm. Isth. 8.5 μm.).

DISTRIBUTION: Florida, Louisiana, Mississippi.

PLATE LXXV, figs. 21–21b; PLATE LXXVI, figs. 2–2b.

91. Euastrum quebecense Irénée-Marie 1938, Flore Desmid. Montréal, p. 132. Pl. 18, Figs. 16, 17.

A medium-sized *Euastrum*, broadly truncate-oval in outline, 1.5 times longer than broad; semicells truncate-pyramidal in outline, the angles of the basal lobes angular, bilobed and furnished with stout mucros, the lateral margins retuse above the upper angle of the basal lobes to an upper, lateral protrusion bearing a stout spine, then sharply retuse to the polar lobe, which has subparallel margins, the apical angles provided with a stout, horizontally directed spine, the apical margin truncate but elevated to the deep median incision, the wall thickened around the opening; apical angles with a subapical tubercle within the margin; face of semicell with a low, granule-bearing protrusion; cell wall with intramarginal granules at all lobes, smooth otherwise; sinus narrow and closed; lateral view elliptic, the poles narrowly rounded, with a subapical tubercle on either side, lateral margins convex to the basal, granular protrusion; vertical view elliptic, the poles sharply rounded and furnished with a blunt spine, with a pair of subapical tubercles on either side, the margins convex with a low, granular protrusion in the midregion. L. 51–58 μm. W. 36–38 μm. Isth. 7–8 μm. Ap. 23–25 μm. Th. 20–22 μm. Zygospore unknown.

This species has a similarity to *Euastrum evolutum* (Nordst.) West & West and its varieties as seen in face view, but the polar lobe is different in lateral view and is readily separated.

DISTRIBUTION: Ohio. Québec.

PLATE LXXVII, figs. 16–16b.

92. **Euastrum rectangulare** Fritsch & Rich 1937, Trans. Roy. Soc. South Africa 25(II): 174. Figs. 5M, 5N.

A small *Euastrum*, quadrangular in outline, length only slightly more than the width; semicells quadrate in outline, basal lobes broadly rounded-quadrate and emarginate, the lateral margins of the semicells deeply retuse to a short polar lobe with subparallel margins (somewhat convex), the angles of the polar lobe furnished with a short spine, the apical margin flat, with a short, broadly open median incision; face of semicell with a broad, smooth median protrusion; wall with a pattern of granules within the apical lobules and a granule within the lobules of the basal lobes, wall otherwise smooth; sinus narrow and closed; lateral view broadly elliptic in outline, the poles rounded and somewhat produced, the margins diverging to a prominent median protrusion; vertical view oval, the poles rounded, the lateral margins expanded to a median, bimammillate protrusion. L. 10–11 μm. W. 9–10 μm. Isth. 3–4 μm. Th. 6–7 μm. Zygospore unknown.

DISTRIBUTION: Florida. Africa, South America.

PLATE LXVIII, figs. 12–12b.

93. **Euastrum rimula** Irénée-Marie 1958, Nat. Canadien 85(5): 138. Fig. 20.

A small *Euastrum*, subrectangular in outline, 1 1/3 times longer than broad; semicells subquadrangular, the basal lobes broadly rounded, the lateral margins retuse to a short, truncate polar lobe which has subparallel margins, the angles rounded, the apical margin flat, with a relatively deep, narrow median incision; face of semicell with a granule-bearing protrusion in the midregion; cell wall smooth; sinus narrow and closed throughout; lateral view elliptic, the poles rounded, the margin smooth except for a prominent, submedian granule; vertical view elliptic, with a low, median swelling with a granule on either side. L. 25.5–26 μm. W. 19–19.5 μm. Zygospore unknown.

This incompletely described species should be compared with *Euastrum subhexalobum* West & West, from which it differs by being smaller, and in having a shorter and differently shaped polar lobe, the median incision being deeper.

DISTRIBUTION: Québec.

PLATE LXXVIII, fig. 13.

94. **Euastrum sibiricum** Boldt 1885, Öfv. Kongl. Vet. Akad. Förhandl. 42(2): 99. Pl. 5, Fig. 2 var. **sibiricum** f. **sibiricum**.

A small *Euastrum*, subrectangular in outline, 1.25 times longer than broad; semicells quadrangular, basal lobes truncate with a basal mucro and a spine at the upper angles, the upper, lateral angle extended further than the base, the lateral margins of the semicell retuse to the polar lobe, which has a stout spine at the angles, the apical margin flat, with a slight retuseness in the midregion that is bordered by a small spine on either side; face of semicell with a prominent, truncate protrusion; wall with a small intramarginal spine at the upper, lateral extension of the basal lobes; sinus narrow and closed; lateral view truncate-oval, the poles flat and with a small spine at either angle and an intramarginal spine in the midregion, the lateral margins extended to a prominent, truncate protrusion, the margins convex below but converging to the base of the semicells; vertical view broadly elliptic, the poles extended into a sharp spine, with a small, subapical spine on either side, the lateral margins convex with a truncate projection in the midregion on either side. L. 15.6–20 μm. W. 11–16 μm. Isth. 3–6 μm. Ap. 10–11 μm. Th. 8–8.4 μm. Zygospore unknown.

DISTRIBUTION: Connecticut, Florida, Massachusetts, Michigan, Montana, New York. Québec. Europe, Asia, Australia, Africa, South America, Arctic, Puerto Rico.

PLATE LXXIII, figs. 1, 1a.

94a. Euastrum sibiricum var. **sibiricum** f. **africanum** Grönblad & Scott, in Grönblad, Prowse & Scott 1958, Acta Bot Fennica 58: 16. Figs. 21–23.

A form similar in shape to the typical, with the spines at the angles much reduced or lacking, and with a much less prominent protrusion in the face of the semicell; lateral view of semicell circular, with a tubercular process at the midregion on either side. L. 17–18 μm. W. 12.5–15 μm. Isth. 5 μm. Th. 11–12 μm.

DISTRIBUTION: Virginia (Rep. without illust.). Africa.

PLATE LXXIII, figs. 2–2b.

94b. Euastrum sibiricum var. **exsectum** (Grönblad) Krieger 1937, Rabenhorst's Kryptogamen-Flora 13(1:3): 566. Pl. 77, Fig. 4 f. **exsectum**.

Euastrum sibiricum f. *exsecta* Grönblad 1921, Acta Soc. Fauna Flora Fennica 49: 13. Pl. 3, Figs. 16, 19.

A variety similar in shape to the typical but with spines at the angles reduced or lacking (as in f. *africanum*) and differing mostly in the apex of the polar lobe, which is broadly retuse at the margin, giving the appearance of 2 lobules, the apical margin with 2 small intramarginal granules on either side of the median invagination. L. 19.8–19.9 μm. W. 15.8–19 μm. Isth. 3.5–6.3 μm. Ap. 12–13 μm. Th. 8.4 μm.

DISTRIBUTION: Georgia, Louisiana, Massachusetts, Michigan, New England States. Québec. Europe, Africa, Iceland, Arctic.

PLATE LXXIII, fig. 3.

94c. Euastrum sibiricum var. **exsectum** f. **bituberculatum** Prescott f. nov.

Forma magnitudine formae typicae similis, differens, autem ut sinus palam apertus atque superficies semicellulae duos tumores tuberculares praebet; semicellula a vertice visa ovata, par tumorum tubercularium utroque in latere in regione media habens. Long. 20 μm. Lat. 18 μm. Lat. ad isthmum 5 μm.

ORIGO: E fossa distante 2 milia passuum (3.2 km.) ad occidentem a loco High Bluff, Franklin Co., Florida dicto.

HOLOTYPUS: Scott Coll. Fla-61.

ICONOTYPUS: Plate LXXIII, figs. 4, 4a.

A form similar in size to the typical but with a definitely open sinus, and with 2 tubercular swellings on the face of the semicell; vertical view oval, with a pair of tubercular swellings on either side in the midregion. L. 20 μm. W. 18 μm. Isth. 5 μm.

DISTRIBUTION: Florida.
PLATE LXXIII, figs. 4, 4a.

94d. Euastrum sibiricum var. reductum Prescott & Scott 1945, Amer. Mid. Nat. 34(1): 243. Pl. 1, Fig. 11.

A variety similar in shape to the typical but with a reduction in the size of the spines at the angles, and with a reduction in the prominence of the facial protrusion; upper, lateral angles of the basal lobes with a pair of spines at the margin; lower angles of the basal lobes without a spine; lateral view of the semicells broadly oval, with a tubercular projection in the midregion, the poles with 3 sharp spines. L. 22 μm. W. 18 μm. Isth. 5 μm. Th. 13 μm.
DISTRIBUTION: Louisiana.
PLATE LXXIII, figs. 5–5b.

95. Euastrum sinuosum Lenormand 1845, Herbarium, ex Krieger 1937, Rabenhorst's Kryptogamen-Flora 13(1:3): 499. Pl. 62, Figs. 9–11 var. **sinuosum.**

Euastrum circulare Hassall ex Ralfs 1848, Brit. Desm., p. 85. Pl. 13, Fig. 5a.

An average-sized *Euastrum*, truncate-oval in outline, 1.5 times longer than broad; semicells truncate-pyramidal, the basal lobes broadly rounded, the lateral margins retuse and slightly converging to upper, lateral lobules, then sharply retuse to the polar lobe with subparallel margins, the apex slightly inflated, the angles broadly rounded, the apical margin flat, the median incision relatively deep and narrow; face of semicell with a median basal protrusion and a protrusion on either side in the basal lobes, a similar protrusion on either side opposite the upper, lateral lobules, with a mucilage pore within each of the protrusions; cell wall scrobiculate and the margins roughened; sinus narrow and closed; lateral view truncate-oval, the poles truncate and slightly retuse, the apex produced, the lateral margins with 2 undulations, a mucilage pore within each; vertical view quadrate, the poles produced and knoblike, the lateral margins with 5 undulations and a mucilage pore within each. L. 52–87 μm. W. 28–50 μm. Isth. 8–15 μm. Ap. 17–26.4 μm. Th. 21–40 μm. Zygospore globose with numerous long spines, 27 μm in diameter without spines.
DISTRIBUTION: Alaska, Connecticut, Florida, Georgia, Indiana, Maine, Massachusetts, Michigan, Minnesota, Mississippi, New Hampshire, New Jersey, North Carolina, Pennsylvania, Rhode Island, Virginia, Washington. Québec. Great Britain, Europe, Asia, Australia, New Zealand, Africa, South Africa, Hawaii, Philippine Islands, Arctic.
PLATE LX, fig. 13.

95a. Euastrum sinuosum var. dideltoides f. glabrum Prescott & Scott 1945, Amer. Mid. Nat. 34(1): 244. Pl. 5, Fig. 2.

A form differing from the typical by being relatively broader in the basal lobes and by lacking a central mucilage pore in the face of the semicell; upper, lateral lobules weakly developed; wall coarsely punctate rather than scrobiculate as in the typical. L. 70 μm. W. 38 μm. Isth. 9.5 μm. Ap. 17.5 μm.
DISTRIBUTION: Louisiana.
PLATE LX, figs. 14, 14a.

95b. Euastrum sinuosum var. gangense (Turner) Krieger 1937, Rabenhorst's Kryptogamen-Flora 13(1:3): 502. Pl. 62, Figs. 20–22.

Euastrum gangense Turner 1892, Kongl. Svenska Vet.-Akad. Handl. 25(5): 87. Pl. 11, Fig. 20.

A variety differing from the typical by being stouter, with a wider and less produced polar lobe, with a central mucilage pore, but without mucilage pores in the protrusions of the semicell; lateral view with the apex truncate and not retuse as in the typical, the polar lobe somewhat produced, the lower, lateral margins biundulate with a single mucilage pore between the swellings. L. 52–67 μm. W. 29–36 μm. Isth. 8–12.5 μm. Th. 20–25.5 μm.

DISTRIBUTION: Mississippi. Asia, South America.

PLATE LXI, figs. 3, 3a.

95c. Euastrum sinuosum var. **germanicum** (Racib.) Krieger 1937, Rabenhorst's Kryptogamen-Flora 13(1:3): 502. Pl. 63. Fig. 1.

Euastrum sinuosum f. *germanica* (Racib.) Lütkemüller 1900, Verh. k.k. Zool.-Bot. Ges. Wien 50: 75.

Euastrum sinuosum var. *jenneri* f. *germanica* Raciborski 1889, Pam. Wydz. Mat.-Przyr. Akad. Umiej. Krakowie 17: 103. Pl. 2, Fig. 10.

A variety differing from the typical by being relatively broader at the base of the semicells, the basal lobes being more narrowly rounded, the upper, lateral lobules less prominent, and the polar lobe more produced, with subparallel margins and not so inflated at the apex; face of semicell with a triangular pattern of 4 mucilage pores in the midregion, the 5 facial protuberances without mucilage pores; cell wall punctate rather than scrobiculate, and rough. L. 64–78 μm. W. 36–44 μm. Isth. 11–12 μm. Ap. 18–21 μm. Th. 21–30 μm.

DISTRIBUTION: Nova Scotia. Europe, Asia, Australia, Caroline Is.

PLATE LXI, figs. 1, 2.

95d. Euastrum sinuosum var. **parallelum** Krieger 1932, Arch. f. Hydrobiol. Suppl. 11: 215. Pl. 22, Fig. 7.

A variety smaller than the typical and relatively narrow, the basal lobes bilobed, the upper lobule extended equal to the basal; polar lobe narrower and somewhat more produced than in the typical; face of semicell with protuberances as in the typical, each with a mucilage pore; wall smooth (or punctate ?). L. 44–57 μm. W. 25–30 μm. Isth. 10 μm. Th. 18 μm.

The American specimens are slightly smaller than described for this variety by Krieger in Sumatra.

DISTRIBUTION: Louisiana, Mississippi. Asia, East Indies.

PLATE LX, fig. 20.

95e. Euastrum sinuosum var. **perforatum** Krieger 1937, Rabenhorst's Kryptogamen-Flora 13(1:3): 503. Pl. 62, Fig. 24.

A variety relatively stouter than the typical; face of semicell with a central mucilage pore, with pores in the protuberances as well as pores in the basal and upper, lateral lobules; wall punctate. L. 56–73 μm. W. 22–45 μm. Isth. 12–15 μm.

DISTRIBUTION: Alaska, Oregon. Europe, Asia, Arctic.

PLATE LX, fig. 17.

95f. Euastrum sinuosum var. **reductum** West & West 1897, Jour. Bot. 35: 83; Krieger 1937, Rabenhorst's Kryptogamen-Flora 13(1:3): 503. Pl. 62, Fig. 25.

A variety averaging slightly smaller than the typical, with a shorter polar lobe that is not inflated at the apex; the upper, lateral lobules less developed; apical angles smoothly rounded; face of semicell with protrusions as in the typical but without mucilage pores; wall smooth. L. 42–74 μm. W. 25–44 μm. Isth. 7–15 μm. Ap. 13–22.5 μm. Th. 12–27 μm.

DISTRIBUTION: Alaska, Connecticut, Georgia, Idaho, Louisiana, Michigan, Montana, New York, North Carolina, Oregon, Washington, Wyoming. British Columbia, Newfoundland, Québec. Great Britain, Europe, Asia, Africa.

PLATE LX, fig. 19.

95g. **Euastrum sinuosum** var. **scrobiculatum** (Nordstedt) Krieger 1937, Rabenhorst's Kryptogamen-Flora 13(1:3): 503. Pl. 63, Figs. 2, 3.

Euastrum sinuosum f. *scrobiculata* Nordstedt 1873, Acta Univ. Lund 9: 9.

A variety with basal lobes and upper, lateral lobules broadly rounded, the polar lobe short and inflated with diverging lateral margins; face of semicell with a pattern of mucilage pores between the protuberances, the protrusions without mucilage pores. L. 38–86 μm. W. 20–47 μm. Isth. 11–15 μm. Ap. 18–21 μm. Th. 21–33 μm.

DISTRIBUTION: Alaska. Québec. Europe, Asia, South America, Arctic.

PLATE LX, fig. 15.

95h. **Euastrum sinuosum** var. **subjenneri** West & West 1902a, Trans. Linn. Soc. London, Bot., II, 6: 148. Pl. 19, Fig. 17.

A variety smaller than the typical and more slender; face of semicell with a pattern of protuberances numbering 11, each with a mucilage pore. L. 48–70 μm. W. 30–42.5 μm. Isth. 9–13 μm. Ap. 15 μm.

DISTRIBUTION: Louisiana, Mississippi. Québec. Asia, South America.

PLATE LX, fig. 18.

96. **Euastrum solidum** West & West 1896a, Trans. Linn. Soc. London, Bot., II, 5: 244. Pl. 14, Fig. 29.

A small *Euastrum*, nearly circular in outline, only slightly longer than broad; semicells pentagonal in outline, basal lobes with 3 angles, the lateral margins of the semicell somewhat diverging from the base to an angle which extends more than the basal angle, the margins then retuse to a third sharply pointed angle, then retuse to the polar lobe, which is short, with subparallel margins, the angles sharply pointed, the apical margin truncate and retuse in the midregion; face of semicell with a truncate, prominent protrusion; wall with an intramarginal granule on either side of the apical notch and with 2 intramarginal granules at the basal lobes; lateral view truncately oval in outline, the poles flat, the margins diverging to the prominent median protrusion; vertical view oval to rectangular in outline, the poles broadly rounded and furnished with 3 small spines, and a subapical, hornlike extension on either side, the lateral margins concave to the median, truncate protrusion. L. 27 μm. W. 23 μm. Isth. 4.5 μm. Ap. 14.5 μm. Th. 13.5 μm.

DISTRIBUTION: North America (West & West), Minnesota.

PLATE LXX, figs. 21–21b, 22, 22a.

97. **Euastrum solum** (Nordstedt) Grönblad & Scott, in Grönblad, Prowse & Scott 1958, Acta Bot. Fennica 58: 16 var. **solum**.

Euastrum cuneatum var. *solum* Nordstedt 1887, Bot. Notiser 1887: 156.

A medium-sized *Euastrum*, elongate-rectangular in outline, about 3 times longer than broad; semicells elongate, truncate-pyramidal, the basal angles rather narrowly rounded, the margins evenly converging to a truncate apex, the apical margin flat, with a narrow but shallow median notch; face of semicell with a single, mammillate protrusion at the isthmus; cell wall smooth (?); lateral view elongate-oval to subrectangular, the poles broadly rounded, the lateral margins slightly diverging to a slight swelling and then subparallel to the base of the semicell; vertical view oval, the poles broadly rounded, the lateral margins with a slight protrusion in the midregion. L. 106–130 μm. W. 34–42 μm. Isth. 16 μm. Ap. 17–22 μm. Th. 27–32 μm. Zygospore unknown.

DISTRIBUTION: Not reported from North America. Australia, New Zealand.

PLATE LXXXV, fig. 6.

97a. Euastrum solum (Nordstedt) Grönblad & Scott var. **africanum** (Fritsch & Rich) Grönblad & Scott, in Grönblad, Prowse & Scott 1958, Acta Bot. Fennica 58: 17. Figs. 13, 31, 32, photo 34.

Euastrum brasiliense var. *africanum*   Fritsch & Rich 1924, Trans. Roy. Soc. South Africa 11: 332. Fig. 9.

A variety more elongate and more slender than the typical; cells narrowly truncate-oval in outline, about 3 times longer than wide; semicells elongate-pyramidal with margins at first parallel and then convex from narrowly rounded and somewhat inflated basal angles; apex truncate with rounded angles, the apical margin flat, with a relatively deep, narrow median incision; face of semicell with a slight central basal inflation immediately above the isthmus; wall smooth; sinus narrow and closed throughout; lateral view narrowly oval, the poles broadly rounded and smooth, lateral margins convex to the base of the semicell. L. 92–101 μm. W. 30–33 μm. Isth. 13 μm. Th. 30 μm.

Var. *angustum* Grönblad & Scott is described and illustrated as being circular in vertical view, the base of the semicell orbicular or broadly oval.

Euastrum solum has been reinstated to replace *Euastrum cuneatum* var. *solum* Nordstedt, which is recognized by Krieger (1937, p. 481). More material must be studied and the zygospores identified before *E. solum* and its varieties can be separated from *E. cuneatum*.

DISTRIBUTION: Virginia (Rep. without illust.). Australia, Africa.

PLATE LVIII, figs. 5, 5a.

98. Euastrum sphyroides Nordstedt 1888, Kongl. Svenska Vet.-Akad. Handl. 22: 32. Pl. 3, Fig. 3 var. **sphyroides** f. **sphyroides**.

A small *Euastrum*, broadly oval in outline, 1.5 times longer than broad; basal lobes broadly rounded with acute granules at the angles, the lateral margins deeply retuse to a collarlike polar lobe, the apical angles rounded and bispinate, the apical margin flat and without a median incision; face of semicell with a median circular pattern of granules, a row of 3 intramarginal granules showing at the apex, and a pattern of sharp granules on the basal lobes; sinus narrow and closed; lateral view narrowly oval, the poles truncate and granular, the lateral margins subparallel and then inflated at the base of the semicell, the wall roughened with granules; vertical view narrowly oval, the poles broadly rounded and furnished with sharp granules, the lateral margins convex with a prominent, median granular protrusion. L. 36–55 μm. W. 26–50 μm. Isth. 11–14 μm. Th. 20 μm. Zygospore unknown.

DISTRIBUTION: Typical form not reported from North America. New Zealand, Africa.

PLATE LXXX, figs. 11-11b, 13-13b.

98a. Euastrum sphyroides var. **sphyroides** f. **granulata** Scott & Prescott 1952, Hydrobiologia 4(4): 391. Pl. 3, Fig. 11.

A form differing from the typical by the possession of 3 rows of granules that encircle the apical lobe (one row being intramarginal), and by having a prominent, supraisthmial granule immediately below the facial protuberance; facial protuberance with a pattern of 4 granules encircled by a ring of granules, the one nearest the isthmus being larger. L. 52-54 μm. W. 43-51 μm. Isth. 10-12 μm. Th. 22-26 μm.

DISTRIBUTION: Florida.

PLATE LXXX, figs. 12, 12a.

98b. Euastrum sphyroides var. **intermedium** Lütkemüller 1900a, Ann. des Nat. Hofmuseums 15: 121. Pl. 6, Fig. 22.

A variety differing from the typical by being relatively wider, scarcely longer than wide, and in having the polar lobe somewhat retuse at the apex, forming 2 apical lobules; granules on the lobes more numerous and acute to short spinelike. L. 42 μm. W. 38 μm. Isth. 11 μm. Ap. 14 μm.

DISTRIBUTION: Québec. Asia.

PLATE LXXX, fig. 14.

99. Euastrum spinulosum Delponte 1876, Mem. R. Acad. Sci. di Torino 28: 95. Pl. 6, Figs. 17, 18 var. **spinulosum.**

A medium-sized *Euastrum*, nearly circular in outline, 1 1/9 to 1 1/5 times longer than broad; semicells broadly oval transversely, the basal angles broadly rounded and produced; margins with a sinus between the basal lobes and equal-sized upper, lateral lobes, with a short, crownlike polar lobe, the angles rounded, the apical margin slightly retuse in the midregion, margins of all lobes with short, sharp spinelike granules and with patterns of similar spines on the face of the lobules; face of semicell with a broad central protuberance bearing a circular pattern of large granules; wall smooth otherwise; sinus narrow and closed within, opening slightly at the outer extremities; lateral view narrowly oval, the poles broadly rounded and spiny, the margins subparallel below the apex and then inflated to form a prominent, median protrusion which bears a circle of granules; vertical view narrowly oval, the poles rounded and furnished with spinelike granules, the lateral margins slightly convex with a prominent median protrusion with granules on the margin. L. 42-80 μm. W. 38-73 μm. Isth. 10-17 μm. Ap. 14-27 μm. Th. 22-42 μm. Zygospore globose with long, fork-tipped spines, conically rounded at the base, 46-50 μm in diameter without spines.

DISTRIBUTION: Iowa, Louisiana, Oklahoma, Virginia. Europe, Asia, Australia, Africa, South America, Celebes, Philippine Is.

PLATE LXXXI, fig. 10.

99a. Euastrum spinulosum var. **inermius** Nordstedt 1880, Acta Univ. Lund 16: 9. Pl. 1, Fig. 9.

A variety circular in outline, the basal lobes more truncately rounded than in the typical, the upper, lateral angles truncate, and the sinus between the upper, lateral angles and the polar lobe very narrow and closed, the polar lobe short and anvil-shaped, the apical margin retuse; cell wall coarsely granular on the margins, the pattern of granules on the median protrusion of the semicell larger than in the typical; lateral view truncately oval, the poles truncate and relatively broad, granular at the angles, the poles somewhat produced, the lateral margins strongly convex in the midregion, the protrusion prominent and granular. L. 49.2–81.9 μm. W. 42–67 μm. Isth. 10–18 μm. Ap. 17–27 μm. Th. 22–36 μm.

DISTRIBUTION: Oregon. Europe, Asia, Australia, Africa, South America.
PLATE LXXXII, figs. 6–6b.

99b. **Euastrum spinulosum** var. **lindae** Grönblad & Scott, in Grönblad, Prowse & Scott 1958, Acta Bot. Fennica 58: 17. Figs. 84–88.

A variety very similar to the typical but with all lobes more extended, the apical lobules more developed (as in var. *henriquesii* Sampaio) and the apical margin more deeply retuse; face of the semicell with a very prominent median protrusion; cell wall with distinct spinelike granulations at the margins and over most of the face, forming rows on all the lobules, being especially larger on the median protrusion, where there may be emarginate teeth, and 2 blunt, toothlike spines, one on either side of the median notch at the apex. L. 95–96 μm, without spines. W. 89–93 μm. Isth. 21–23 μm. Th. 51 μm.

DISTRIBUTION: Virginia. South Africa,
PLATE LXXXII, fig. 8.

100. **Euastrum subalpinum** Messikommer 1935, Jahrb. St. Gall. Naturw. Ges. 67: 120. Pl. 1, Fig. 5 f. **subalpinum.**

A small *Euastrum*, quadrate in outline, 1 1/3 times longer than broad; semicells quadrate; basal lobes truncately bilobed at the margin, the lateral margins retuse to the polar lobe, which has diverging margins to sharp angles of the apex, the apical margin deeply retuse and broadly V-shaped, the angles furnished with a large mucro; face of semicell with a low, smooth protrusion; cell wall with a pair of intramarginal granules on either side of the median notch, and with a pair of granules at the margin of the basal lobes; lateral view narrowly oval, the poles smoothly rounded, the margin inflated at the base of the semicell; vertical view oval, the poles broadly rounded with a small apical mucro, the lateral margins with a low, truncate protrusion in the midregion. L. 17.5–18.5 μm. W. 13.5–14.6 μm. Isth. 4–4.6 μm. Ap. 10.7 μm.

DISTRIBUTION: Alaska. Labrador, Europe, Asia, Africa, Arctic.
PLATE LXXIII, figs. 8–8b.

100a. **Euastrum subalpinum** f. **crassum** (Messikommer) Prescott stat. nov.

Euastrum subalpinum var. *crassum* Messikommer 1942, Beitr. z. Geobot. Landes. der Schweiz 24: 140. Pl. 3, Fig. 9.

A form differing from the typical only in being thicker as seen in lateral view, the facial protrusion more extended and truncate; in vertical view with a pattern of granules at the poles. L. 16–21 μm. W. 13–17 μm. Isth. 4.5–6.5 μm. Th. 8–13 μm.

DISTRIBUTION: Labrador. Europe.
PLATE LXXIII, figs. 7–7b.

101. **Euastrum subhexalobum** West & West 1898, Linn. Soc. Jour. Bot. 33: 287. Pl. 16, Fig. 7 var. **subhexalobum.**

A small *Euastrum*, broadly truncate-oval in outline, 1 2/3 times longer than broad; semicells truncate-pyramidal, the basal lobes angular at the extremities with diverging margins, the upper angle extending further than the basal angle; lateral margins of the semicell converging and concave to the polar lobe, which has rounded angles, a nearly flat apex and a short, narrow median notch; face of semicell with 22 prominent, smooth protrusions; cell wall smooth; sinus narrow and closed; lateral view truncate-oval in outline, the poles broadly rounded, the sides subparallel and then diverging to a prominent, median protrusion; vertical view oval, the poles broadly rounded and smooth, somewhat produced, the margins with 2 prominent median inflations or convolutions. L. 38–42 μm. W. 23–26 μm. Isth. 6–10 μm. Ap. 13–15.3 μm. Th. 15 μm. Zygospore unknown.

DISTRIBUTION: Florida, Louisiana, North Carolina. Newfoundland, Québec. Asia (Ceylon).

PLATE LXV, figs. 13–13b.

101a. Euastrum subhexalobum var. scrobiculatum Grönblad 1945, Acta Soc. Sci. Fennicae, II, B, II(6): 14.

Euastrum subhexalobum var. *poriferum* Prescott & Scott 1945, Amer. Mid. Nat. 34(1): 245. Pl. 5, Fig. 10.

A variety about the same size as the typical but with basal angles less sharp, and with a median mucilage pore in the face of the semicell; wall sparsely but roughly scrobiculate. L. 39–42.5 μm. W. 24–28 μm. Isth. 8.5–9 μm. Th. 15 μm.

DISTRIBUTION: Louisiana. Africa, South America.

PLATE LXV, figs. 15, 15a.

102. Euastrum subinerme West & West 1897, Jour. Bot. 35: 84. Pl. 366, Fig. 18 f. subinerme.

A small *Euastrum*, broadly oval in outline, 1.5 times longer than broad; semicells pyramidal in outline, the basal lobes rounded, the margins of the semicells retuse to upper, lateral lobules, then retuse to form a very short polar lobe which is equally bilobed, with a narrow, closed median incision, the apical margin smooth; face of semicell with 3 low protrusions across the base and a pair of protrusions, one on either side of the midregion, with a pattern of mucilage pores both within and between the facial protrusions; vertical view narrowly oval, the poles narrowly rounded and somewhat produced, with 5 undulations on either side. L. 35 μm. W. 12 μm. Isth. 8 μm. Ap. 11 μm. Zygospore unknown.

DISTRIBUTION: Typical form not reported from North America. Africa.

PLATE LXI, fig. 10.

102a. Euastrum subinerme f. glabrum Prescott f. nov.

Forma magnitudine figuraque formae typicae quasi similis, membrana, autem, sine poris mucosis, et solum 3 protuberationes, superficiales, in parte basali semicellulae habens; lobi laterales inferiores latiores quam in forma typica atque paululum bilobati. Long. 30.5 μm. Lat. 21.5 μm. Lat. ad isthmum 4.5 μm.

ORIGO: E palude Sphagno pleno in loco Isle Royale, Michigan dicto.

HOLOTYPUS: GWP Coll. Mich. A1-8.

ICONOTYPUS: Plate LXI, fig. 6.

A form about the same size and shape as the typical, but without mucilage pores in the wall, and with only 3 facial protuberances in the basal part of the semicell; the lower, lateral lobes broader than in the typical and slightly bilobed. L. 30.5 μm. W. 21.5 μm. Isth: 4.5 μm.

DISTRIBUTION: Michigan.
PLATE LXI, fig. 6.

103. **Euastrum sublobatum** Brébisson, ex Ralfs 1848, Brit. Desm., p. 91. Pl. 32, Fig. 4 var. **sublobatum** f. **sublobatum.**

A small *Euastrum*, simple, *Cosmarium*-like, cells quadrate in outline; semicells quadrate in outline, the basal angles broadly rounded and smooth, retuse to the broad polar lobe, which has broadly rounded angles, a flat but slightly concave apical margin; face of semicell with a low, smooth protuberance; cell wall smooth; sinus closed; lateral view truncate-oval, the poles truncately rounded, the lateral margins subparallel to the slight basal inflation; vertical view quadrate, the poles narrowly rounded and somewhat produced, with a subapical inflation and a median swelling on either side. L. 18–48 μm. W. (13)20–39 μm. Isth. 5.2–12.5 μm. Ap. 10–26 μm. Th. 13–21 μm. Zygospore unknown.

This plain *Euastrum* varies greatly in dimensions according to various authors. It seems likely that some species of *Cosmarium* have been confused with it, although the vertical view clearly exhibits *Euastrum*-like characters.

DISTRIBUTION: Florida, Georgia, Louisiana, Michigan, Oregon, Rhode Island. British Columbia, Québec. Europe, Asia, New Zealand, Africa, South America.
PLATE LXII, fig. 6; PLATE LXIX, figs. 10–10b.

103a. **Euastrum sublobatum** var. **kriegeri** Grönblad 1936, Soc. Sci. Fennica. Comm. Biol. 5(6): 3. Pl. 2, Fig. 35 f. **tuberculatum** Prescott f. nov.

Forma relative longior angustiorque quam forma typica, angulis loborum basalium acutius rotundatis (nostra forma cum, forma typical sine, mucrone); lobus polaris aliquantulum magis productus, angulis abrupte rotundatis aut acutis; superficies semicellularum proiectionem parvam tuberculiformem media in regione praebens; semicellulae a latere visae latiores, polis latius rotundatis, et proiectionem umbonatam in media regione utroque in latere habentes. Long. 24.5 μm. Lat. 16.5 μm. Lat. ad isthmum 4.5 μm.

ORIGO: E fossa distante 3 milia passuum (4.8 km.) ad orientam a loco Rigoleta, Louisiana dicto.

HOLOTYPUS: Scott Coll. La-84.

ICONOTYPUS: Plate LXX, figs. 4, 4a.

A form relatively longer and narrower than the typical, the angles of the basal lobes more sharply rounded (and in ours with a mucro lacking in the typical variety); polar lobe somewhat more produced, with sharply rounded or acute angles; face of semicells with a small tubercelike projection in the midregion; in lateral view the semicells wider, the poles more broadly rounded, and with a knoblike projection in the midregion on either side. L. 24.5 μm. W. 16.5 μm. Isth. 4.5 μm.

DISTRIBUTION: Louisiana.
PLATE LXX, figs. 4, 4a.

103b. **Euastrum sublobatum** var. **notatum** Grönblad 1945, Acta Soc. Sci. Fennicae, II, B, II(6): 14. Pl. 4, Fig. 71.

Euastrum sublobatum var. *obtusatum* f. Scott & Prescott 1958, Rec. America-Australian Sci. Exped. Arnhem Land, 3: 38. Fig. 6.

A variety with the polar lobe possessing subparallel margins, relatively narrower than in the typical; the face of the semicell with a knoblike protrusion (as in var. *tuberculatum*). L. 21 μm. W. 11–18 μm. Isth. 3.5–4.5 μm. Th. 10 μm.

DISTRIBUTION: Florida. Australia. South America.

PLATE LXIX, figs. 14–14b.

103c. **Euastrum sublobatum** var. **obtusatum** (Gutwinski) Krieger 1937, Rabenhorst's Kryptogamen-Flora 13(1:3): 545. Pl. 74, Figs. 9–11.

Cosmarium angustatum f. *obtusata* Gutwinski 1902, Bull. Inter. Acad. Sci. crakovie 1902: 601. Pl. 39, Fig. 53.

A variety narrower and relatively longer than the typical, the basal lobes with obliquely truncate apices; polar lobe with the angles sharper and slightly recurved; face of semicell without a median swelling; lateral view narrowly oval, the semicells truncately cone-shaped. L. 18–26 μm. W. 8–17 μm. Isth. (3)3.8–5 μm. Ap. 8.4 μm. Th. 8–17 μm.

DISTRIBUTION: Louisiana, Michigan. Europe, Asia, East Indies, Australia, South America.

PLATE LXIX, figs. 11, 15–15b.

104. **Euastrum subornatum** West & West 1896a, Trans. Linn. Soc. London, Bot., II, 5: 243. Pl. 14, Fig. 30 var. **subornatum**.

A small *Euastrum*, transversely rectangular in outline, *Cosmarium*-like, about as broad as long or slightly broader; semicells transversely oval in outline, the basal lobes broadly rounded at the angles and granular, the lateral margins of the semicell deeply retuse to the very short polar lobe, which has acute angles, the apical margin truncate but retuse in the midregion; face of semicell with a prominent protrusion bearing a pattern of large granules; wall with granules within the apical margin and over the extremities of the basal lobes; sinus narrow and closed within but opening outwardly; lateral view broadly truncate-oval, the poles truncate and granular, the margins diverging to extended, median, granular swellings at the base of the semicell; vertical view narrowly oval, the poles rounded and granular, margins with a pronounced granular protrusion in the midregion on either side. L. 17–18 μm. W. 20–22 μm. Isth. 6–9 μm. Ap. 8–10 μm. Th. 14 μm. Zygospore unknown.

DISTRIBUTION: Kansas, Virginia, North America (West & West).

PLATE LXXI, fig. 18–18b.

104a. **Euastrum subornatum** var. **apertum** Scott & Grönblad 1957, Acta Soc. Sci. Fennicae, II, B, II(8): 15. Pl. 3, Fig. 12.

A variety in which the sinus is more open and with the angles of the polar lobe more rounded. L. 20 μm. W. 20 μm. Isth. 6.5 μm. Th. 14 μm.

DISTRIBUTION: Florida.

PLATE LXXI, figs. 16, 16a.

105. **Euastrum tetralobum** Nordstedt 1875. Öfv. Kongl. Vet. Akad. Förhandl. 1875: 30. Pl. 8, Fig. 30.

A medium-sized *Euastrum*, broadly oval in outline, 1.5 times longer than broad; semicells semicircular in outline, the lateral margins almost equally divided into 4 lobes, which are again divided to form 2 emarginate lobules each by V-shaped incisions, the sinuses between the lobes narrow and the upper, lateral sinuses closed; face of semicell with a broad, bandlike protrusion near the base, and with a more reduced swelling across the face of the semicell at the base of the polar lobe; wall with verrucae on the facial swelling and 3 pairs of verrucae on each of the lobes; sinus narrow and closed; lateral view elliptic, the poles bluntly pointed, the lateral margins convex to form cone-shaped semicells, roughened at the margin with granular verrucae immediately below the apex and more prominent in the upper half of the semicell and at the base; vertical view elliptic, the poles bluntly pointed, the margins symmetrically undulate. L. 74–88 μm. W. 48–64 μm. Isth. 19.2–28.8 μm. Ap. 24–31.2 μm. Th. 27.6–33 μm. Zygospore unknown.

DISTRIBUTION: Alaska. Europe, Arctic.
PLATE LXXIII, figs. 20–20b.

106. Euastrum tribulbosum Silva 1951, Dissert., p. 349. Pl. 26, Fig. 9; 1953, Bull. Torr. Bot. Club 80(4): 349. Fig. 10.

A medium-sized *Euastrum*, broadly truncate-oval in outline, 1.5 times longer than broad; semicells truncate-pyramidal in outline, the basal angles rounded, the lateral margins of the semicells concave and evenly converging to a polar lobe which has broadly rounded angles, the apical margin deeply retuse; face of semicell with a protrusion in the median area of each basal lobe and one in the midregion of the polar lobe; wall smooth; sinus closed within but opening outwardly; lateral view narrowly oval, the semicells cone-shaped with broadly rounded apices and biundulate margins; vertical view narrowly oval, the poles rounded, the margins triundulate. L. 60 μm. W. 40 μm. Isth. 10 μm. Th. 20 μm. Zygospore unknown.

DISTRIBUTION: Tennessee.
PLATE LXXI, figs. 15–15b.

107. Euastrum trigibberum West & West 1895, Trans. Linn. Soc. London, Bot., II, 5: 53. Pl. 6, Fig. 22.

Euastrum denticulatum var. *trigibberum* (West & West) Schmidle 1897, Nuova Notarisia 1897(8): 65.

A small *Euastrum*, broadly truncate-oval in outline, 1 1/5 times longer than broad; semicells truncate-pyramidal in outline, the basal lobes with broadly rounded, granular apices, a deep, U-shaped sinus between the basal lobes and the polar lobe, which bears a sharp spine at the angles, and with a bigranular protrusion immediately below the polar angles, the apical margin truncate, flat but undulate, with a U-shaped median incision; face of semicell with a basal, median protrusion furnished with a circular pattern of granules; wall with a pattern of granules within the polar lobules and a cluster of granules on each of the basal lobes, with a row of intramarginal granules; sinus narrow and closed; lateral view narrowly oval, the poles truncate with a sharp spine at the apex, the polar region slightly swollen, retuse between the apical lobe and the prominent, truncate swelling at the base; vertical view oval in outline, the poles rounded and granular, with a subapical granular protrusion, the margins deeply retuse between the subapical protrusion and a prominent, truncate, granular protrusion in the midregion. L. 19.5–30 μm. W. 18.5–30 μm. Isth. 6–7 μm. Ap. 14–15.5 μm. Th. 13.5 μm. Zygospore unknown.

DISTRIBUTION: Colorado, Georgia, Kansas, Michigan, Mississippi, Wisconsin. Asia, Africa, Arctic.
PLATE LXXV, fig. 11.

108. **Euastrum turneri** West 1892, Linn. Soc. Jour. Bot. 29: 141. Pl. 20, Fig. 18 var. **turneri** f. **turneri**.

A small *Euastrum*, truncate-oval in outline, 1.5 times longer than broad; semicells truncate-pyramidal in outline, the basal angles biundulate at the margins, the margins of the semicells retuse and converging to upper, lateral lobules, a U-shaped sinus between the upper, lateral lobules and the polar lobe, which has a small protrusion immediately below the apical angles where there is a short, stout spine; the apical margin flat but undulate, with a short, open median incision; face of semicell with a median protrusion bearing a circle of large granules, and with a pair of mucilage pores above, one on either side; wall with a scattering of granules in the apical lobules and over the face of the basal lobes; lateral view elliptic, the poles pointed with a mucro, the apical margin undulate, with an upper, low protrusion and a prominent, granular protrusion at the base. L. 33–50 μm. W. 26–33 μm. Isth. 6–11 μm. Ap. 18–26.4 μm. Th. 17–23 μm. Zygospore unknown.

Krieger (1937, p. 589) places *E. denticulatum* var. *stictum* Börges. in synonymy with *E. turneri* West. But we agree with Scott & Prescott (1952, p. 392) in the assignment of the former taxon to *E. turneri* var. *stictum*.

DISTRIBUTION: Alaska, Colorado, Florida, Illinois, Kentucky, Michigan, New Hampshire, New York, North Carolina, Vermont, Québec. Great Britain, Europe, Asia, Australia, New Zealand, Africa, South America, Arctic.

PLATE LXXV, fig. 16.

108a. **Euastrum turneri** var. **turneri** f. **laeve** Irénée-Marie 1958, Nat. Canadien 85(5): 143. Fig. 24.

A form very similar to the typical but with a narrower, deeper sinus between the upper, lateral lobules and the polar lobe; marginal granules reduced in number; face of semicell without mucilage pores. L. 42–45 μm. W. 31.5–34.2 μm. Isth. 8–9.7 μm.

DISTRIBUTION: Québec.

PLATE LXXV, fig. 19.

108b. **Euastrum turneri** var. **bohemicum** (Lütkemüller) Krieger 1937, Rabenhorst's Kryptogamen-Flora 13(1:3): 590. Pl. 82, Fig. 22.

Euastrum turneri f. *bohemica* Lütkemüller 1910, Verh. k.k. Zool.-Bot. Ges. Wien 60: 483. Pl. 2, Fig. 2.

Euastrum turneri West f. Taylor 1935, Pap. Michigan Acad. Sci., Arts & Lettr. 20: 209. Pl. 39, Fig. 8.

A variety with semicells nearly quadrate to semicircular rather than truncate-pyramidal as in the typical; basal lobes more definitely bimarginate and furnished with a pair of spines, as are the upper, lateral lobes and the angles of the polar lobe; face of semicell with 4 granules and 4 (or 2) mucilage pores immediately above. L. 36–40 μm. W. 27–30 μm. Isth. 6–10 μm. Th. 16–20 μm. Zygospore globose with slender spines, conical at the base, 30 μm in diameter without spines.

DISTRIBUTION: Newfoundland. Europe, South America, Arctic.

PLATE LXXV, figs. 17, 20.

108c. **Euastrum turneri** var. **poriferum** (Scott & Prescott) Prescott stat. nov.

Euastrum turneri f. *poriferum* Scott & Prescott 1952, Hydrobiologia 4(4): 392. Pl. 3, Fig. 6.

A variety in which the invaginations and sinuses between the lobes and lobules are much reduced, the apices of the lobes rounded and bearing minute granules; polar incision deeper and narrower than in the typical; face of semicell with a pair of mucilage pores immediately above the facial swelling, one on either side. L. 46–51 μm. W. 31–36 μm. Isth. 7–9 μm. Th. 19 μm.

DISTRIBUTION: Florida.

PLATE LXXV, figs. 23, 23a.

108d. Euastrum turneri var. **stictum** (Börgesen) Scott & Prescott 1952, Hydrobiologia 4(4): 392. Pl. I, Fig. 9.

Euastrum denticulatum var. *stictum* Börgesen 1890, Vid. Medd. Naturh. Foren. Kjöbenhavn 1890: 939. Pl. III, Fig. 18.

A variety differing from the typical by having less produced lobes and lobules, the subapical protrusions on the margin of the polar lobe scarcely evident; apical notch deep and V-shaped; lateral view showing a greater extension of the median protrusion; in vertical view broadly oval, the poles narrowly rounded and with 2 mammillate protrusions on either side in the midregion. L. 36–37 μm. W. 25–26 μm. Isth. 7 μm. Th. 18 μm.

DISTRIBUTION: Florida, Michigan. Australia, South America.

PLATE LXXV, fig. 18.

109. Euastrum umbonatum (West & West) Schmidle 1896, Flora 82(3): 310; Krieger 1937, Rabenhorst's Kryptogamen-Flora 13(1:3): 607. Pl. 84, Figs. 16–18 var. **umbonatum.**

Euastrum rostratum subsp. *umbonatum* West & West 1895, Trans. Linn. Soc. London, Bot., II, 5: 52. Pl. 6, Fig. 16.
Euastrum rostratum subsp. *umbonatum* var. *ornatum* West & West 1895, Trans. Linn. Soc. London, Bot., II, 5: 52. Pl. 6, Fig. 17.

A medium-sized *Euastrum*, broadly truncate-oval in outline, 1.5 times longer than broad; semicells quadrate in outline, the basal lobes broadly rounded and tipped with 3 spines, the margins deeply retuse to upper, lateral lobules which extend nearly as far as the basal lobes, the lateral margins concave to a short polar lobe which has parallel margins, the angles extended into horizontally directed spines, the apical margin elevated to the short, open median notch; wall with a pair of prominent granules on either side of the midline of the polar lobe, a granule on the upper, lateral lobules and a pattern of granules on the outer part of the basal lobes; face of semicell with a prominent, smooth median protrusion; sinus narrow and opening slightly at the apex and at the outer limits of the basal lobes; lateral view broadly elliptic, the poles narrowly rounded, with a sharp, hornlike protrusion immediately below the apex, and a prominent, smooth protrusion at the base of the semicell; vertical view oval, the poles rounded and bearing 4 small, sharp granules, the margins convex but with a large, smooth swelling on either side in the midregion. L. 36.5–56 μm. W. 26–34.5 μm. Isth. 7–11 μm. Th. 17–24.5 μm.

DISTRIBUTION: Colorado, Kansas. Africa.

PLATE LXXVIII, fig. 20.

109a. Euastrum umbonatum var. **ceylanicum** Krieger 1937, Rabenhorst's Kryptogamen-Flora 13(1:3): 608. Pl. 84, Figs. 19, 20.

Euastrum rostratum f. West & West 1902a, Trans. Linn. Soc. London, Bot., II, 6: 154. Pl. 20, Fig. 12.

A variety shorter and relatively wider than the typical, and with a flat apical margin; the lobes and lobules of the semicells nearly smooth or with greatly reduced granules rather than short spines; face of semicell with a large granular projection on the extremities of the lower basal lobes, immediately above the sinus; vertical view oval, the poles narrowly rounded and produced, the lateral margins triundulate. L. 37–47 μm. W. 26–35 μm. Isth. 6–7.5 μm. Th. 18 μm.

DISTRIBUTION: Kansas. Asia (Ceylon).

PLATE LXXVIII, figs. 17, 17a.

110. Euastrum undulatum Prescott sp. nov.

Euastrum parvum, subquadrangulare, ca. aeque longum ac latum; semicellulae transverse rectangulares, lobis basalibus truncatis, bigranulatis, margines semicellularum retusi ad lobum spueriorem lateralem triangulatum qui tam longe quam lobi basales inferiores extendit, margine deinde concavo ad lobum polarem brevem latumque, in margine apicali 5-undulatum, formandum; incisura apicalis deest; superficies semicellulae protrusionem prominentum quae circulum granulorum fert praebens; membrana ordinem duplicem granulorum trans partes exteriores semicellulae habens, in vicinitate protrusionis mediae levis. Long. 21.7 μm. Lat. 21.7 μm. Lat. ad isthmum 6.5 μm. Zygospora ignota.

ORIGO: E flumine Baptisme dicto, Minnesota.

HOLOTYPUS: GWP Coll. Minn-75.

ICONOTYPUS: Plate LXXXIV, fig. 12.

A small Euastrum, subquadrangular in outline, about as long as broad; semicells transversely rectangular in outline, the basal lobes truncate, bigranulate; the margins of the semicells retuse to an upper, lateral, triangular lobule which extends as far as the lower basal lobes, the margins of the semicells then concave to form a short, broad polar lobe which is 5-undulate on the apical margin, apical incision not present; face of semicell with a prominent protrusion bearing a circle of granules; cell wall with a double row of granules over the outer portions of the semicell, smooth in the vicinity of the median protrusion. L. 21.7 μm. W. 21.7 μm. Isth. 6.5 μm. Zygospore unknown.

DISTRIBUTION: Minnesota.

PLATE LXXXIV, fig. 12.

111. Euastrum urnaforme Wolle 1892, Desm. U.S., p. 111. Pl. 63, Figs. 11, 12 f. urnaforme.

A large Euastrum, truncate-oval in outline, 1 1/3 times longer than broad; the semicells truncate-pyramidal in outline, the basal lobes produced, truncate, bilobed at the apices, the lateral margins abruptly retuse to form broad, shoulder-like upper, lateral lobules, a deep, U-shaped incison to form a polar lobe which is broadly anvil-shaped, the angles narrowly rounded, the apical margin truncate but slightly retuse in the midregion so that 2 prominent apical lobules are produced; face of semicell with 2 widely separated supraisthmial protrusions, and with a low protrusion within each of the apical lobules, a protrusion on either side of the midregion of the semicell base, and one at the base of the polar lobe; cell wall scrobiculate (?); vertical view rectangular in outline, the poles rounded and produced, with a subapical protrusion on either side of the midregion which is smooth and straight. L. 50–76 μm. W. 54–75 μm. Isth. 13–15 μm. Ap. 35.4–36 μm. Zygospore unknown.

Krieger (1937, p. 608) gives this species a questionable status, but the records of Irénée-Marie (1939, p. 137) indicate that this is an identifiable taxon.

It resembles *Euastrum pectinatum* Bréb. in some aspects, but the polar lobe is distinctly different, as are the upper, lateral lobules of the semicell.

DISTRIBUTION: Maine, New Jersey. Québec.

PLATE LXIII, fig. 9.

111a. Euastrum urnaforme f. rostrata Irénée-Marie 1958, Nat. Canadien 85(5): 144. Fig. 25.

A form larger than the typical and different by having a notably enlongated spur on the lower angle of the basal lobes. L. 77.3–77.5 µm. W. 59.6–60 µm. Isth. 23.7–24.5 µm.

DISTRIBUTION: Québec.

PLATE LXIII, fig. 12.

112. Euastrum validum West & West 1896a. Trans. Linn. Soc. London, Bot., II, 5: 245. Pl. 14, Fig. 32 var. **validum.**

A small *Euastrum*, broadly truncate-oval in outline, 1 1/3 times longer than broad; semicells truncate-pyramidal, the basal angles broadly rounded and plain, the lateral margins of the semicells converging and concave to form a short, slightly inflated polar lobe, the polar angles rather sharply rounded and the wall thickened, the apical margin biconvex, with a slight retuseness in the midregion; face of semicell with a low, median, tubercular protrusion immediately above the isthmus, with a mucilage pore in the center of the semicell; wall with a tubercular granule within the margin of the basal lobe angles; sinus narrow and closed; lateral view narrowly oval, the poles rounded, the margins symmetrically convex, with a tubercular granule at the base of the semicell; vertical view narrowly oval, the poles rounded, with a subapical granular thickening on either side. L. 20–30 µm. W. 16.8–22 µm. Isth. 3.6–9 µm. Ap. 11–14.5 µm. Th. 7–11 µm. Zygospore unknown.

DISTRIBUTION: Alaska, Florida, Kentucky, Louisiana, Massachusetts, Michigan, Minnesota, Washington. British Columbia, Québec. Great Britain, Europe, Asia, Australia, Africa, South America, Arctic.

PLATE LXX, figs. 1–1b.

112a. Euastrum validum var. **glabrum** Krieger 1937, Rabenhorst's Kryptogamen-Flora 13(1:3): 547. Pl. 72, Figs. 21, 22 f. **glabrum.**

Euastrum validum f. West & West 1907, Ann. Roy.Bot. Garden Calcutta 6(2): 199. Pl. 1, Fig. 4.

A variety differing from the typical by having no tubercular granules on the apices of the basal lobes, which are more broadly rounded; polar lobe smooth at the apex and less inflated, without a thickening of the wall so that there is no appearance of the angles being slightly recurved as in the typical. L. 25–29 µm. W. 16–21 µm. Isth. 4.8–6.5 µm. Ap. 14.4–15.6 µm. Th. 11–11.4 µm.

DISTRIBUTION: Louisiana, Michigan, Mississippi. Europe, Asia, South America.

PLATE LXX, figs. 2–2b.

112b. Euastrum validum var. **glabrum** f. **inflatum** Prescott f. nov.

Forma a forma typica differens ut multo crassior a latere verticeque visa, semicellula a latere visa fere circularis, indentatione subapicali parva sed distincta quae porum mucosum habet, praedita; semicellula a vertice visa late symmetrique ovata, membrana porum mucosum utroque in latere media in regione praebens. Long. 31 μm. Lat. 21 μm. Lat. ad isthmum 7.5 μm.

ORIGO: E fossa e qua solum exemptum est prope pontem in loco Sun, Louisiana dicto.

HOLOTYPUS: Scott Coll. La-(1156).

ICONOTYPUS: Plate LXX, figs. 3–3b.

A form differing from the typical variety by being much thicker when seen in lateral and vertical views, the semicells nearly circular in lateral view, and with a small but distinct subapical indentation where there is a mucilage pore; vertical view broadly symmetrically oval, a mucilage pore showing in the wall on either side in the midregion. L. 31 μm. W. 21 μm. Isth. 7.5 μm.

DISTRIBUTION: Louisiana.

PLATE LXX, figs. 3–3b.

113. **Euastrum ventricosum** Lundell 1871, Nova Acta Reg. Soc. Sci. Upsal., III, 8(2): 18. Pl. 2, Fig. 2 var. **ventricosum.**

A large *Euastrum*, broadly truncate-oval in outline, 1.6 times longer than broad; semicells broadly truncate-pyramidal, the basal lobes broadly rounded, the lateral margins of the semicells concave and converging to upper, lateral lobules which are upwardly directed, a narrow, closed sinus between the upper, lateral lobules and the short, capitate, anvil-shaped polar lobe, the angles obtusely rounded, the apical margin elevated and then flat to the deep, narrow median incision; face of semicell with a supraisthmial, prominent protrusion in the midregion, a similar protrusion on either side of the basal lobes, and a protrusion on either side of the midline of the semicell, with a mucilage pore midway between the 2 swellings; wall coarsely scrobiculate; sinus narrow and closed; lateral view truncate-oval, the poles broad and flat, slightly retuse in the midregion, the margins at first subparallel and then inflated to form a rectangular base with parallel margins, but with a protrusion in the midregion on either side; vertical view broadly elliptic, the poles narrowly rounded and somewhat produced, the lateral margins convex and 5-undulate, with a mucilage pore showing in the wall in the midregion. L. 80–175 μm. W. 49–103 μm. Isth. 14–28 μm. Ap. 32–35 μm. Th. 33–48 μm. Zygospore globose, beset with scattered, conical processes, 110 μm in diameter including processes.

DISTRIBUTION: Florida, Louisiana, Massachusetts, Michigan, Mississippi. Newfoundland. Great Britain, Europe, Asia, South America.

PLATE LXII, fig. 11.

113a. **Euastrum ventricosum** var. **floridanum** (Turner) Turner 1892, Kongl. Svenska Vet.-Akad. Handl. 25(5): 78. Pl. 11, Fig. 19.

Euastrum floridanum Turner 1885, Jour. Roy. Microsc. Soc., II, 5: 935. Fig. 7.

A variety differing from the typical by lacking the supraisthmial protuberances and by having prominent protuberances on the apical lobules; lateral view rectangular in outline, the poles convex with subapical, lateral protuberances giving the polar lobe a distinct capitate appearance, the margins subparallel below the protuberances and then abruptly inflated in the midregion, where there is a lobelike extension, with the semicell base showing a less produced inflation. L. 96 μm. W. 54 μm. Isth. 14 μm.

DISTRIBUTION: Florida. Asia.

PLATE LXII, figs. 12, 12a.

113b. **Euastrum ventricosum** var. **glabrum** Prescott & Scott 1945, Amer. Mid. Nat. 34(1): 246. Pl. 4, Fig. 4.

A variety differing from the typical by having a single, low protuberance at the isthmus in the midregion and without a mucilage pore in the midregion of the semicell; basal angles extended and sharply rounded, the sinus between the upper, lateral lobules more widely open, the polar lobe distinctly anvil-shaped; wall smooth or finely punctate. L. 154 μm. W. 94 μm. Isth. 18 μm.

DISTRIBUTION: Louisiana.

PLATE LXII, fig. 9.

113c. **Euastrum ventricosum** var. **mucronatum** Scott & Grönblad 1957, Acta Soc. Sci. Fennicae, II, B, II(8): 15. Pl. 3, Fig. 3.

A variety similar in shape to var. *glabrum* Presc. & Scott but with all lobes bearing a pronounced mucro at the apices rather than being broadly rounded; polar lobe with sharply pointed, mucronate angles. L. 138-144 μm. W. 106-109 μm. Isth. 22-23 μm.

This variety is distinct in some respects, but probably should be assigned to a form of var. *glabrum*.

DISTRIBUTION: Florida.

PLATE LXII, fig. 10.

113d. **Euastrum ventricosum** var. **rectangulare** Prescott & Scott 1945, Amer. Mid. Nat. 34(1): 246. Pl. 4, Fig. 5.

A variety differing from the typical by having angular basal lobes, the lateral margins of the semicells at first subparallel and then deeply retuse to form the somewhat longer, upper lateral lobules; facial protrusions and mucilage pore as in the typical, but with the 2 lateral basal protrusions slightly emarginate. L. 104 μm. W. 55 μm. Isth. 16 μm.

DISTRIBUTION: Louisiana.

PLATE LXII, figs. 8, 8a.

113e. **Euastrum ventricosum** var. **sopchoppiense** Scott & Prescott 1952, Hydrobiologia 4(4): 393. Pl. 2, Fig. 2.

A variety larger than the typical, 1.8-2 times longer than broad; basal lobes sharply rounded at the angles and extended so that they overlap the lobes of the opposite semicell, the sinus accordingly closed, deep and enlarged at the apex; lateral margins of the semicells broadly concave and then diverging to form the upper, lateral lobules, which are narrower than in the typical and directed sharply upward, a deep sinus between the upper, lateral lobules and the polar lobe, which is broadly cuneate, the apical margin convex with a narrow median incision; face of the semicell with a prominent, supraisthmial protuberance and with a smaller intramarginal protuberance on either side at the base of the semicell; lacking a central mucilage pore; a low protrusion in the midregion of the polar lobules; wall coarsely scrobiculate; lateral view of semicell truncate-pyramidal with margins symmetrically converging to an inflated apex that is 3-lobed; vertical view elliptic in outline with a protuberance on either side of the sharply rounded poles, and a prominent lateral protrusion in the midregion. L. 169-180 μm. W. 90-99 μm. Isth. 23-25 μm. Th. 58-64 μm.

DISTRIBUTION: Florida. Ontario.

PLATE LXII, figs. 7-7b.

114. **Euastrum verrucosum** Ehrenb. ex Ralfs 1848, Brit. Desm., p. 79. Pl. XI, Figs. 2a–2d var. **verrucosum** f. **verrucosum**.

Euastrum verrucosum Ehrenberg 1835, Phys. Abh. K. Acad. Wiss. Berlin 1833: 247.
Euastrum intermedium Delponte 1876, Mem. R. Acad. Sci. Torino, II, 28: 97. Pl. 6, Figs. 1–23.

A large *Euastrum*, approximately circular in outline, 1 1/8 times longer than broad; semicells broadly truncate-pyramidal, the basal lobes extended and rounded, the lateral margins of the semicells converging and concave to form short, upwardly directed upper, lateral lobules, a U-shaped sinus between the upper, lateral lobules and the short, broad polar lobe, with rounded, inflated angles, the apical margin retuse, without a median incision; face of semicell with a large central protrusion, decorated with concentric circles of large granules, and with a smaller protrusion on either side in the midregion of the basal lobes; wall coarsely granular throughout, the granules prominent on the margins of all lobes and lobules; sinus narrow and closed throughout; lateral view subrectangular, the poles truncate and bilobed, retuse in the midregion, the margins concave to the large basal inflation of the semicell; vertical view broadly oval in outline, the poles produced and narrowly rounded, the lateral margins equally triundulate with granular protrusions. L. 70–118 μm. W. 63–108 μm. Isth. 19–23 μm. Ap. 25–42 μm. Th. 42–57 μm. Zygospore unknown.

It is noteworthy that, as far as known, this large species has not been found in conjugation. Collections often show numerous individuals to be present, indicating ample opportunities for zygospore formation. It is one of the most variable of all the *Euastrum*.

DISTRIBUTION: Widely distributed throughout the United States and Canada. Cosmopolitan.

PLATE LXXXII, fig. 9.

114a. **Euastrum verrucosum** var. **alatum** Wolle 1884a, Desm. U.S., p. 101. Pl. 26, Fig. 4 f. **alatum**.

Euastrum verrucosum var. *alatum* f. *minus* (Lobik) Kossinskaja 1936b, Sovietskaja Bot. 1936: 110. Pl. 1, Fig. 5.

A variety differing from the typical by having narrower, more produced basal lobes, with the sinus closed within but opening widely for most of its length; polar lobe somewhat more produced and cuneate. L. 72–110 μm. W. 60–108 μm. Isth. 14–22.5 μm. Ap. 22–41 μm. Th. 32–43 μm.

DISTRIBUTION: Widely distributed throughout the United States. Newfoundland, Québec. Great Britain, Europe, Asia, Africa, South America, Arctic.

PLATE LXXXII, fig. 13; PLATE LXXXIII, fig. 1.

114b. **Euastrum verrucosum** var. **alatum** f. **cyclops** Jackson 1971, Dissert., p. 165. Pl. 40, Fig. 4.

A form smaller than the typical, with a large facial protuberance ornamented with especially large granules, the lateral facial protuberances much reduced. L. 79–86 μm. W. 66–71 μm. Isth. 18–22 μm. Th. 37.5–45 μm.

DISTRIBUTION: Montana.

PLATE LXXXII, figs. 10–10b.

114c. **Euastrum verrucosum** var. **alatum** f. **extensum** Prescott & Scott 1952, Hydrobiologia 4(4): 394. Pl. 2, Fig. 6.

A form differing from the typical by having longer and narrower upper, lateral lobules; basal lobes widely extended as in the typical; with a definite row of prominent granules extending across the base of the semicell; in lateral view the polar lobe narrower, bilobed at the apex, the margins subparallel above and then broadly inflated in the basal region; vertical view broadly elliptic, the poles extended into cone-shaped lobes with narrowly rounded apices, with a subapical lobe on either side, and a prominent rotund swelling in the midregion bearing rows of granules. L. 85–87 μm. W. 79–88 μm. Isth. 18 μm. Th. 42 μm.

DISTRIBUTION: Florida, Montana.

PLATE LXXXIII, figs. 2–2b.

114d. Euastrum verrucosum var. **alatum** f. **floridense** Prescott f. nov.

Forma a forma typica differens quod lobi superiores laterales longiores angustioresque, sinu inter lobos superiores laterales et lobum polarem lato; lobus polaris magis productus, marginibus parallelibus. Long. 85 μm. Lat. 79 μm. Lat. ad isthmum 18 μm. Crass. 42 μm.

ORIGO: E palude distante 9 milia passuum (14.5 km.) a urbe Naples, Florida dicta.

HOLOTYPUS: Scott Coll. Fla-80.

ICONOTYPUS: Plate LXXXV, figs. 3–3b.

A form differing from the typical variety by having longer and narrower upper, lateral lobules, the sinus between the upper, lateral lobules and the polar lobe broad; polar lobe more produced, with parallel margins. L. 85 μm. W. 79 μm. Isth. 18 μm. Th. 42 μm.

DISTRIBUTION: Florida.

PLATE LXXXV, figs. 3–3b.

114e. Euastrum verrucosum var. **alatum** f. **rostrata** Irénée-Marie 1947, Nat. Canadien 74(3/4): 123. Pl. II, Fig. 7.

A form differing from the typical by having the angles of the basal lobes extended and curved so that the apices of the lobes of the 2 semicells overlap, closing the sinus outwardly. L. 90.2–92 μm. W. 77.3–79 μm. Isth. 19.3–20 μm. Ap. 38.6 μm.

DISTRIBUTION: Québec.

PLATE LXXXIII, figs. 9, 9a.

114f. Euastrum verrucosum var. **alpinum** (Huber-Pestalozzi) Krieger 1937, Rabenhorst's Kryptogamen-Flora 13(1:3): 646. Pl. 94, Fig. 14.

A variety smaller than the typical and smaller than var. *alatum*, 1 1/5 to 1 1/3 times longer than broad; the basal lobes stout, the upper, lateral lobules short and reduced. L. 130 μm.

DISTRIBUTION: Colorado, Minnesota. Europe.

PLATE LXXXII, figs. 11 (f.), 12.

114g. Euastrum verrucosum var. **apiculatum** Istvanffi 1887, Notarisia 5: 235; Krieger 1937, Rabenhorst's Kryptogamen-Flora 13(1:3): 646. Pl. 94, Fig. 6.

A variety differing from the typical by having basal and upper, lateral lobules narrower and tipped with long, spinelike granules; the sinus widely open as in var. *alatum*; facial protuberances, especially the lateral ones, reduced. L. 90–106 μm. W. 86–103 μm. Isth. 13 μm.

Among the many variations of *Euastrum verrucosum* it seems likely that this one cannot be separated taxonomically from var. *alatum* f. *extensum* Presc.
DISTRIBUTION: Québec. Europe, Arctic.
PLATE LXXXV, fig. 1.

114h. Euastrum verrucosum var. **californicum** Prescott var. nov.

Varietas a varietate typica differens quod lobus polaris incisuram mediam profundam angustamque praebet, atque lobi laterales superiores minus producuntur, atque sinus vel incisura inter lobos superiores laterales et basales deest; lobi basales ad extremitates angusti atque aliquantulum producti; superficies semicellulae 2 protrusiones contra basim loborum lateralium superiorum habens; sinus intus inapertus, extrinsecus, autem, aperiens; membrana omnio dense granulosa. Long. 115.4 µm. Lat. 104.5 µm. Lat. ad isthmum 24 µm.
ORIGO: E loco Lily Lake, north Warner Mountains, California dicto.
HOLOTYPUS: GWP Coll. Cal-5.
ICONOTYPUS: Plate LXXXIII, fig. 5.

A variety differing from the typical by having a polar lobe with a deep, narrow median incision; the upper lateral lobules less produced, and with no sinus or incision between the upper lateral lobules and the basal lobes, which are narrow at the extremities and somewhat produced (and have a spiny process at their extremities); face of semicell with 2 protrusions opposite the base of the upper, lateral lobules; sinus closed within but opening outward; cell wall densely granular throughout. L. 115.4 µm. W. 104.5 µm. Isth. 24 µm.

This is an anomalous variety, possessing some of the characteristics of *Euastrum spinulosum* var. *orbiculare* (Wallich) Wildeman, especially in reference to the incision in the polar lobe, but with basal lobes and small upper, lateral lobules similar to *Euastrum verrucosum* var. *pterygoideum* Huber-Pestalozzi. The California plant might be assignable to the latter, but the polar lobe is more like var. *orbiculare* of *E. spinulosum*. The integradation is exemplified by Wallich (1860, p. 282), who describes and illustrates what he identifies as a form of *Euastrum verrucosum* (Pl. 14, Figs. 5–7) and also a new species, *E. orbiculare* (Pl. XIV, Figs. 8–11). Krieger (1937, p. 637, Pl. 93, Fig. 14) uses Wallich's illustration of *Euastrum verrucosum* to illustrate *Euastrum spinulosum* var. *orbiculare* (Wallich) Wildeman. A monographer of the genus *Euastrum* probably will wish to separate *E. orbiculare* from *E. spinulosum* and give it a new, legal epithet.
DISTRIBUTION: California.
PLATE LXXXIII, fig. 5.

114i. Euastrum verrucosum var. **coarctatum** Delponte 1873, Mem. R. Acad. Sci. Torino 28: 95. Pl. 6, Fig. 16.

Euastrum verrucosum var. *reductum* Nordstedt 1880. Acta Univ. Lund 16: 9. Pl. 1, Fig. 14.

A variety averaging smaller than the typical, stout, with the basal lobes and the upper, lateral lobules reduced, the margins between the basal angles and the upper, lateral lobules shallow-concave, the sinus between the upper, lateral lobules and the polar lobe forming a shoulder. L. 69–117 µm. W. 64–100 µm. Isth. 14.8–23 µm. Ap. 23–45 µm. Th. 33–62 µm.
DISTRIBUTION: California, Colorado, Iowa, Maine, Mississippi, New Hampshire, New York, North Carolina, Oregon, Washington, Wisconsin. British Columbia, Québec. Great Britain, Europe, Asia, South America, Arctic.
PLATE LXXXIII, fig. 3.

114j. **Euastrum verrucosum** var. **crassum** (Irénée-Marie) Prescott comb. nov. f. **crassum.**

Euastrum verrucosum var. *alatum* f. Irénée-Marie 1952, Hydrobiologia 4(1/2): 176. Pl. 18. Fig. 2.

A variety differing from the typical by having the basal lobes high and broadly convex, the upper, lateral lobules relatively short, the sinus between the upper, lateral lobules and the polar lobe narrower, the margins of the polar lobe parallel, producing a quadrate apex, the angles narrowly rounded; facial ornamentation as in the typical; sinus narrow and closed.

DISTRIBUTION: Québec.

PLATE LXXXII, fig. 14; Plate LXXXIII, fig. 4.

114k. **Euastrum verrucosum** var. **crassum** f. **angustum** (Irénée-Marie) Prescott comb. nov.

Euastrum verrucosum var. *alatum* f. Irénée-Marie 1952, Hydrobiologia 4(1/2): 176. Pl. 19, Fig. 2.

A form differing from the typical by having narrower, downwardly directed basal lobes, and with the sinus narrow within, open for a distance and then closed at the extremities of the basal lobes; polar lobe with subparallel margins, produced and quadrate; upper, lateral lobules less produced, appearing as low elevations on the shoulder of the basal lobes.

DISTRIBUTION: Québec.

PLATE LXXXIV. fig. 15.

114l. **Euastrum verrucosum** var. **crux-africanum** Wolle 1884a, Desm. U.S., p. 101. Pl. 26, Fig. 2.

A variety differing from the typical by being simpler in morphology, the semicells broadly truncate-pyramidal or rhomboidal, the basal angles narrowly rounded, with a shallow, smooth concavity between the basal lobes and the upper, lateral lobules, which are slightly produced, but which are separated from the polar lobe by a narrow, long sinus; the polar lobe short and cuneate, the angles narrowly rounded, the apical margin slightly retuse; granules restricted to the facial protuberances and to the apices of the lobes, with a low protuberance in the midregion of each polar lobule; sinus narrowly open throughout. L. 80–100 μm. W. 65–80 μm.

DISTRIBUTION: Connecticut, Nebraska, North Carolina, Pennsylvania, Washington, Wyoming. New Brunswick.

PLATE LXXXIII, fig. 6.

114m. **Euastrum verrucosum** var. **dalbisii** Laporte 1931, Encyclop. Biol. 9: 88. Pl. 9, Fig. 94; Krieger 1937, Rabenhorst's Kryptogamen-Flora 13(1:3): 648. Pl. 95, Fig. 2 f. **dalbisii.**

A variety relatively more narrow than the typical and more compact; basal lobes broadly rounded and higher than the typical, the upper, lateral lobules less produced, the sinus between the upper, lateral lobules and the polar lobe acute at the apex but broadly open; polar lobe short, the angles broadly rounded and not produced as much as in the typical; sinus narrow and closed. L. 96 μm. W. 70 μm. Isth. 20 μm.

DISTRIBUTION: Oregon, New England States. Europe.

PLATE LXXXIII, figs. 8, 8a; PLATE LXXXIV, figs. 14, 14a.

114n. Euastrum verrucosum var. **dalbisii** f. **minus** Prescott & Scott 1945, Amer. Mid. Nat. 34(1): 247. Pl. 6, Fig. 4.

A form differing from the typical by its smaller size and by the less developed upper, lateral lobules. L. 54–62.4 µm. W. 48.3–51.5 µm. Isth. 13.5–16 µm. Ap. 25.8–27 µm. Th. 28.5 µm.

DISTRIBUTION: Mississippi. Québec.

PLATE LXXXIII, figs. 7–7b.

114o. Euastrum verrucosum var. **ornatum** Woodhead & Tweed 1960, Hydrobiologia 15(4): 324. Fig. 19.

A variety differing from the typical by being smaller and by having broadly rounded, extended basal lobes; the upper, lateral lobules truncate and upwardly projecting, bearing a pair of short, sharp spines, the sinus between the upper, lateral lobules and the polar lobe broad; margins of the polar lobe subparallel and extending to form an upwardly projecting, truncate lobule bearing 3 sharp spines, the apical margin between the lobules flat; sinus closed within but opening widely with the shape of the basal lobes. L. 68 µm. W. 61 µm. Ap. 30 µm.

This incompletely described variety differs primarily in the shape of the basal lobes, the shape of the polar lobules, and in the presence of spines at the apices of the upper, lateral lobules.

DISTRIBUTION: Newfoundland.

PLATE LXXXV, fig. 2.

114p. Euastrum verrucosum var. **perforatum** Grönblad 1920, Acta Soc. Fauna Flora Fennica 47(4): 32. Pl. 6, Figs. 40, 41.

A variety differing from the typical by having truncate-pyramidal or rhomboidal semicells; basal lobes extended and narrowly rounded, the lateral margins of the semicells converging gently and slightly convex rather than concave, so that a high basal lobe is formed, the upper, lateral lobules small and little-produced, the lateral margins forming a shoulderlike invagination to the polar lobe, which is very short, with converging margins and broadly rounded angles, the apical margin slightly retuse; facial protrusion with pores interspersed among the granules; sinus closed within but opening widely. L. 114–120 µm. W. 105–114 µm. Isth. 30 µm. Th. 57 µm.

DISTRIBUTION: California. Europe.

PLATE LXXXIV, figs. 1, 2.

114q. Euastrum verrucosum var. **planctonicum** West & West 1903, Linn. Soc. Jour. Bot. 35: 537. Pl. 15, Fig. 4.

A simplified variety in which the basal lobes are narrow and much extended, the apices narrowly rounded; upper, lateral lobules appearing only as slight swelling on the dorsal margin of the basal lobes; the polar lobe short with diverging margins, the angles narrowly rounded, the apical margin retuse, the polar lobules more extended than in the typical. L. 90 µm. W. 91 µm. Isth. 19.5 µm.

DISTRIBUTION: Florida, Michigan, New York. British Columbia. Great Britain.

PLATE LXXXII, fig. 15; PLATE LXXXIV, fig. 3.

114r. Euastrum verrucosum var. **pterygoideum** (Huber-Pestalozzi) Krieger 1937, Rabenhorst's Kryptogamen-Flora 13(1:3): 650. Pl. 96, Fig. 1.

Euastrum verrucosum var. *rhomboideum* f. *pterygoideum* Huber-Pestalozzi 1931, Arch. f. Hydrobiol. 22: 444. Textfig. 37.

A variety with much extended basal lobes, narrowly rounded at the apices; lateral margins straight and converging to upper, lateral lobules, which are small and upwardly directed; polar lobe relatively broad and quadrate with subparallel lateral margins; sinus closed within but opening widely. L. 96–104 μm. W. 86–91 μm.

This variety is similar in several respects to var. *lundellii* Krieger, but is differentiated by the more developed upper, lateral lobules and the relatively wider polar lobe. It perhaps should be assigned as a form of that variety.

DISTRIBUTION: Colorado, Montana. Europe, South America, Arctic.

PLATE LXXXIV, figs. 4, 5 (f.).

114s. **Euastrum verrucosum** var. **rhomboideum** Lundell 1871, Nova Acta Reg. Soc. Sci. Upsal., III, 8: 16; Krieger 1937, Rabenhorst's Kryptogamen-Flora 13(1:3): 650. Pl. 96, Fig. 2.

Euastrum verrucosum var. *rhomboideum* f. *levanderi* Roll 1928, Arch. Russ. Protistol. 7: 133. Pl. 3, Fig. 6.

A variety in which the basal lobes are much extended and narrowly rounded, their dorsal margins straight and converging to the upper, lateral lobules, which are only slightly produced and narrow; the polar lobe with converging lateral margins, the polar lobules only slightly produced, with the apical margin retuse; sinus closed within but widely open throughout most of its length. L. 88–135 μm. W. 76–119 μm. Isth. 19–26 μm. Ap. 32–36 μm. Th. 57–62 μm.

The chief difference between this variety and var. *pterygoideum* (Huber-Pestalozzi) Krieger is the shape of the basal lobes, which are higher (thicker vertically) and more rounded at their apices.

DISTRIBUTION: Alaska, California, Colorado, Idaho, Oklahoma, Utah. Ontario. Europe, South America, Arctic.

PLATE LXXXIV, fig. 8.

114t. **Euastrum verrucosum** var. **ricardii** Irénée-Marie 1947, Nat. Canadien 74: 124. Pl. II, Fig. 10.

A variety with subquadrate cells, with a broad, quadrate polar lobe; semicells transversely quadrate, the basal lobes broadly rounded, the lateral margins of the semicells concave above the basal lobes to the equal-sized upper, lateral lobules, which extend as far as the basal lobes; the upper, lateral lobules forming a right-angled shoulder below the quadrate polar lobe, which has parallel margins, the angles rounded but very slightly enlarged to form polar lobules, the apical margin convex in the midregion; sinus narrow and closed within, opening widely in the outer extremities. L. 88.5–106 μm. W. 67.5–74 μm. Ap. 27–32.4 μm. Isth. 16.1–19.3 μm.

This apparently is a variety which exhibits many forms. Irénée-Marie (*l.c.*, Pl. II, Fig. 11) illustrates an expression which (if constant) should be described as a form of his var. *ricardii*.

DISTRIBUTION: Québec.

PLATE LXXXIV, fig. 13.

114u. **Euastrum verrucosum** var. **schoenavii** Kaiser 1919, Krypt. Forsch. 1: 221. Textfig. 9.

A simplified and compact variety in which the basal lobes are broadly rounded and not produced laterally, the lateral margins of the semicells concave and converging to shoulderlike upper, lateral lobules, which are only slightly produced, being only swellings; the polar lobe short, broad, and quadrate, the angles rounded and the apical margin retuse. L. 84–104 μm. W. 67–84 μm. Isth. 20–22 μm. Ap. 34–38 μm. Th. 49 μm.

DISTRIBUTION: Montana. Europe, Arctic.

PLATE LXXXIV, figs. 6–6b.

114v. **Euastrum verrucosum** var. **subalatum** Huber-Pestalozzi 1931, Arch. f. Hydrobiol. 22: 431. Textfig. 17.

A compact, stout variety; the semicells truncate-pyramidal in outline; basal lobes rounded at the apices and convex on the upper margins at such an angle as to give the appearance of being somewhat downward projecting, the lateral margins of the semicells becoming vertical to form upward projecting lateral lobules, the sinus between the upper, lateral lobules and the polar lobe narrow, the polar lobe with diverging lateral margins to broadly rounded apical angles, the apical margin retusc, forming prominent apical lobules; sinus narrow and closed throughout, or sometimes closed within, then opening and closed again by the downward projecting basal lobes. L. 86–100 μm. W. 72–85 μm. Th. 53–59 μm.

DISTRIBUTION: Michigan, New York. Europe.

PLATE LXXXIV, fig. 7.

114w. **Euastrum verrucosum** var. **valesiacum** Viret 1909, Bull. Soc. Bot. Genève 1: 259, Pl. 3, Fig. 7.

Euastrum verrucosum var. *sublongum* Ackley 1929, Trans. Amer. Microsc. Soc. 48(3): 303. Figs. 9–11.

A variety relatively short, stout, and somewhat compact, with the basal lobes thick (vertically) and broadly rounded; lateral margins of the semicells concave to slightly produced and short, upper, lateral lobules, the sinus between the upper, lateral lobules and the polar lobe broadly U-shaped; polar lobe very short and quadrate, the lateral apical angles scarcely extended to form lobules; sinus narrow and closed throughout except at the extreme outer position. L. 73–88 μm. W. 57–69 μm. Isth. 12–15 μm. Th. 30–35 μm.

DISTRIBUTION: California, Michigan, Europe, Asia.

PLATE LXXXIV, figs. 9–9b.

114x. **Euastrum verrucosum** var. **wailesii** Croasdale, in Croasdale & Grönblad 1964, Trans. Amer. Microsc. Soc. 83(2): 172. Pl. 8, Figs. 6–8.

Euastrum verrucosum f. Wailes 1933, Vancouver City Mus. Bull., March, 1933: 1, 2. Figs. 7, 8.

A variety very similar in shape to the typical, but with the sinus widely open in the outer extremities, and with a prominent row of large granules across the base of the semicell, and with granules on the polar lobules in definite vertical and horizontal series. L. 88–96 μm. W. 68–78 μm. Isth. 19 μm. Th. 40–44 μm.

DISTRIBUTION: British Columbia, Labrador.

PLATE LXXXIV, figs. 10, 10a.

114y. **Euastrum verrucosum** var. **willei** Irénée-Marie 1956, Rev. Algol., 2(1/2): 120. Pl. 1, Fig. 15.

A relatively large variety with rhomboid semicells; the basal lobes with a downward projecting lobule at the lower angle, then sinuate to form a horizontally projecting lobule, thus producing a bilobed basal lobe, the lateral margins of the semicells converging straight to slightly elevated protrusions rather than upper, lateral lobules, the upper margins shoulderlike below the polar lobe, which is short with subparallel or slightly diverging lateral margins, the apical angles produced to form truncate lobules, the apical margin of the lobes straight; all lobes coarsely punctate; sinus widely open within, closed at the outer extremities by the basal lobes, L. 128–130 μm. W. 100–104 μm. Isth. 32–34 μm. Ap. 41.5–42.5 μm.

DISTRIBUTION: Québec.
PLATE LXXXIV, figs. 11, 11a.

115. Euastrum vinyardii Prescott sp. nov.

Euastrum sp. nov. Vinyard 1951, Thesis, p. 32, Pl. IV, Fig. 1.

Euastrum mediocre, late ovatum, c. 1 1.3 plo longius quam latum; semi-cellulae semicirculares, lobis basalibus ad angulos biundulatis, protrusione superiore tam longe aut longius quam lobus basalis extendente; margines semi-cellularum concavi ad lobum superiorem lateralem latum emarginatumque, sinu angusto profundoque inter lobos superiores laterales et lobum polarem parvum cuneatum; margo apicalis convexus, incisuram median angustam inapertamque habens; superficies semicellulae ut videtur levis, protrusionem haud altam prae-bens. Long. 41 μm. Lat. 25.7 μm. Lat. ad isthmum 8 μm.

ORIGO: E stagno super cinerem volcanicum superiacente in loco Mt. Jeffer-son Park, Cascade Mountains, Oregon dicto.

HOLOTYPUS: GWP Coll. OI-66.

ICONOTYPUS: Plate LXVIII, fig. 9; PLATE LXXXV, fig. 5.

A medium-sized *Euastrum* broadly oval in outline, about 1 1/3 times longer than broad; semicells semicircular in outline, the basal lobes biundulate at the angles, the upper protrusion extending as far or further than the basal lobule; margins of semicells concave to an upper, broad, emarginate lateral lobule, a narrow, deep sinus between the upper, lateral lobule and the small, cuneate polar lobe; apical margin convex with a narrow and closed median notch; face of semicell apparently smooth and with a very low protrusion. L. 41 μm. W. 25.7 μm. Isth. 8 μm.

This incompletely known species is included for the record because of its unique features. Lateral and vertical views are not recorded, hence additional observations are needed to complete the diagnosis.

DISTRIBUTION: Oregon.
PLATE LXVIII, fig. 9; PLATE LXXXV, fig. 5.

116. Euastrum wollei Lagerheim 1885, Öfv. Kongl. Vet.-Akad. Förhandl. 1885(7): 233; Wolle 1892, Desm. U.S., p. 108. Pl. 33, Figs. 1, 2; Krieger 1937, Rabenhorst's Kryptogamen-Flora 13(1:3): 531. Pl. 72, Figs. 1–3 var. **wollei**.

Euastrum intermedium var. *cuspidatum* Wolle 1884a, Desm. U.S., p. 103. Pl. 29, Figs. 3–5.
Euastrum wollei var. *quadrigibberum* Lagerheim 1885, Öfv. Kongl. Vet.-Akad. Förhandl. 1885(7): 233. Pl. 27, Fig. 6.
Euastrum wollei var. *cuspidatum* De Toni 1889, Sylloge Algar., p. 1104.

A large *Euastrum*, truncate-oval in outline, 1.5 times longer than broad; semicells distinctly 3-lobed, the basal lobes transversely cone-shaped, extended to form sharply rounded or pointed angles, their dorsal margins converging symmet-rically to form a broad polar lobe with parallel margins, the apex 4-lobed (2 lobules in view), the apical margin flat and elevated to a narrow median notch;

face of semicell with 2 horizontally disposed protrusions furnished with a circular pattern of granules; wall coarsely punctate; sinus acute at the apex, opening widely to form the bordering convex lower margins of the basal lobes; lateral view subrectangular in outline, the poles truncate with 2 cone-shaped lobules, the apical margin convex medianly, lateral margins deeply concave to a prominent, broadly conical protrusion in the midregion; vertical view broadly elliptic, the poles produced and narrowly rounded, conical, with a subapical swelling on the margin and 2 prominent, mammillate protrusions in the midregion on either side. L. 155–180 μm. W. 102–120 μm. Isth. 21–30 μm. Ap. 60–62 μm. Th. 66–72 μm. Zygospore unknown.

DISTRIBUTION: Alabama, Florida, Georgia, Louisiana, Massachusetts, Mississippi, New Hampshire, New Jersey, North Carolina. Québec.

PLATE LXVI, figs. 7–7b, 9, 9a, 10, 10a.

116a. **Euastrum wollei** var. **pearlingtonense** Prescott & Scott 1942, Trans. Amer. Microsc. Soc. 61(1): 9. Pl. 1, Fig. 22.

A variety with the basal lobes less extended, the apices broadly rounded but mucronate; the lateral margins of the semicells converging evenly to an upper, lateral lobule, then retuse to the polar lobe, which has subparallel margins, the poles broadly 4-lobed (2 lobules visible); face of semicell with a supraisthmial protrusion, an emarginate protrusion on either side in the basal lobes, and a pair of projections, one on either side of the midline immediately above the basal protrusion; wall coarsely punctate. L. 120–126 μm. W. 79–85 μm. Isth. 24–27 μm. Ap. 41–43 μm.

DISTRIBUTION: Louisiana, Mississippi. Québec.

PLATE LXVI, figs. 5–5b.

EUASTRUM: North American Taxa Rejected or in Synonym

Cosmarium angustatum (Wittr.) Nordstedt f. *obtusata* Gutwinski 1902, p. 215. Pl. 39, Fig. 53 = *Euastrum sublobatum* var. *obtusatum* (Gutw.) Krieger 1937, p. 545. Pl. 74, Figs. 9–11.
Cosmarium crassum Brébisson, in: Meneghini 1840, p. 222 = *Euastrum crassum* (Bréb.) Ralfs 1848, p. 81. Pl. XI, Figs. 3a–3f.
Cosmarium elegans Brébisson, in: Meneghini 1840, p. 222 = *Euastrum elegans* (Bréb.) Ralfs 1848, p. 89. Pl. XIV, Figs. 7a–7l.
Cosmarium gemmatum Brébisson, in: Meneghini 1840, p. 221 = *Euastrum gemmatum* (Bréb.) Ralfs 1848, p. 87. Pl. XIV, Figs. 4a–4e.
Cosmarium pectinatum Brébisson, in: Meneghini 1840, p. 222 = *Euastrum pectinatum* (Bréb.) Ralfs 1848, p. 86. Pl. XIV, Figs. 5a–5f.
Echinella oblonga Greville, in: Hooker 1830, p. 398 = *Euastrum oblongum* (Grev.) Ralfs 1848, p. 80. Pl. XII.
Euastrum abruptum Nordstedt var. *evolutum* Nordstedt 1878, p. 21. Pl. 2, Fig. 7 = *Euastrum evolutum* (Nordst.) West & West 1896, p. 243. Pl. 14, Fig. 22.
Euastrum ampullaceum Ralfs f. *scrobiculatum* Nordstedt 1873, p. 8 = *Euastrum ampullaceum* Ralfs 1848, p. 83. Pl. XIII, Figs. 4a–4d.
Euastrum ansatum Ehrenberg (p.p.), in: Hylander 1928, p. 80. Pl. 10, Fig. 11 = *Euastrum ansatum* Ehrenb. ex Ralfs var. *concavum* Krieger 1937, p. 487. Pl. 58, Fig. 12.
Euastrum ansatum (Ehrenb.) Ralfs var. *turgidum* Börgesen 1891, p. 938. Pl. 3, Fig. 15 = *Euastrum brasiliense* Borge 1903, p. 112. Pl. 5, Fig. 1.
Euastrum ansatum Ehrenb. ex Ralfs var. *valleculatum* Schaarschmidt 1882, p. 263. Fig. 1 = *Euastrum ansatum* Ehrenb. ex Ralfs 1848, p. 85. Pl. XIV, Figs. 2a–2f.
Euastrum attenuatum Wolle var. *brasiliense* Grönblad 1945, p. 12. Pl. 3, Fig. 39 = *Euastrum attenuatum* Wolle var. *hastingsii* (Wolle) Prescott comb. nov.
Euastrum attenuatum Wolle var. *brasiliense* Grönblad (p.p.), in: Förster 1972, p. 533. Pl. 6, Fig. 12 = *Euastrum attenuatum* Wolle var. *splendens* (Fritsch & Rich) Grönblad & Scott f. *foersteri* Prescott stat. nov.

Euastrum attenuatum Wolle var. *pulchellum* (Presc. & Scott) Bourrelly 1966, p. 88. Pl. 1, Fig. 3 = *Euastrum attenuatum* Wolle var. *lithuanicum* Woloszynska f. *pulchellum* Prescott & Scott 1945, p. 233. Pl. 1, Fig. 3; Pl. 4, Fig. 10.

Euastrum binale (Turp.) Ehrenberg (1840)1841, p. 208 = *Euastrum binale* (Turp.) Ralfs 1848, p. 90. Pl. XIV, Figs 8a–8c, 8g–8h.

Euastrum binale (Turp.) Ralfs f. *gutwinskii* Schmidle 1893, p. 552 = *Euastrum binale* (Turp.) Ehrenberg var. *gutwinski* (Schm.) Krieger 1937, p. 551. Pl. 75, Figs. 13–15.

Euastrum binale (Turp.) Ehrenberg f. *hians* West & West 1892, p. 140. Pl. 20, Fig. 14 = *Euastrum binale* (Turp.) Ralfs var. *hians* (West) Krieger 1937, p. 551. Pl. 75, Fig. 16.

Euastrum binale (Turp.) Ehrenberg var. *dissimile* Nordstedt 1875, p. 31. Pl. 8, Fig. 21 = *Euastrum dissimile* (Nordst.) Schmidle 1898, p. 45.

Euastrum binale (Turp.) Ehrenberg var. *elobatum* Lundell 1871, p. 22. Pl. 2, Fig. 7 = *Euastrum elobatum* (Lund.) Roy & Bissett 1893, p. 176.

Euastrum binale (Turp.) Ehrenberg var. *hians* W. West (p.p.), in Conn & Webster 1908, p. 62, Fig. 105 = *Euastrum integrum* (Wolle) Wolle 1892, p. 117. Pl. 31, Figs. 18–22.

Euastrum binale (Turp.) Ehrenberg var. *insulare* Wittrock 1872, p. 49. Pl. 4, Fig. 7 = *Euastrum insulare* (Wittr.) Roy 1877, p. 68. Pl. 40, Figs. 11–13.

Euastrum binale (Turp.) Ehrenberg var. *lacustre* Messikommer 1927, p. 98. Pl. 1, Fig. 16 = *Euastrum insulare* var. *lacustre* (Messik.) Krieger 1937, p. 556. Pl. 75, Figs. 15–18.

Euastrum binale (Turp.) Ehrenberg var. *lagoense* Nordstedt 1869, p. 218. Pl. 2, Fig. 4 = *Euastrum abruptum* var. *lagoense* (Nordst.) Krieger 1937, p. 606. Pl. 83, Figs. 4–6.

Euastrum binale (Turp.) Ehrenberg var. *minus* (W. West) Krieger 1937, p. 552. Pl. 75, Fig. 17 = *Euastrum binale* (Turp.) Ralfs f. *minor* West 1888, p. 340.

Euastrum binale (Turp.) Ralfs var. *subelobatum* West 1892, p. 140. Pl. 20, Fig. 15 = *Euastrum elobatum* (Lund.) Roy & Bissett 1893, p. 176.

Euastrum boldtii Schmidle var. *isthmochondrum* Grönblad 1921, p. 10. Pl. 3, Figs. 1–4 = *Euastrum boldtii* Schmidle 1896, p. 27. Pl. 16, Fig. 5.

Euastrum brasiliense Borge f. Fritsch & Rich 1924, p. 330. Textfigs. 8E, 8F = *Euastrum brasiliense* Borge var. *convergens* Krieger 1937, p. 484. Pl. 57, Fig. 17.

Euastrum brasiliense Borge var. *africanum* Fritsch & Rich 1924, p. 332. Fig. 9 = *Euastrum solum* (Nordst.) Grönblad & Scott var. *africanum* (Fritsch & Rich) Grönblad & Scott, in Grönblad, Prowse & Scott 1958, p. 17. Figs. 13, 31, 32.

Euastrum candianum Delponte 1876, p. 107. Pl. 6, Figs. 11, 12 = *Euastrum divaricatum* Lundell 1871, p. 21. Pl. 21, Fig. 5.

Euastrum candianum Delponte var. *minutum* Turner f. *candiana* Irénée-Marie 1958, p. 116. Fig. 7 = *Euastrum divaricatum* Lundell var. *spinosum* (Irénée-Marie) Prescott nom. nov.

Euastrum chickeringii Prescott, in Prescott & Magnotta 1935, p. 164. Pl. 25, Figs. 15, 16 = *Euastrum crassicolle* Lundell var. *bicrenatum* DeToni 1889, p. 1073.

Euastrum circulare Hassall var. *ralfsii* Brébisson 1856, p. 122 = *Euastrum jenneri* Archer 1861, p. 730.

Euastrum circulare Hassall ex Ralfs 1848, p. 85. Pl. XIII, Fig. 5a = *Euastrum sinuosum* Lenormand 1845, ex Krieger 1937, p. 499. Pl. 62, Figs. 9–11.

Euastrum compactum Wolle 1884a, p. 15 = *Euastrum elegans* var. *compactum* (Wolle) Krieger 1937, p. 593. Pl. 82, Figs. 1, 2.

Euastrum coralloides Joshua var. *subintegrum* West & West f. Scott & Prescott 1961, p. 24, in Sieminska 1965, p. 121. Pl. 8, Figs. 23, 24 = *Euastrum deuticulatum* (Kirch.) Gay 1884, p. 335.

Euastrum coralloides Joshua var. *subintegrum* West & West 1907, p. 198. Pl. 14, Fig. 8 = *Euastrum octogibbosum* Krieger 1937, p. 586. Pl. 80, Figs. 27–29.

Euastrum coronatum Turner 1885, p. 935. Pl. 15, Fig. 9 = *Euastrum evolutum* (Nordst.) West & West 1890, p. 243. Pl. 14, Fig. 22.

Euastrum cosmarioides West & West 1895, p. 54. Pl. 6, Fig. 23 = *Euastrum binale* var. *cosmarioides* (West & West) Krieger 1937, p. 550. Pl. 75, Figs. 6, 8.

Euastrum crameri Raciborski f. Krieger 1932, p. 211. Pl. 20, Fig. 12 = *Euastrum kriegeri* Prescott nom. nov.

Euastrum crameri Raciborski var. *tropicum* Krieger 1937, p. 620. Pl. 90, Figs. 4–6 = *Euastrum kriegeri* Prescott nom. nov.

Euastrum crassangulatum Börgesen var. *carniolicum* Lütkemüller 1900, p. 73. Pl. 1, Figs. 20–22 = *Euastrum luetkemuelleri* var. *carniolicum* (Luetk.) Krieger 1937, p. 561. Pl. 80, Figs. 5–7.

Euastrum crassum (Bréb.) Kützing (p.p.) in Smith 1924, p. 22. Pl. 56, Fig. 1 = *Euastrum crassum* (Bréb.) Ralfs var. *microcephalum* Krieger 1937, p. 512. Pl. 65, Fig. 6.

Euastrum crassum (Bréb.) Kützing var. *ornatum* (Wood) Hansgirg 1888, p. 205 = *Euastrum crassum* (Bréb.) Ralfs 1848, p. 81, Pl. XI, Fig. 3.

Euastrum crassum (Bréb.) Ralfs var. *pulchrum* Cushman 1905, p. 251. Pl. 64, Fig. 1 = *Euastrum crassum* (Bréb.) Ralfs var. *scrobiculatum* Lundell 1871, p. 18. Pl. 2, Fig. 1.

Euastrum crassum (Bréb.) Ralfs var. *taturnii* West & West f. *allorgei* Laporte 1931, p. 82. Pl. 10, Figs. 107, 108 = *Euastrum crassum* (Bréb.) Ralfs var. *scrobiculatum* Lundell 1871, p. 18. Pl. 2, Fig. 1.

Euastrum crassum (Bréb.) Ralfs var. *tumidum* Okada (p.p.), in Jackson 1971, p. 164. Pl. 32, Fig. 2 = *Euastrum crassum* (Bréb.) Ralfs var. *tumidum* Okada f. *undulata* Prescott comb. nov.

Euastrum crassum (Bréb.) Ralfs var. *tumidum* Okada f. *suboblongum* Jackson 1971, p. 164. Pl. 32, Fig. 3 = *Euastrum crassum* (Bréb.) Ralfs var. *suboblongum* (Jackson) Prescott comb. nov.

Euastrum cuneatum Jenner var. *granulatum* Cushman 1905, p. 251 = *Euastrum cuneatum* Jenner ex Ralfs 1848, p. 90. Pl. 32, Fig. 3.

Euastrum cuneatum Jenner var. *solum* Nordstedt 1887, p. 156 = *Euastrum solum* (Nordst.) Grönblad & Scott comb. nov., in Grönblad, Prowse & Scott 1958, p. 16.

Euastrum denticulatum (Kirch.) Gay f. Krieger 1932, p. 212. Pl. 20, Fig. 14 = *Euastrum denticulatum* (Kirch.) Gay var. *quadrifarium* Krieger 1937, p. 585. Pl. 80, Figs. 20, 21.

Euastrum denticulatum (Kirch.) Gay f. Boldt 1888, p. 6. Pl. 1, Fig. 9 = *Euastrum boldtii* Schmidle var. *groenbladii* Krieger 1937, p. 564. Pl. 77, Fig. 8.

Euastrum denticulatum (Kirch.) Gay var. *stictum* Börgesen 1890, p. 939. Pl. 3, Fig. 18 = *Euastrum turneri* West var. *stictum* (Börges.) Scott & Prescott 1952, p. 392. Pl. 1, Fig. 9.

Euastrum denticulatum (Kirch.) Gay var. *trigibberum* (West & West) Schmidle 1897, p. 65 = *Euastrum trigibberum* West & West 1895, p. 53. Pl. 6, Fig. 22.

Euastrum didelta (Turp.) Ralfs f. *ansatiforme* Schmidle 1898, p. 46. Pl. II, Fig. 28 = *Euastrum didelta* (Turp.) Ralfs 1848, p. 84, Pl. XIV, Fig. 1.

Euastrum didelta (Turp.) Ralfs var. *ansatiforme* (Schmidle) Ducellier 1915, p. 84. Fig. 9 = *Euastrum didelta* (Turp.) Ralfs 1848, p. 84, Pl. XIV, Fig. 1.

Euastrum didelta (Turp.) Ralfs var. *ansatiforme* Schmidle (p.p.), in Irénée-Marie 1938, p. 124. Pl. 16, Fig. 4 = *Euastrum didelta* (Turp.) Ralfs 1848, p. 84, Pl. XIV, Fig. 2.

Euastrum didelta (Turp.) Ralfs f. *longicolle* Irénée-Marie 1958, p. 123. Fig. 13 = *Euastrum ansatum* Ehrenberg ex Ralfs 1848, p. 85, Pl. XIV, Fig. 1.

Euastrum didelta (Turp.) Ralfs var. *intermedium* Ducellier, in Irénée-Marie 1958, p. 122 = *Euastrum ansatum* Ehrenberg ex Ralfs var. *subconcavum* Prescott nom. nov.

Euastrum didelta (Turp.) Ralfs var. *scrobiculata* (Nordst.) Irénée-Marie 1956, p. 114. Fig. 13 = *Euastrum didelta* (Turp.) Ralfs f. *scrobiculata* Nordstedt 1873, p. 9.

Euastrum dideltoides (Racib.) West & West f. *javanica* Gutwinski 1902, p. 602. Pl. 39, Fig. 5 = *Euastrum ansatum* Ehrenberg ex Ralfs var. *javanicum* (Gutw.) Krieger 1937, p. 488. Pl. 58, Fig. 9.

Euastrum divaricatum Lundell var. *inerme* Irénée-Marie 1958, p. 123. Fig. 14 = *Euastrum divaricatum* Lundell var. *elevatum* (Irénée-Marie) Prescott nom. nov.

Euastrum dubium Nägeli f. *mauritiana* Irénée-Marie 1958, p. 124 = *Euastrum lapponicum* Schmidle var. *mauritianum* (Irénée-Marie) Prescott comb. nov.

Euastrum dubium Nägeli f. *scrobiculata* Lütkemüller 1910 p. 482. Pl. 2, Fig. 1 = *Euastrum dubium* Nägeli var. *scrobiculatum* (Lütkem.) Krieger 1937, p. 573. Pl. 79, Fig. 13.

Euastrum dubium Nägeli var. *anglicum* (Turner) West & West 1905a, p. 44, Pl. 38, Fig. 9 = *Euastrum dubium* Nägeli 1849, p. 122. Pl. 70, Fig. 2.

Euastrum elegans (Bréb.) Kützing var. *bidentatum* (Nägeli) Jacobsen 1875, p. 191 = *Euastrum elegans* (Bréb.) Ralfs 1848, p. 89, Pl. XIV, Fig. 7.

Euastrum elegans (Bréb.) Ralfs var. *compactum* f. Prescott & Scott 1945, p. 238. Pl. 2, Fig. 9 = *Euastrum elegans* (Bréb.) Ralfs f. *Prescottii* Förster 1972, p. 536. Pl. 8, Fig. 3.

Euastrum elegans (Bréb.) Ralfs var. *inerme* Ralfs 1848, p. 89. Pl. XVII, Fig. 7e = *Euastrum inerme* (Ralfs) Lundell 1871, p. 20. Pl. 2, Fig. 3.

Euastrum elegans (Bréb.) Ralfs var. *medianum* Nordstedt 1888, p. 34 = *Euastrum cornubiense* West & West var. *medianum* (Nordst.) Krieger 1937, p. 574. Pl. 78, Figs. 7–9.

Euastrum elegans (Bréb.) Ralfs var. *oculatum* Istvànffi 1887, p. 236 = *Euastrum bidentatum* var. *oculatum* (Istv.) Krieger 1937, p. 603. Pl. 85, Fig. 2.

Euastrum elegans (Bréb.) Ralfs var. *poljanae* Pevalek 1925, p. 289. Fig. 1 = *Euastrum elegans* (Bréb.) Ralfs 1848, p. 89. Pl. XIV, Fig. 7.

Euastrum elegans (Bréb.) Ralfs var. *quebecense* Irénée-Marie 1956, p. 115. Fig. 7 = *Euastrum elegans* (Bréb.) Ralfs var. *pseudelegans* (Turner) West & West f. *quebecense* (Irénée-Marie) Prescott stat. nov.

Euastrum elegans (Bréb.) Ralfs var. *speciosum* Boldt 1888, p. 9. Pl. 1, Figs. 10, 11 = *Euastrum bidentatum* Nägeli var. *speciosum* (Boldt) Schmidle 1898, p. 47.

Euastrum everettense Wolle var. *crassum* (Presc. & Scott) Jackson 1971, p. 170. Pl. 34, Fig. 5 = *Euastrum didelta* var. *crassum* (Presc. & Scott) Förster 1972, p. 535. Pl. 6, Fig. 7.

Euastrum evolutum (Nordst.) West & West f. *minor* West & West 1898, p. 292. Fig. 1c = *Euastrum evolutum* (Nordst.) West & West var. *guianense* (Racib.) West & West 1898, p. 292.

Euastrum floridanum Turner 1885, p. 935. Fig. 7 = *Euastrum ventricosum* Lundell var. *floridanum* (Turner) Turner 1892, p. 78. Pl. 11, Fig. 19.

Euastrum formosum Gay 1884, p. 55. Pl. 1, Fig. 9 = *Euastrum gayanum* DeToni 1889, p. 1075.

Euastrum gangense Turner 1892, p. 87. Pl. 11, Fig. 20 = *Euastrum sinuosum* Lenormand var. *gangense* (Turner) Krieger 1937, p. 502. Pl. 62, Figs. 20–22.

Euastrum gemmatum (Bréb) Ralfs f. Borge 1918, p. 61. Pl. 5, Fig. 8 = *Euastrum gemmatum* (Bréb.) Ralfs var. *tenuis* Krieger 1937, p. 641. Pl. 92, Figs. 12, 13.

Euastrum gemmatum (Bréb) Ralfs f. *latior* Grönblad 1945, p. 13. Fig. 57 = *Euastrum gemmatum* (Bréb) Ralfs var. *taftii* Prescott nom. nov.

Euastrum gemmatum (Bréb.) Ralfs var. *alatum* Taft 1949, p. 214. Pl. 2, Fig. 6 = *Euastrum gemmatum* (Bréb.) Ralfs var. *taftii* Prescott nom. nov.

Euastrum giganteum (Wood) Nordstedt var. *latum* Taylor 1935, p. 203. Pl. 41, Figs 8, 8a = *Euastrum giganteum* (Wood) Nordstedt 1889, in DeToni 1889, p. 1106.

Euastrum glaziovii Börgesen 1890, p. 941. Pl. 3, Fig. 23 = *Euastrum evolutum* (Nordst.) West & West 1898, p. 292.

Euastrum hastingsii Wolle 1892, p. 113. Pl. 42, Figs. 16, 17 = *Euastrum attenuatum* Wolle var. *hastingsii* (Wolle) Prescott comb. nov.

Euastrum humerosum Ralfs f. Prescott 1935, p. 24. Pl. 325, Figs. 11, 12 = *Euastrum humerosum* Ralfs var. *evolutum* Krieger 1937, p. 525. Pl. 70, Figs. 1, 2.

Euastrum humerosum Ralfs f. *scrobiculata* Nordstedt 1875, p. 8 = *Euastrum humerosum* Ralfs 1848, p. 82. Pl. XIII, Fig. 2.

Euastrum humerosum Ralfs var. *affine* (Ralfs) Raciborski 1885, p. 93 = *Euastrum affine* Ralfs 1848, p. 82. Pl. XIII, Fig. 3.

Euastrum incudiforme Börgesen 1890, p. 940. Pl. 3, Fig. 22 = *Euastrum evolutum* (Nordst.) West & West var. *incudiforme* (Börges.) West & West 1890, p. 292.

Euastrum incurvatum Turner 1892, p. 85. Pl. 11, Fig. 1 = *Euastrum elegans* (Bréb.) Ralfs var. *pseudelegans* (Turner) West & West 1905, p. 49. Pl. 38, Figs. 22, 23.

Euastrum incurvatum Turner 1892, p. 85. Pl. 11, Fig. 1 = *Euastrum elegans* (Bréb.) Ralfs var. *pseudelegans* (Turner) West & West f. *quebecense* (Irénée-Marie) Prescott stat. nov.

Euastrum insigne Hassall (*p.p.*), in Prescott 1935, p. 37. Pl. 325, Figs. 9, 10 = *Euastrum insigne* Hassall ex Ralfs var. *lobulatum* Presc. & Scott 1945, p. 249. Pl. 2, Fig. 16.

Euastrum insigne Hassall ex Ralfs f. Grönblad 1921, p. 35. Pl. 3, Fig. 35 = *Euastrum insigne* Hassall ex Ralfs var. *lobulatum* Presc. & Scott 1945, p. 240. Pl. 2, Fig. 16.

Euastrum insigne Hassall ex Ralfs f. Taylor 1935, p. 205. Pl. 45, Fig. 8 = *Euastrum insigne* Hassall ex Ralfs var. *lobulatum* Presc. & Scott f. *taylorii* Prescott & Scott 1945, p. 241. Pl. 7, Fig. 6.

Euastrum insulare (Wittr.) Roy f. *silesiaca* Grönblad 1926, p. 13. Pl. 1, Figs. 26, 27 = *Euastrum insulare* (Wittr.) Roy var. *silesiacum* (Grönblad) Krieger 1937, p. 557. Pl. 76, Figs. 19–21.

Euastrum intermedium (Cleve) West & West 1896a, p. 242. Pl. 14, Figs. 18, 19 = *Euastrum intermedium* (Cleve) West & West var. *longicolle* Borge 1925, p. 21. Pl. 2, Figs. 30–32.

Euastrum intermedium Delponte 1876, p. 97. Pl. 6, Figs. 21–23 = *Euastrum verrucosum* Ehrenberg ex Ralfs 1848, p. 79. Pl. XI, Fig. 2.

Euastrum intermedium (Cleve) West & West f. *scrobiculata* Schmidle 1896, p. 46. Pl. 2, Fig. 27 = *Euastrum intermedium* (Cleve) West & West var. *scrobiculatum* (Schmidle) Krieger 1937, p. 533. Pl. 71, Fig. 13.

Euastrum intermedium Cleve var. *cuspidatum* Wolle 1884a, p. 103. Pl. 29, Figs. 3–5 = *Euastrum wollei* Lagerheim 1885, p. 233.

Euastrum lapponicum Schmidle f. Prescott & Scott 1945, p. 242. Pl. 3, Fig. 10 = *Euastrum lapponicum* Schmidle var. *protuberans* Prescott var. nov.

Euastrum latipes Nordstedt f. *evoluta* Grönblad 1945, p. 13. Fig. 58 = *Euastrum latipes* Nordstedt var. *evoluta* (Grönblad) Taft 1949, p. 214. Pl. 1, Fig. 1.

Euastrum longicolle Nordstedt var. *capitatum* West & West f. Scott & Grönblad 1957, p. 13. Pl.

3, Fig. 1 = *Euastrum longicolle* Nordstedt var. *capitatum* West & West f. *extensum* (Scott & Grönblad) Prescott f. nov.

Euastrum magnificum Wolle var. *crassioides* Hastings, in Wolle 1892, p. 108 = *Euastrum crassum* (Bréb.) ex Ralfs 1848, p. 81. Pl. XI, Fig. 3.

Euastrum mammillosum Wolle 1883, p. 13. Pl. 27, Fig. 21 = *Euastrum insigne* Hassall ex Ralfs 1848, p. 83, Pl. XIII, Fig. 6.

Euastrum monticulosum Taylor 1935a, p. 774. Pl. 3, Fig. 2 = *Euastrum evolutum* (Nordst.) West & West var. *monticulosum* (Taylor) Krieger 1937, p. 616. Pl. 88, Figs. 7, 8.

Euastrum multilobatum Wood 1870, p. 16 = *Euastrum pinnatum* Ralfs 1848, p. 81. Pl. 13, Fig. 1.

Euastrum muricatum Bailey 1846, p. 126. Figs. 1, 2 = *Micrasterias muricata* (Bailey) Ralfs 1848, p. 210.

Euastrum nordstedtianum Wolle 1884a, p. 105. Pl. 26, Figs. 7, 9–15. Pl. 52, Figs. 13–15 = *Euastrum evolutum* (Nordst.) West & West 1890, p. 243. Pl. 14, Fig. 22.

Euastrum nordstedtianum Wolle 1884a, (p.p.), p. 105. Pl. 26, Fig. 8 = *Euastrum evolutum* var. *guianense* (Racib.) West & West 1898, p. 292.

Euastrum obesum Joshua f. Irénée-Marie 1938, p. 125. Pl. 8, Fig. 9 = *Euastrum obesum* Joshua var. *subangulare* West & West 1895, p. 50. Pl. 6, Fig. 15.

Euastrum obesum Joshua var. *crassum* Prescott & Scott 1942, p. 9. Pl. 1, Fig. 23 = *Euastrum didelta* (Turp.) Ralfs var. *crassum* (Presc. & Scott) Förster 1972, p. 535. Pl. 6, Fig. 7.

Euastrum oblongum (Grev.) Ralfs f. *elliptica* Irénée-Marie 1947, p. 120. Pl. II, Fig. 5 = *Euastrum oblongum* (Grev.) Ralfs var. *ellipticum* (Irénée-Marie) Irénée-Marie 1958, p. 133.

Euastrum oblongum (Grev.) Ralfs f. *scrobiculata* Nordstedt 1875, p. 7 = *Euastrum oblongum* (Grev.) Ralfs 1848, p. 80. Pl. XII.

Euastrum oculatum Börgesen var. *tonsum* West & West f. Prescott & Scott 1942, p. 9. Pl. 1, Fig. 27 = *Euastrum oculatum* Börges var. *prescottii* Förster 1972, p. 539. Pl. 8, Figs. 1, 2.

Euastrum ohioense Taft 1945, p. 190. Pl. 2, Fig. 17 = *Euastrum hypochondrum* Nordstedt var. *ohioense* (Taft) Prescott comb. nov.

Euastrum ornatum Wood 1874, p. 137. Pl. 21, Fig. 12 = *Euastrum crassum* (Bréb.) Ralfs 1848, p. 81. Pl. XI, Fig. 3.

Euastrum papilio Raciborski 1885, p. 39. Pl. 13, Fig. 9 = *Euastrum bidentatum* Nägeli var. *speciosum* (Boldt) Schmidle 1898, p. 47.

Euastrum pecten Ehrenberg 1838, p. 162. Pl. 12, Fig. IV = *Euastrum oblongum* (Grev.) Ralfs 1848, p. 80. Pl. XII.

Euastrum pectinatum Brébisson ex Ralfs var., Wolle 1887, p. 36. Pl. 5, Figs. 10–12 = *Euastrum pectinatum* (Bréb.) Ralfs var. *inevolutum* West & West 1905, p. 17. Pl. 1, Figs. 13, 14.

Euastrum pectinatum Brébisson ex Ralfs var. *brachylobum* Wittrock f. *rostrata* Taylor 1935, p. 206. Pl. 41, Fig. 4 = *Euastrum pectinatum* (Bréb.) Ralfs var. *rostratum* (Taylor) Krieger 1937, p. 539. Pl. 73, Figs. 4, 5.

Euastrum pictum Börgesen var. *subrectangulare* West & West 1898, p. 293. Textfig. 1f = *Euastrum bidentatum* Nägeli 1849, p. 122. Pl. 7D, Figs. 1a–1f.

Euastrum pictum Börgesen var. *subrectangulare* West & West, in Taylor, 1935, p. 207. Pl. 40, Fig. 12 = *Euastrum pictum* Börgesen var. *subrectangulare* West & West.

Euastrum pinnatum Ralfs f. Prescott & Scott 1945, p. 243. Pl. 7, Figs. 7–10 = *Euastrum pinnatum* Ralfs var. *pres-scottii* Irénée-Marie 1958, p. 136. Figs. 18, 19.

Euastrum porrectum Borge 1903, p. 115. Pl. 5, Fig. 8 = *Euastrum johnsonii* West & West var. *porrectum* (Borge) Irénée-Marie 1947, p. 118. Pl. 2, Fig. 2.

Euastrum pseudelegans Turner 1885, p. 935. Pl. 15, Fig. 8 = *Euastrum elegans* (Bréb.) Kützing var. *pseudelegans* (Turner) West & West 1905a, p. 49. Pl. 38, Figs. 22, 23.

Euastrum pseudotuddalense Messikommer 1942, p. 141. Pl. 4, Fig. 5 = *Euastrum bipartitum* Krieger var. *pseudotuddalense* (Messikommer) Prescott comb. nov.

Euastrum pulchellum Brébisson var. *abruptum* Grönblad, Miscr. Prescott & Scott 1942, p. 9. Pl. 4, Fig. 3 = *Euastrum pulchellum* Brébisson var. *subabruptum* Grönblad 1921, p. 12. Pl. 3, Figs. 27, 28.

Euastrum purum Wolle 1887, p. 37. Pl. 58, Figs. 9–11 = *Euastrum intermedium* Cleve 1864, p. 484. Pl. 4, Fig. 1.

Euastrum quadriceps Nordstedt 1870, p. 216. Pl. 2, Fig. 5 = *Euastrum didelta* Ralfs var. *quadriceps* (Nordst.) Krieger 1937, p. 520. Pl. 67, Figs. 8–10.

Euastrum ralfsii Rabenhorst 1868, p. 184 = *Euastrum ansatum* Ehrenberg ex Ralfs 1848, p. 85, Pl. XIV, Fig. 2.

Euastrum ralfsii Rabenhorst (p.p.) in Wood 1874, p. 139. Pl. 13, Fig. 1 = *Euastrum obesum* Joshua 1886, p. 638. Pl. 23, Figs. 19, 20.

Euastrum rostratum Ralfs 1844, p. 192 = *Euastrum bidentatum* Nägeli 1849, p. 122. Pl. 7D, Figs. 1a–1f.

Euastrum rostratum Ralfs f. West & West 1902a, p. 154. Pl. 20, Fig. 12 = *Euastrum umbonatum* (West & West) Schmidle var. *ceylanicum* Krieger 1937, p. 608. Pl. 84, Figs. 19, 20.

Euastrum rostratum Ralfs subsp. *umbonatum* West & West 1895, p. 51. Pl. 6, Fig. 16 = *Euastrum umbonatum* (West & West) Schmidle 1896, p. 310.

Euastrum rostratum Ralfs subsp. *umbonatum* West & West var. *ornatum* West & West 1895, p. 52. Pl. 6, Fig. 17 = *Euastrum umbonatum* (West & West) Schmidle 1896, p. 310.

Euastrum scaphaephorum Skuja 1934, p. 95. Fig. 95 = *Euastrum luetkemuelleri* Ducellier var. *carniolicum* (Lütkm.) Krieger 1937, p. 561. Pl. 80, Figs. 5–7.

Euastrum sexspinatum Irénée-Marie & Hilliard 1963, p. 107. Pl. 2, Fig. 14 = *Micrasterias sexspinata* (Irénée-Marie & Hilliard) Croasdale & Prescott comb. nov.

Euastrum sibiricum Boldt f. *exsecta* Grönblad 1921, p. 13. Pl. 3, Figs. 16, 19 = *Euastrum sibiricum* Boldt var. *exsectum* (Grönblad) Krieger 1937, p. 566. Pl. 77, Fig. 4.

Euastrum sibiricum Boldt f. *fennica* Grönblad 1921, p. 13. Pl. 13, Figs. 17, 18 = *Euastrum gayanum* DeToni f. *fennicum* (Grönblad) Krieger 1937, p. 566. Pl. 77, Figs. 12, 13.

Euastrum simplex Wolle 1884, p. 106. Pl. 27, Figs. 18–22 = *Euastrum integrum* (Wolle) Wolle 1892, p. 117. Pl. 31, Figs. 18–22.

Euastrum sinuosum Lenormand f. *germanica* (Racib.) Lütkemüller 1900, p. 75 = *Euastrum sinuosum* Lenormand var. *germanicum* (Racib.) Krieger 1937, p. 502. Pl. 63, Fig. 1.

Euastrum sinuosum Lenormand f. *scrobiculata* Nordstedt 1873, p. 9 = *Euastrum sinuosum* Lenormand var. *scrobiculatum* (Nordst.) Krieger 1937, p. 503. Pl. 62, Figs. 2, 3.

Euastrum sinuosum Lenormand var. *aboense* (Elfv.) Cedergren 1932, p. 35 = *Euastrum aboense* Elfving 1881, p. 7.

Euastrum sinuosum Lenormand var. *bidentatum* Irénée-Marie 1958, p. 140. Fig. 22 = *Euastrum denticulatum* (Kirch.) Gay var. *bidentatum* (Irénée-Marie) Prescott comb. nov.

Euastrum sinuosum Lenormand var. *jenneri* (Arch.) Raciborski 1889, p. 103. Pl. 2, Fig. 10 = *Euastrum sinuosum* Lenormand var. *germanicum* (Racib.) Krieger 1937, p. 502. Pl. 63, Fig. 1.

Euastrum snowdoniense Turner 1893, p. 343. Fig. 2 = *Euastrum dubium* Nägeli var. *snowdoniense* (Turner) West & West 1905a, p. 45. Pl. 38, Fig. 11.

Euastrum spicatum Turner var. *ceylanicum* (West & West) Krieger 1932, p. 215. Pl. 21, Fig. 5 = *Euastrum ceylanicum* (West & West) Krieger 1937, p. 627. Pl. 90, Figs. 16, 17.

Euastrum spinosum Ralfs 1844, p. 193. Pl. 7, Fig. 6 = *Euastrum elegans* (Bréb.) Ralfs 1848, p. 89. Pl. XIV, Fig. 7.

Euastrum spinosum Ralfs, in Wolle 1884, p. 106. Pl. 27, Figs. 4, 5 = *Euastrum evolutum* (Nordst.) West & West var. *integrius* West & West 1896a, p. 244. Pl. 14, Figs. 23–25.

Euastrum spinulosum Delponte subsp. *inermius* Nordst. var. *laticeps* Borge 1903, p. 113. Pl. 5, Fig. 4 = *Euastrum gemmatum* (Bréb.) var. *monocyclum* Nordstedt 1880, p. 8. Pl. 1, Fig. 13.

Euastrum splendens Fritsch & Rich 1937, p. 175. Figs. 9a–9c = *Euastrum attenuatum* Wolle var. *splendens* (Fritsch & Rich) Grönblad & Scott, in Grönblad, Prowse & Scott 1958, p. 12. Figs. 33, 34.

Euastrum subalpinum Messikommer var. *crassum* Messikommer 1942, p. 140. Pl. 3, Fig. 9 = *Euastrum subalpinum* Messikommer f. *crassum* (Messikommer) Prescott stat. nov.

Euastrum subglaziovii Borge 1899, p. 25. Pl. 1, Fig. 29 = *Euastrum abruptum* Nordstedt 1870, p. 217.

Euastrum subglaziovii Borge var. *minor* Borge 1903, p. 111. Pl. 4, Fig. 26 = *Euastrum abruptum* Nordstedt var. *subglaziovii* Krieger 1937, p. 606. Pl. 83, Figs. 7, 8.

Euastrum subhexalobum West & West var. *poriferum* Prescott & Scott 1945, p. 245. Pl. 5, Fig. 10 = *Euastrum subhexalobum* West & West var. *scrobiculatum* Grönblad 1945, p. 14.

Euastrum sublobatum Brébisson var. *crispulum* Nordstedt 1873, p. 10. Pl. 1, Fig. 9 = *Euastrum crispulum* (Nordst.) West & West 1905a, p. 72. Pl. 40, Figs. 15–18.

Euastrum sublobatum Brébisson var. *obtusatum* (Gutw.) Krieger f. Prescott & Scott 1958, p. 38. Fig. 6 = *Euastrum sublobatum* Bréb. ex Ralfs var. *notatum* Grönblad 1945, p. 14. Pl. 4, Fig. 71.

Euastrum tuddalense Ström var. *novae-terrae* Taylor 1935, p. 209. Pl. 41, Fig. 5 = *Euastrum denticulatum* (Kirch.) Gay var. *angusticeps* Grönblad 1921, p. 13. Pl. 3, Figs. 10, 11.

Euastrum turneri West f. Taylor 1935, p. 209. Pl. 39, Fig. 8 = *Euastrum turneri* West var. *bohemicum* (Lütkem.) Krieger 1937, p. 590. Pl. 82, Fig. 22.

Euastrum turneri West f. *bohemica* Lütkmüller 1910, p. 483. Pl. 2, Fig. 2 = *Euastrum turneri* West var. *bohemicum* (Lütkm.) Krieger 1937, p. 590. Pl. 82, Fig. 22.

Euastrum turneri West f. *poriferum* Scott & Prescott 1952, p. 392. Pl. 3, Fig. 6 = *Euastrum turneri* West var. *poriferum* (Scott & Presc.) Prescott stat. nov.

Euastrum turneri West var. *poriferum* Prescott & Scott in Jackson 1971, p. 174. Pl. 37, Figs. 7, 8 = *Euastrum evolutum* (Nordst.) West & West var. *monticulosum* (Taylor) Krieger f. *poriferum* Scott & Prescott 1952, p. 381. Pl. 3, Fig. 2.

Euastrum validum f. West & West 1907, p. 199. Pl. 1, Fig. 4 = *Euastrum validum* West & West var. *glabrum* Krieger 1937, p. 547. Pl. 72, Figs. 21, 22.

Euastrum verrucosum Ehrenberg f. Wailes 1933, p. 1, Figs. 7, 8 = *Euastrum verrucosum* Ehrenb. ex Ralfs var. *wailesii* Croasdale, in Croasdale & Grönblad 1964, p. 172. Pl. 8, Figs. 6–8.

Euastrum verrucosum Ehrenberg var. *alatum* Wolle f. Irénée-Marie 1952, p. 176. Pl. 18, Fig. 2 = *Euastrum verrucosum* Ehrenb. ex Ralfs var. *crassum* (Irénée-Marie) Prescott comb. nov.

Euastrum verrucosum Ehrenberg var. *alatum* Wolle f. Irénée-Marie 1952, 176. Pl. 19, Fig. 2 = *Euastrum verrucosum* Ehrenb. ex Ralfs var. *crassum* Irénée-Marie f. *angustum* (Irénée-Marie) Prescott comb. nov.

Euastrum verrucosum Ehrenberg var. *alatum* Wolle f. *minus* (Lobik) Kossinskaja 1936a, p. 110. Pl. 1, Fig. 5 = *Euastrum verrucosum* Ehrenb. ex Ralfs var. *alatum* Wolle 1884a, p. 101. Pl. 26, Fig. 4.

Euastrum verrucosum Ehrenberg var. *alatum* Wolle f. *alpina* Huber-Pestalozzi 1931, p. 429. Fig. 14 = *Euastrum verrucosum* Ehrenberg var. *alpinum* (Huber-Pestalozzi) Krieger 1937, p. 646. Pl. 94, Fig. 5.

Euastrum verrucosum Ehrenberg var. *reductum* Nordstedt 1880, p. 9. Pl. 1, Fig. 14 = *Euastrum verrucosum* Ehrenb. ex Ralfs var. *coarctatum* Delponte 1873, p. 95. Pl. 6, Fig. 16.

Euastrum verrucosum Ehrenberg var. *rhomboideum* Lundell f. *levanderi* Roll 1928, p. 133. Pl. 3, Fig. 6 = *Euastrum verrucosum* Ehrenb. ex Ralfs var. *rhomboideum* Lundell 1871, p. 16.

Euastrum verrucosum Ehrenberg var. *rhomboideum* Lundell f. *pterygoideum* (Huber-Pestalozzi) Krieger 1937, p. 650. Pl. 96, Fig. 1 = *Euastrum verrucosum* Ehrenberg ex Ralfs var. *pterygoideum* (Huber-Pestalozzi) Krieger 1937, p. 650. Pl. 96, Fig. 1.

Euastrum verrucosum Ehrenberg var. *suboblongum* Ackley 1929, p. 303. Figs. 9–11 = *Euastrum verrucosum* Ehrenb. ex Ralfs var. *valesiacum* Viret 1909, p. 259. Pl. 3, Fig. 7.

Euastrum wollei Lagerheim var. *cuspidatum* DeToni 1889, p. 1104 = *Euastrum wollei* Lagerheim 1885, p. 233.

Euastrum wollei Lagerheim var. *quadrigibberum* Lagerheim 1885, p. 233. Pl. 27, Fig. 6 = *Euastrum wollei* Lagerheim 1885, p. 233.

Euastrum sp. nov. Vinyard 1951, p. 32. Pl. IV, Fig. 1 = *Euastrum vinyardii* Prescott sp. nov.

Heterocarpella binalis Turpin 1820, Fig. 14 = *Euastrum binale* (Turp.) Ralfs 1844, p. 193. Pl. 7, Fig. 7 = *Euastrum binale* (Turp.) Ehrenberg 1841, p. 208.

Heterocarpella didelta Turpin 1828, p. 315 - *Euastrum didelta* (Turp.) Ralfs 1848, p. 84, Pl. XIV, Fig. 1.

Tetmemorus giganteus Wood 1869, p. 19 = *Euastrum giganteum* (Wood) Nordstedt 1889, in: DeToni 1889, p. 1106.

MICRASTERIAS Agardh ex Ralfs 1848, Brit. Desm., p. 68.

Micrasterias is one of the largest and most beautiful of all the desmids. The genus is characterized in general by its relatively flat cells and radiating, often deeply dissected lobes. While its affinities with other genera are not clear, some species such as *M. suboblonga* Nordst. and *M. americana* (Ehrenb.) Ralfs show a certain likeness to some species of *Euastrum*, and highly ornamented species such as *M. anomala* Turn. seem somewhat similar to *Xanthidium*.

The basic structural plan in the genus is a single cell which is a flattened disc, showing bilateral and bipolar symmetry. A deep, usually narrow median sinus divides the cell into 2 semicells, which are again divided into 3 lobes: a polar lobe and 2 lateral lobes. The polar lobe forms the apex of the cell and may be at its summit, flat, slightly rounded, depressed, or shallowly notched, but it is not again divided, although its upper lateral extremities may be variously drawn out into processes which are sometimes referred to as polar lobes or lobules. It is now believed that the form and structure of the polar lobe is fundamental in classification within the genus.

The lateral lobes are in most species divided into lobules, and vary in the angle and number of incisions. In Section I (Holocystis) and in part of Section III (Schizocystis) the incisions are horizontal, parallel to the sinus, but in Section II (Actinocystis), which contains the majority of the species, the incisions are radial. The first division of a lateral lobe results in an upper and a lower lateral lobule. If this is the only division, the lobe is described as being "once divided" or "divided to the first order" (Pls. XCIV, XCV). In many species these lobules are again divided, by a shallower incision, and the lobe is then described as being "twice divided" or "divided to the second order" (Pls. CVIII, CXI). Further divisions may bring it to the third order (Pls. CXXXIV, CXXXV) or the fourth order (Pl. CXXXIII, Figs. 1, 9) or even the fifth order (Pl. CXXX, Fig. 2). In some species the upper lateral lobule is larger and undergoes one more division than the lower lateral lobule (Pl. CXXXIV, Figs. 1–4). The extremities of the ultimate divisions are usually excavate or bidentate, although other conditions exist.

A single-lobed chloroplast, controlled by the nucleus, conforms to the shape of the semicell and shows several to many pyrenoids. The cell walls are always thin, of two layers, and always punctate or scrobiculate in the secondary wall; in some instances sharp granules or spines are present, usually within the margins of the major incisions, and in a few species the surface is ornamented with circles of granules or with verrucae, or even short processes. The semicells in vertical view are for the most part narrowly fusiform, sometimes showing a small central protuberance; in lateral view they are usually narrowly tapering from a slightly swollen base. Zygospores are known from only about one-sixth of the species and are generally formed singly between 2 cells; they are usually spherical with long simple or forked spines. The initial product of germination is very simple in form, but at the first vegetative division normal cells are formed.

The single, large, symmetrical cell of *Micrasterias*, which grows well in culture, has been a favorable subject for various kinds of research, particularly electron microscopy, scanning electron microscopy, cytophysiology, and irradiation. From this research have arisen important concepts, specific for *Micrasterias*, but some possibly applicable for broader interpretation.

From the work of Kiermayer (1970), Kiermayer & Dobberstein (1973), Lacalli & Acton (1974), Pickett-Heaps (1975), Tippit & Pickett-Heaps (1974), Waddington (1963), and others we have a well-established theory that the bilateral symmetry characteristic of the normal cells of *Micrasterias* is determined by a cytoplasmic framework comprising some basic units which in vegetative cell division are transmitted from the parental to the new semicell, and are thus responsible for the continuity of symmetry in the offspring. The composition and intracellular position of these basic units are still matters of conflicting opinion.

The nucleus, while not controlling the symmetry directly, has an influence on the morphology of the developing semicells: the extent of the lobes and how far subdivision proceeds. It can affect the tendency to vary in wing number (degree of radiation), and controls the chloroplast. The nucleus is striking for having, in most species investigated, large number of very small chromosomes, variable in number sometimes even within the same clone, this variability being attributed to ease of fragmentation. Some important papers in this field are: Brandham (1965), Kallio (1951, 1968), Kiermayer (1970), Selman (1966), and Waris & Kallio (1964).

The formation and structure of the primary and secondary wall, as well as the septum between the 2 semicells, have been the object of much study by electron microscopy and cytophysiological methods. It has been shown that the septum forms in the cortical cytoplasm inside the isthmus, presumably by the action of microtubules, and independent of the position of the mitotic spindle. It seems that in the growing semicell the maximum wall extension is concentrated in the tips and, as in fungal hyphae, depends on the 2 competing processes of wall extension by incorporation of new material and wall hardening. A new type of Golgi vesicles (F-vesicles) was found, probably formed from the distal poles of dictyosomes, whose membranes possibly function as templates for the highly patterned secondary wall. Outstanding contributions in this field are by Kiermayer (1968, 1970), Kiermayer & Dobberstein (1973), and Lacalli & Acton (1974).

Summarizing our present-day knowledge, and adding a new outlook gained from the scanning electron microscope are two important works: Tippit & Pickett-Heaps (1974) and Pickett-Heaps (1974, 1975).

Experiments on polyploidy have shown that the balance between nuclear mass and cell size is preserved to a large degree in ploidization. The diploid to polyploid cell may respond in 2 ways: its lobes may increase in size and degree of division, or, often after a delay of a few generations, they may increase in number (degree of radiation). See Kallio (1951, 1968), and others.

Since size of cell and degree of incision of the lateral lobes have been considered important for a long time in the determination of species and varieties, these findings cast serious doubt on the validity of certain taxa, some of which may be merely haploid or diploid phases of another taxon. It is recommended that more emphasis be placed on the morphology of the polar lobe, which appears to be the more stable element in the cell.

Key to the North American Species of *Micrasterias*

1. Polar lobe without horizontally furcate or accessory processes.
  2. Polar lobe very broad, without a well-marked median incision; the incisions which separate the lateral lobules horizontal (Section I, Holocystis).
    3. Lateral lobes undivided.
      4. Angles of polar lobe extended laterally as far as the lateral lobes.
        5. Subpolar incision narrow, much less than the thickness at base of the lateral lobes.
          6. Lateral lobes tapering toward their extremities.   19. *M. laticeps.*
          6. Lateral lobes subrectangular, not tapering toward their extremities.
                33. *M. suboblonga.*
        5. Subpolar incision a sinus, equal to or wider than the thickness at base of the lateral lobes.
          7. Sinus no wider than width of lateral lobe at base.   21. *M. muricata* var. *simplex.*
          7. Sinus much wider than width of lateral lobe at base.   6c. *M. arcuata* var. *gracilis.*
      4. Angles of polar lobe not extended as far as the lateral lobes.
        8. Subpolar incision (sinus) much wider than the breadth at base of the lateral lobes.   6. *M. arcuata.*
        8. Subpolar incision (sinus) not wider than the breadth at base of the lateral lobes.
          9. Lateral lobes subquadrate with paired teeth at upper and lower extremities, lateral margins straight.   27. *M. piquata.*
          9. Lateral lobes tapering outwardly, ending in 1 or 2 teeth, or furcate.
            10. Polar lobe broad at base and little extended horizontally, broadly curved or flattened.
              11. Cells longer than broad; polar lobe rounded, teeth at angles not upwardly directed.   24. *M. oscitans.*
              11. Cells broader than long; polar lobe with median notch, teeth at angles upwardly directed.   32. *M. sexspinata.*
            10. Polar lobe narrow and horizontally extended.
              12. All margins smooth; apex flat or concave.   26. *M. pinnatifida.*

  12. All margins rough; apex retuse.   38. *M. tropica.*
 3. Lateral lobes divided into 2 lobules, which may be again divided.
  13. Incisions between lobules as broad as lobules. 21c. *M. muricata* var. *simplex.*
  13. Incisions between lobules much narrower than lobules.
   14. Cells circular or broader than long, if longer than broad subpolar incision
    narrow.
    15. Polar lobe at base nearly as broad as lateral lobe, subpolar incision widely
     open; upper margin of lateral lobe horizontal, cells less than 75 μm long.
                     11. *M. decemdentata.*
    15. Polar lobe at base narrower than lateral lobe, subpolar incision closed or
     narrow, or if widely open the upper margin of lateral lobes not horizontal.
     16. Lateral extremities of polar lobe bidentate and extended beyond the
      flattened apex; cells less than 70 μm. long.   1. *M. abrupta.*
     16. Lateral extremities of polar lobe not exserted beyond the apex; cells
      75–144 μm long.   39. *M. truncata.*
   14. Cells longer than broad; subpolar incision large with broadly rounded interior.
                     13. *M. depauperata.*
 2. Polar lobe with median incision of variable depth; incisions between lateral lobules ±
  radially disposed (Section II, Actinocystis).
  17. Lateral lobes no more than twice divided (to the second order), rarely undivided.
   18. Angles of polar lobe extended into long, narrow processes.
    19. Dorsal margin of upper lateral lobule erect and paralleling polar lobe for
     about half its length.   2. *M. alata.*
    19. Dorsal margin of upper lateral lobules extending radially, forming a broad
     angle with polar lobe.
     20. Ultimate lateral lobules very slender.   29. *M. radiata.*
     20. Ultimate lateral lobules stout.
      21. Incisions shallow, ultimate lobules as broad as long. 10. *M. crux-melitensis* p.p.
      21. Incisions deep, ultimate lobules longer than broad. 28. *M. radians.*
   18. Angles of polar lobe not extended into long, narrow processes.
    22. Division of lateral lobes shallow, extending less than halfway to center of
     semicell.
    23. Sinus and/or infrapolar incisions closed; cells longer than broad. 17. *M. jenneri.*
    23. Sinus and/or infrapolar incisions open.
     24. Interlobular incisions very shallow; teeth terminating processes of polar
      lobes and lower lateral lobules long and curved. 3. *M. ambadiensis.*
     24. Interlobular incisions extending nearly halfway to center of semicell; teeth
      terminating processes of polar lobe and lower lateral lobules short.
      25. Polar lobe stout with lateral margins parallel or slightly convex, its apex
       bearing tumors and/or spines, standing apart from the lateral lobes,
       which are very shallowly divided.   8. *M. brachyptera.*
      25. Polar lobe diverging toward its extremities, if standing apart from lateral
       lobes it is slender and long-extended; apical margin of polar lobe
       without tumors or spines.   10. *M. crux-melitensis* p.p.
    22. Division of lateral lobes deep, extending more than halfway to center of
     semicell.
    26. Angles of polar lobe ending in 2 teeth.   28. *M. radians.*
    26. Angles of polar lobe ending in 1 tooth or mucro.
     27. Apex of polar lobe nearly flat, with shallow median depression and very
      short tooth or mucro at angle.   14. *M. floridensis.*
     27. Apex of polar lobe deeply depressed, angles ending in long tooth.
                    18. *M. johnsonii.*
  17. Lateral lobes, at least in upper portion, more than twice divided (to the third or
   forth order).
   28. Lateral lobes, at least in upper portion, 3 times divided.
    29. Incision between polar lobe and upper lateral lobules opening widely.
     30. Cells square in outline, united by apical processes to form a filament.
                   15. *M. foliacea.*
    30. ,Cells ± circular, not united into a filament.
     31. Polar lobe expanded to apex, much shorter than long. 37. *M. triangularis.*
     31. Polar lobe longer than broad.

32. Surface of semicell sparsely covered with spines; cells mostly more than 150 μm long.   5. *M. apiculata* p.p.

32. Surface of semicell not covered with spines.

    33. Polar lobes spreading broadly at apex; cells mostly less than 150 μm long.   23. *M. novae-terrae.*

    33. Polar lobe not spreading broadly at apex; cells mostly more than 150 μm long.

        34. Lateral lobules not tapering to extremities.   5. *M. apiculata* p.p.

        34. Lateral lobules tapering to extremities.   36. *M. torreyi* p.p.

29. Incision between polar lobe and upper lateral lobules narrow or closed.

    35. Ultimate lateral lobules tapering toward extremities.   36. *M. torreyi* p.p.

    35. Ultimate lateral lobules not tapering.

        36. Polar lobe subcuneate, apex convex, although retuse in middle, angles reflexed.   9. *M. conferta.*

        36. Polar lobe at apex concave, angles not reflexed.

            37. Angles of polar lobe ending in paired teeth; with subapical and/or intramarginal spines.   25. *M. papillifera.*

            37. Angles of polar lobe ending in mucro; subapical and intramarginal spines lacking.   31c. *M. rotata* var. *japonica.*

28. Lateral lobes 4 times divided, at least in upper portion.

    38. Surface of semicell not ornamented with conical teeth or protuberances.

        39. Extremities of lobules bidentate or excavate, with sharp corners.

            40. Incisions not very deep, primary division of lateral lobe extending half to two-thirds of way to center of semicell; rarely with intramarginal spines.

                41. Apex of polar lobe convex with median depression; extremities ending in mucro.   34. *M. tetraptera.*

                41. Apex of polar lobe flat or concave; extremities bidentate.

                    42. Extremities long-toothed.   5. *M. apiculata* p.p.

                    42. Extremities short-toothed or excavate to truncate.

                        43. Polar lobe exserted.   31. *M. rotata.*

                        43. Polar lobe not exserted.

                          44. Polar lobe flat, without tooth at notch.   12. *M. denticulata* var. *intermedia.*

                          44. Polar lobe concave, with tooth at notch.   35a. *M. thomasiana* var. *notata.*

            40. Incisions very deep, extending about 3/4 way to center of semicell.

                45. Surface without protuberances and without spines except along incisions.   30. *M. radiosa.*

                45. Whole surface often covered with spines; semicell sometimes with supraisthmial protuberance.   7. *M. borgei.*

        39. Extremities of lobules rounded-truncate, or excavate with rounded angles; semicell sometimes with supraisthmial protuberances.

            46. Surface of semicell smooth.   12. *M. denticulata.*

            46. Surface of semicell with a few long, stout curving spines.   16. *M. horrida.*

    38. Surface of semicell ornamented with conical teeth or protuberances.

        47. Granular circular protuberances scattered over surface.   40. *M. verrucosa.*

        47. Conical teeth on surface at base of lateral lobules, and 3 large curving protuberances above isthmus.   35. *M. thomasiana.*

1. Polar lobe horizontally furcate or with accessory processes. (Section III, Schizocystis).

    48. Margins of lobules irregular, rough.

        49. Lateral lobes once divided.   4. *M. americana.*

        49. Lateral lobes twice divided.   20. *M. mahabuleshwarensis.*

    48. Margins of lobules smooth.

        50. Cells with the 2 lateral lobules equal in length, horizontally furcate, separated by wide-open sinus.   21. *M. muricata.*

        50. Cells with vertically formed lateral lobe and short lateral protuberance on long, erect portion of polar lobe.   22. *M. nordstedtiana.*

1. **Micrasterias abrupta** West & West 1896a, Trans. Linn. Soc. London, Bot., II, 5(5): 241. Pl. 14, Figs. 13–16.

Micrasterias truncata (Corda) Ralfs 1848, Brit. Desm., p. 75. Pl. VIII, Fig. 4; Pl. 5, Fig. 10 (*p.p.*) in Wolle 1892, Desm. U.S., p. 126. Pl. 44, Fig. 9.

Cells small, about as broad as long; lateral lobes twice divided, although incisions are shallow and second division may be slight and variable; polar lobe flat with very short, diverging bidentate lateral extensions; sinus soon open broadly due to slight divergence of lateral lobes; subpolar incision broad, rounded in interior; semicells in vertical view subangular-fusiform with acute poles, in lateral view angular-ovate. L. 40–67 μm. W. 42–65 μm. I. 7.5–13 μm. W. at apex 25–30 μm. Zygospore unknown.

Förster (1969, p. 38), with good reason, reduced Krieger's var. *borgei* to the type. The minimal difference between them is covered by Wests' description and figures of the type.

DISTRIBUTION: Florida, Louisiana, Michigan, Pennsylvania, Wisconsin. British Columbia, Ontario, Québec. Africa, South America.

PLATE XCIV, figs. 1–6.

2. **Micrasterias alata** Wallich 1860, Ann. Mag. Nat. Hist., III, 5: 279. Pl. 13, Fig. 11.

Cells medium-sized, about 1 1/5 times longer than broad; sinus closed in inner quarter, then widely open; lateral lobes once divided, the upper member again divided, the lower member simple, all divisions deep; lobules swollen at base, then abruptly and narrowly tapering to tridentate extremities, the upper margin of the upper member extending vertically, parallel to the polar lobe, separated from it by a deep and narrow incision; polar lobe slender with parallel sides, abruptly branching at apex into 2 slender diverging processes, extreme apex sometimes rather coarsely punctate. L. 152–254 μm. W. 125–228 μm. W. at apex to 125 μm. I. 14–21 μm. Th. c. 28 μm. Zygospore spherical with relatively few strong, sometimes bent spines. Diameter excluding spines 58–60 μm, including spines 98–100 μm.

Scott (Prescott & Scott 1952, p. 230, Pl. 6, Fig. 6) noted on specimens with partially dry surface a few bright refractive spots caused by refraction from small swellings on the wall. Somewhat the same feature was seen by Grönblad (1945, p. 14, Pl. 4, Fig. 75) in specimens from Brazil.

DISTRIBUTION: Florida. Tropical Asia, Japan, Australia, Africa, Brazil, Cuba.

PLATE CXII, figs. 3–8.

3. **Micrasterias ambadiensis** (Grönbl. & Scott) Thomasson 1960, Nova Acta Reg. Soc. Sci. Upsaliensis, IV, 17(2): 22. Pl. 4, Fig. 10; Pl. 10, Fig. 4.

Micrasterias radians Turn. var. *ambadiensis* Grönbl. & Scott in Grönblad, Prowse & Scott 1958, Acta Bot. Fennica 58: 21. Pl. 11, Fig. 119; Pl. 26. Fig. 363.

Cells 1 1/4 or more times longer than broad; lateral lobes stout, shallowly twice divided, the second division, however, reduced and irregular; polar lobe stout, strongly exserted, only slightly extended at the extremities, which end in long, paired, diverging teeth; all incisions usually open, especially the subpolar one; semicells in lateral view tumid, broadest near the base. L. 120–175 μm. W. 108–137 μm. I. 18–22 μm. Th. 50 μm. Zygospore unknown.

Cells in outline suggesting *M. brachyptera* Lund. but smaller, and without surface ornamentation. From *M. brachyptera* f. *glabriuscula* (Nordst.) Homfeld (1929, p. 34) it differs in its polar lobe not being swollen at base and its lateral lobes being less divided. When Scott & Grönblad (*l.c.*) named this as a new variety of *M. radians* they suggested that it was perhaps a new species, and we support Thomasson's (*l.c.*) decision to make it so. It is perhaps linked to *M. radians* by its form *M. ambadiensis* f. *latiloba* (Grönbl. & Scott) Thomasson (1960, p. 22).

DISTRIBUTION: Virginia. North and South Africa.

PLATE CXVI, figs. 1, 2.

4. Micrasterias americana (Ehrenb.) Ralfs 1848, Brit. Desm., p. xix, 74. Pl. 10, Fig. 1 a-d var. americana f. americana.

Micrasterias americana var. *wollei* Raciborski 1889, Pamiet. Wydz. Mat.-Przyr. Akad. Umiej. Krakow 17: 106, Pl. 7, Fig. 3.
Micrasterias americana var. *spinulifera* Cushman 1904a, Bull. Torrey Bot. Club 31: 395.

Cells of medium size, a little longer than broad, sinus opening moderately from a narrow interior; polar lobe large, outstanding, cuneate from the base or rarely cylindrical in the basal part, apex slightly retuse, each angle extended into a blunt diverging process, denticulate at the end, with a usually shorter and more erect accessory process arising near its origin, the accessory process of one angle on one side of the polar lobe, the other on the other side; lateral lobes cuneate, shallowly twice divided, the lobules of the first order so divided that the 2 proximate lobules are shorter than the two distal ones, their extremities bearing 2 to 4 teeth; surface of semicell with irregular granules on the lobes, with sometimes a circle of granules within the apical margin of the polar lobe; semicells in vertical view rhomboid-fusiform, the angles produced and truncate-dentate, the polar lobe showing the asymmetrical disposition of the accessory processes, in lateral view ovate-pyramidal, showing the central granules and asymmetrically disposed accessory processes. L. 100–160 μm. W. 83–145 μm. W. at apex 50–81 μm. I. 17–36 μm. Th. 41–50 μm.

Cushman's (*l.c.*) plant, described without an illustration, seems rather similar to the typical and perhaps is better included in it.

DISTRIBUTION: Widely distributed in U.S. and Canada. Great Britain, Europe, Asia, Africa, South America, Arctic.

PLATE CXL, figs. 5–9.

Key to the North American Varieties and Forms of *Micrasterias americana*

1. Polar lobe with evident accessory processes; apex usually concave.
  2. Polar lobe slender, its breadth at base less than 1/4 the breadth of the semicell.
    4. var. americana f. americana.
  2. Polar lobe stout, its breadth at base usually more than 1/3 the breadth of the semicell.
    3. With a semicircle of verrucae crowning the polar lobe below the processes.
      4e. var. americana f. spinosa.
    3. Without a semicircle of verrucae crowning the polar lobe.
      4. With a spur from the upper margin of the upper lateral lobule crossing the subpolar incision.   4b. var. *americana* f. *calcarata*.
      4. Without such a spur.
        5. Lateral extensions of polar lobe nearly erect; isthmus opening widely then closing at exterior.   4f. var. *westii*.
        5. Lateral extensions of polar lobe spreading; isthmus opening gradually toward exterior.   4c. var. *americana* f. *gaspensis*.
1. Accessory processes much reduced or lacking; apex usually flat.
  6. Accessory processes reduced to small rounded protuberances; cells more than 100 μm.   4a. var. *americana* f. *boldtii*.
  6. Accessory processes lacking; cells usually less than 110 μm long.
    4d. var. americana f. lewisiana.

4a. Micrasterias americana var. **americana** f. **boldtii** (Gutw.) Croasdale stat. nov.

Micrasterias americana var. *boldtii* Gutwinski 1890, Bot. Centralbl. 43: 73; (1891)1892 Spraw.
    Kom. Fizyjogr. Akad. Umiej. 27: 74. Pl. 3, Fig. 27.
Micrasterias americana var. *recta* Wolle 1876, Bull. Torrey Bot. Club 6: 122; 1892, Desm. U.S.,
    p. 124. Pl. 36, Fig. 3.

A form with the polar lobe reduced, with angles less extended and accessory
processes much reduced or absent; lateral lobes less regularly dentate, with shallower
incisions. L. 110–143 µm. W. 91–128 µm. W. at apex 53–83 µm. I. 19–28 µm. Th.
c. 48 µm.

This is a doubtful form, possibly teratological, the 2 semicells often differing
in degree of reduction, even in Boldt's form (1888, 5. Pl. 1, Fig. 1), on which
Gutwinski (*l.c.*) based his variety. Var. *recta* Wolle (*l.c.*), portrayed very differently by Wolle and by West & West (1905a, 119. Pl. 54, Fig. 4) and other
authors, and f. *lewisiana* West (1890, 286. Pl. 5, Fig. 13) probably should both be
included here, but the latter is kept separate provisionally because of its small size
and more compact appearance.

DISTRIBUTION: Alaska, Colorado, Connecticut, Louisiana, Michigan, New
Hampshire, New Jersey, North Carolina, Pennsylvania, South Carolina, Virginia.
British Columbia, New Brunswick, Québec. Greenland, Great Britain, Europe,
Arctic.

PLATE CXLI, figs. 1–4.

4b. Micrasterias americana var. **americana** f. **calcarata** Croasdale nom. nov.

Micrasterias americana f. *taylorii* Irénée-Marie 1949, Nat. Canadien 76(1/2): 18. Pl. 1, Fig. 2.
Micrasterias americana in Taylor 1935, Michigan. Acad. Sci. Arts & Lettr. 20: 211. Pl. 45, Fig.
    5.

A form differing from the typical by its shorter, broader polar lobe, by its
stouter lateral lobes with less deep incisions, and by the presence of a spurlike
process on the middle of the upper margin of the upper lateral lobule, which
extends into the subpolar incision and sometimes even overlaps the polar lobe;
papillae at base of subpolar incision generally lacking. L. 140–150 µm. W.
115–120 µm. W. at apex 65–67 µm. I. 23–24 µm.

The varietal epithet "*taylorii*" for this species was preempted by Krieger
(1939, p. 48).

DISTRIBUTION: Québec. Newfoundland.

PLATE CXLII, figs. 1, 5.

4c. Micrasterias americana var. **americana** f. **gaspensis** Irénée-Marie 1958a, Hydrobiologia 12: 124. Fig. 13.

A form distinguished by having the polar lobe more massive, the accessory
apical processes shorter, also by the absence of papillae at the base of the sinus,
and by the smooth wall. L. 120–131 µm. W. 99–129 µm. W. at apex 52–70 µm.
I. 27–28.5 µm.

DISTRIBUTION: Québec.

PLATE CXLII, fig. 6.

4d. Micrasterias americana var. **americana** f. **lewisiana** West 1890, Jour. Roy.
Microsc. Soc. 1890: 286. Pl. 5, Fig. 13.

A small, reduced form, the polar lobe with apex nearly flat and without
obvious accessory processes, its angles rounded; cell margins crenate, cell surface

granular in outer part; semicell in vertical view broad, with rounded poles. L. 80–113 µm. W. 70–103.5 µm. W. at apex 47.5–54 µm. I. 16–31 µm.

DISTRIBUTION: Alaska, Arkansas, Connecticut, New York, South Carolina, Virginia. Québec. Great Britain, Europe, Corsica.

PLATE CXLI, figs. 5, 6.

4e. Micrasterias americana var. americana f. spinosa (Turn.) Croasdale stat. nov.

Micrasterias americana var. *spinosa* Turner 1885, Jour. Roy. Microsc. Soc., II, 5(6): 936. Pl. 15, Fig. 13.

A small, compressed form; surface of semicell smooth in central area but ornamented with short, stout spines over the lobes; the polar lobe bearing in addition, near its extremity, a semicircle of truncate spines or verrucae. L. 136 µm. W. 112 µm.

Turner (*l.c.*), without mentioning it in his description, shows the angles of the polar lobe and the accessory processes as relatively long and stout. Krieger (1939, 45) includes this form under the typical.

DISTRIBUTION: Nova Scotia (Turner's record).

PLATE CXLII, fig. 2.

4f. Micrasterias americana var. westii (Roll) Krieger 1939, Rabenhorst's Krypto-gamen-Flora 13(1:2): 48. Pl. 109, Fig. 6 forma.

Micrasterias westii Roll 1925, Arch. Russ. Protistol. 4: 244, 251. Pl. 11, Fig. 5.

Cells small, otherwise quite like the typical; polar lobe low and broad, little narrowed at base, accessory processes as large as one at angles; subpolar incision widely open; sinus very deep, widely open in middle, closed at outside; isthmus very narrow; cell margins dentate but surface wholly smooth. L. 117 µm. W. 101 µm. W. at apex 42–53 µm. I. c. 7 µm.

Our form is much smaller than the dimensions given by Roll (but Krieger (*l.c.*) states that Roll's plant, by plate dimensions, might be much smaller (120 µm x 83 µm) than given in the text (180–190 µm x 135–140 µm). Except for the question of size and open isthmus, our plant agrees very well with Roll's, especially in the strikingly narrow isthmus (which Krieger does not portray well in his copy of Roll's figure).

DISTRIBUTION: Washington (the typical variety only from North Russia).

PLATE CXLII, figs. 3, 4.

5. Micrasterias apiculata (Ehrenb.) Ralfs 1848, Brit. Desm. p. 209 var. apiculata f. apiculata.

Cells large, mostly somewhat broader than long, sinus open; polar lobe exserted, separated from the upper lateral lobes by an incision opening widely from a sharp-angled interior, polar lobe with sides at first parallel, then diverging, its angles slightly produced, bearing 2 (3) curved spines, apex retuse-emarginate, with a curved spine close to each angle and another on either side of the median notch; lateral lobes 3 times divided, the ultimate lobes very short, bispinate; surface bearing spines in irregular radial rows, sparser toward the center of the semicell; in the middle above the isthmus a group of 3 to 7 stouter spines, often on a central protuberance. L. (124) 192–294 (303) µm. W. (120) 140–250 (272) µm. W. at apex 21–96 µm. I. 21–56 µm. Th. 46–90 µm. Zygospore unknown. Delponte (1873. Pl. 5, Fig. 5) figures one, spherical with long, sharp stout spines, but one cannot be certain that it belongs to this species.

DISTRIBUTION: Widely distributed in U.S. and Canada. Great Britain, Europe, Asia, Africa, South America, Arctic.

PLATE CXX, fig. 6; PLATE CXXI, figs. 1, 3.

Key to the North American Varieties and Forms of *Micrasterias apiculata*

1. Polar lobe shallowly retuse at apex, bearing apical spines as well as teeth at angles; spines scattered over surface of semicell.
  2. Lobules ending in sharp teeth; surface spines sharp.   5. var. *apiculata* f. *apiculata*.
  2. Lobules with rounded extremities; surface spines blunt or mammillate.
                              5a. var. *apiculata* f. *mamillata*.
1. Polar lobe rather deeply notched at apex; surface spines absent or only along principal incisions.
  3. Spines along incisions sparse or lacking.
    4. Cells less than 340 μm long.
      5. Angles of polar lobe ending in paired teeth; surface of semicell wholly without spines.   5b. var. *fimbriata* f. *fimbriata*.
      5. Angles of polar lobe bearing one stout incurved spine; surface with a few blunt spines.   5c. var. *fimbriata* f. *depauperata*.
    4. Cells more than 340 μm long.   5d. var. *fimbriata* f. *elephanta*.
  3. Intramarginal spines along principal incisions and across base of semicell.
                              5e. var. *fimbriata* f. *spinosa*.

5a. Micrasterias apiculata var. **apiculata** f. **mamillata** (Turn.) Croasdale comb. nov.

Micrasterias mamillata Turner 1885, Jour. Roy. Microsc. Soc., II, 5(6): 936. Pl. 16, Fig. 12.

A form differing in having the extremities of all its lobes blunt, the surface spines thick and mammillate, the median process broadly furcate. L. of semicell 114 μm. W. 198 μm. I. 23 μm.

The one semicell seen and illustrated by Turner (*l.c.*) is quite possibly a teratological form of *M. apiculata* var. *apiculata*, with which it occurred, but *M. mamillata* has been reported, without illustration, from 3 other places in North America and from many places abroad.

DISTRIBUTION: Pennsylvania, Wisconsin. New Brunswick, Nova Scotia. Great Britain, Europe, Asia, Africa, South America.

PLATE CXXI, fig. 2.

5b. Micrasterias apiculata var. (subsp.) **fimbriata** (Ralfs) Nordstedt 1888, Vidensk. Meddel. Naturh. Foren. Kjöbenhavn 1888: 187. Pl. 6, Figs. 1,2 f. **fimbriata**.

Micrasterias fimbriata Ralfs 1848, Brit. Desm., p. 71. Pl. 8, Fig. 2.
Micrasterias fimbriata var. *simplex* Wolle 1892, Desm. U.S., p. 121. Pl. 40, Fig. 8.

A variety which differs chiefly in its polar lobe, which is narrower and lacks the extramarginal spine on the apex and sometimes also the spines on either side of the apical notch; surface spines are also lacking except for an occasional few along the incisions in forms approaching f. *spinosa*; without median protuberance; lower lateral lobes narrower than the upper and less divided, the lowermost lobule sometimes exserted. L. 196-315 μm. W. 180-295 μm. W. at apex 40-68 μm. I. 25-40 μm. Zygospores are described without illustration by Wolle (1892, p. 121): "The zygospores are orbicular, spinulose; spines rather slender, elongate, scattered, mostly furcate at the ends, and sometimes notched below the middle; punctate ends, with tips recurved."

This variety, which is commoner than var. *apiculata*, has variously been considered a variety of *M. apiculata*, a subspecies of it, and a species in its own right. There is justification for each, but the discovery of forms intermediate between it and *M. apiculata* indicate that it might best be considered a variety of the latter, as Nordstedt (*l.c.*) placed it. Compare also *M. rotata* f. *evoluta* Turn.

DISTRIBUTION: Widely distributed in U.S. and Canada. Great Britain, Europe, Arctic, Panama.

PLATE CXXII, figs. 1, 2.

5c. Micrasterias apiculata var. **fimbriata** f. **depauperata** Irénée-Marie 1951, Nat. Canadien 78(7/8): 179. Pl. 1, Fig. 3.

A form distinguished by its polar lobe, which bears 1 stout incurved spine at each extremity, and has the spine on either side of the apical notch reduced to a mucro; surface spines very few. L. 278–287 μm. W. 253–258 μm. W. at apex 52–60 μm. I. 31–34 μm.

This form perhaps might better be included under the typical.

DISTRIBUTION: Québec.

PLATE CXXI, fig. 6.

5d. Micrasterias apiculata var. **fimbriata** f. **elephanta** (Wolle) Croasdale comb. nov.

Micrasterias fimbriata Ralfs f. *elephanta* Wolle 1884a, Desm. U.S., p. 110. Pl. 36, Fig. 3. (f. *elephantina* Wolle 1892, Desm. U.S., p. 121).
Micrasterias fimbriata var. *elephanta* (Wolle) Krieger 1939, Rabenhorst's Kryptogamen-Flora 13(1:2): 82. Pl. 125, Fig. 2.

A form differing in its larger size and complete lack of surface spines. L. 345–400 μm. W. 347–400 μm.

Irénée-Marie & Hilliard (1963, p. 109) report this form from Alaska with considerably smaller dimensions (L. 300–317 μm, W. 257–305 μm). Since no figure is given, this record is questionable.

DISTRIBUTION: ?Alaska, Louisiana, Massachusetts.

PLATE CXXII, figs. 3, 4.

5e. Micrasterias apiculata var. **fimbriata** f. **spinosa** (Biss.) West & West 1905a, Monogr. II. 100. Pl. 47 Fig. 5.

Micrasterias fimbriata Ralfs var. *spinosa* Biss. in Roy & Bissett 1893, Ann. Scottish Nat. Hist. 3: 173. Pl. 1(4), Fig. 3.
Micrasterias rotata (Grev.) Ralfs var. *papillifera* Raciborski 1895, Flora 81: 34. Pl. 4, Fig. 17.

A form differing by having small surface spines, distributed especially along the major incisions. L. 198–286 μm. W. 169–267 (300) μm. W. at apex 50–70 μm. I. 25–39 μm.

This form, in which the spines, though smaller, are often irregular in distribution; seems to connect var. *fimbriata* with var. *apiculata*.

DISTRIBUTION: Widely distributed in U.S. and Canada. Great Britain, Europe, South America.

PLATE CXXI, figs. 4, 5.

6. Micrasterias arcuata Bailey 1851, Smithson. Contrib. Knowledge 2 (Art. 8): 37. Pl. 1, Fig. 6 var. **arcuata**.

Micrasterias arcuata var. *expansa* (Bail.) Nordst. f. *intermedia* Nordstedt (p.p.) 1878, Öfv. Kongl. Vet.-Akad. Förhandl. 1877 (3): 23. Textfig. II: 3 on p. 23.

Cells of medium size, about as long as broad; the sinus acute within, opening widely; erect portion of polar lobe narrow with parallel sides, dorsal margin flattened, slightly retuse, often somewhat elevated on either side of the middle;

lateral extensions of polar lobe reaching nearly to the extent of the lateral lobes, ending in a short tooth; lateral lobes undivided, narrow, conical, and much extended, curving upward, ending in a short tooth; in vertical view the lateral lobes and the extensions of the polar lobe evenly and sharply pointed; in side view the semicell rectangular with rounded ends and projecting apex. L. 55–130 μm. W. 45–120 μm. I. 9–20 μm. Th. c. 25 μm. Zygospore of type unknown.

Förster (1969, p. 39) describes a zygospore from a small plant he designates as "morpha minor" (L. 47–63 μm. W. 48–73 μm). The zygospore is ellipsoid, smooth-walled, 28–29 μm. x 22.5–24 μm.

DISTRIBUTION: Alabama, Florida, Georgia, Massachusetts, Michigan, Mississippi, New Jersey, North Carolina, Pennsylvania, Wisconsin. Newfoundland, Québec. Europe, Asia, West Africa, South America, Cuba.

PLATE LXXXVI, figs. 1, 2, 4.

Key to the North American Varieties and Forms of *Micrasterias arcuata*

1. Cells as long as broad or longer.
   2. Lateral extensions of polar lobe slender, their length more than 3 times their width at base; dorsal margin of lateral lobe concave.
      3. Lateral extensions of polar lobe horizontal, more than 2/3 the length of the lateral lobes.   6. var. *arcuata.*
      3. Lateral extensions of polar lobe diverging, less than 2/3 the length of the lateral lobes.
         4. Tips of lateral lobes not upturned.   6a. var. *expansa* f. *expansa.*
         4. Tips of lateral lobes upturned.   6b. var. *expansa* f. *recurvata.*
   2. Lateral extensions of polar lobe stout, their length no more than 3 times their width at base; dorsal margin of lateral lobes horizontal or convex.
      5. Semicells without prominent median protuberance.   6e. var. *robusta* f. *robusta.*
      5. Lateral and vertical view of semicell showing prominent median protuberances.   6f. var. *robusta* f. *protuberans.*
1. Cells broader than long.
   6. Lateral lobes very narrow, terminating in a single tooth or mucro.   6c. var. *gracilis.*
   6. Lateral lobes broad at base, ending in paired teeth.   6d. var. *lata.*

6a. **Micrasterias arcuata** var. **expansa** (Bail.) Nordstedt 1878, Öfv. Kongl. Vet.-Akad. Förhandl. 1877(3): 22. Textfigs. II:5.6b on p. 23 f. **expansa.**

Micrasterias expansa Bailey 1851, Smithson. Contrib. Knowledge 2 (Art. 8): 37. Pl. 1, Fig. 7;
     Taylor 1935, Pap. Michigan Acad. Sci., Arts & Lettr. 20: 212. Pl. 44, Fig. 4; Pl. 47, Fig. 1.
Micrasterias arcuata var. *expansa* f. *intermedia* Nordstedt (p.p.) 1878, Öfv. Kongl. Vet.-Akad.
     Förhandl. 1877(3): 22. Textfig. II:4 on p. 23.

Apical portion of polar lobe narrower than in the type; lateral extensions shorter, directed obliquely upward, often with undulating margins, or somewhat swollen at the ends. L. 49–105 μm. W. 47–103 μm. W. at apex 31–42 μm. I. 8–13 μm. Th. c. 23 μm.

Wolle (1884a, p. 117), Taylor (l.c.) and Irénée-Marie (1952, p. 146) prefer to keep this as *M. expansa* Bail. Although generally characterized by the smaller apical portion and stiffly diverging lobes and polar lobe extensions, it is very variable and difficult to define exactly.

DISTRIBUTION: Florida, Louisiana, Massachusetts, Michigan, Minnesota, Mississippi, New Jersey, North Carolina, Pennsylvania, Wisconsin. Newfoundland, Nova Scotia, Québec. Asia, Africa, South America, Cuba.

PLATE LXXXVI, figs. 3, 5.

6b. **Micrasterias arcuata** var. **expansa** f. **recurvata** (Presc. & Scott) Prescott comb. nov.

Micrasterias arcuata var. *robusta* Borge f. *recurvata* Prescott & Scott 1952, Trans. Amer. Microsc. Soc. 71(3): 232. Pl. 5, Fig. 5.

A smaller form with the lateral extensions of the polar lobe diverging upward as in f. *expansa,* but somewhat longer, and with upturned ends; the lateral lobes with horizontal upper margins and upturned ends. L. 49–57 μm. W. 39–49 μm. I. 8–9 μm.

This form seems intermediate between var. *expansa* and var. *robusta.*
DISTRIBUTION: Mississippi, North Carolina.
PLATE LXXXVI, fig. 6.

6c. **Micrasterias arcuata** var. **gracilis** (Bail.) West & West 1896a. Trans. Linn. Soc. London, Bot., II, 5: 238. Pl. 13, Fig. 27.

Cells 1 1/6 to 1 1/3 times broader than long; lateral lobes and lateral extensions of polar lobe very slender; lateral lobes arching upward, polar lobe extensions slightly bent downward. L. (53)65–100 μm. W. 81–133 μm. W. at apex 77–125 μm. I. 9–12 μm.

Scott (unpublished) observed from Florida a form intermediate between var. *gracilis* and the typical. (Pl. LXXXVI, Fig. 7).
DISTRIBUTION: Alabama, Florida, Massachusetts, Mississippi, North Carolina, Pennsylvania, Wisconsin. Québec. Africa, South America.
PLATE LXXXVI, figs. 7, 8; PLATE LXXXVII, fig. 6.

6d. **Micrasterias arcuata** var. **lata** Prescott & Scott 1943, Pap. Michigan Acad. Sci., Arts & Lettr. 28: 70. Pl. 1, Fig. 14; Pl. 5, Fig. 2.

Cells very broad; apex broadly convex, slightly retuse in the middle portion, laterally extended above into long, gradually tapering, slightly recurved arms which are tipped with a stout blunt tooth; lateral lobes disposed horizontally, very stout at the base, tapering gradually to bifurcate tips, which normally slightly exceed in length the converging polar lobe extensions. L. 88–95.5 μm. W. 133–155 μm. I. 14.4 μm.

This variety suggests certain varieties of *M. pinnatifida* (Kütz.) Ralfs, but the cells are much broader, and the tapering, slightly recurved polar lobe extensions with a single terminal tooth are similar to those of *M. arcuata.* Pl. LXXXVII, Fig. 3 shows an abnormal form.
DISTRIBUTION: Florida.
PLATE LXXXVII, figs. 1, 3.

6e. **Micrasterias arcuata** var. **robusta** Borge 1899, Bih. Kongl. Svenska Vet.-Akad. Handl. 24: 3(12): 27. Pl. 2, Figs. 38, 39 f. **robusta.**

Micrasterias expansa Bail. var. *robusta* Borge (wrong cite) in Irénée-Marie 1949, Nat. Canadien 76(1/2): 22. Pl. 1, Fig. 5.

Cells always longer than broad, 1 1/10 to 1 1/4 times longer; lateral lobes broadly conical, mostly extending horizontally; lateral extensions of polar lobe likewise broadly conical, erect portion broader than typical, apex slightly retuse; isthmus broad. L. 45–65 μm. W. 35–59 μm. W. at apex 23–33 μm. I. 7–17 μm.

DISTRIBUTION: Florida, Louisiana, Minnesota, Michigan, Mississippi, North Carolina, Oregon, Wisconsin. Newfoundland, Québec. Europe, South America, Cuba.
PLATE LXXXVII, figs. 2, 4.

6f. Micrasterias arcuata var. **robusta** f. **protuberans** Prescott & Scott 1952, Trans. Amer. Microsc. Soc. 71(3): 232. Pl. 5, Fig. 6.

A form differing from the typical by having a prominent facial protuberance in the median part of the semicell; semicell in lateral view turgid, elongate-ellipsoid, showing 2 swellings on either side above the median incision; in vertical view rhombic, with rounded protuberances and apiculate ends. L. 49–57 μm. W. 39–43 μm. I. 11–12 μm. Th. 17–19 μm.

DISTRIBUTION: Florida.

PLATE LXXXVII, figs. 5, 7.

7. Micrasterias borgei Krieger 1939, Rabenhorst's Kryptogamen-Flora 13(1:2): 86. Pl. 128, Figs. 1–4.

Cells large, 1 1/9 to 1 1/5 times longer than broad, with deep incisions; sinus opening from a closed interior, then tending to close again; polar lobe diverging moderately from a sometimes slightly swollen base with an unarmed tumor on either side of the median incision, the extremities diverging, exserted, bidentate, or with short curving spines; lateral lobes unequal, the upper twice as broad as the lower and once more divided, the ultimate lobules all distinct, bidentate; semicell with a tumor above the isthmus bearing a number of short, stout spines; surface showing small spines along the incisions or in irregular radiate rows over the whole surface. L. (198)221–325 μm. W. 169–265 μm. W. at apex 57–74 μm. I. (26)29–37 μm. Zygospore unknown.

The species differs from *M. apiculata* chiefly in its deeper incision. Our single record from Virginia, since it is unaccompanied by a figure, is somewhat doubtful.

DISTRIBUTION: Virginia. Asia, South America.

PLATE CXXII, figs. 5, 6.

8. Micrasterias brachyptera Lundell 1871, Nova Acta Reg. Soc. Sci. Upsal., III, 8(2): 12. Pl. 1, Fig. 4.

Micrasterias apiculata (Ehrenb.) Menegh. var. *brachyptera* (Lund.) West & West 1905a, Monogr. II: 101. Pl. 46, Fig. 5; Pl. 47, Fig. 6.
Micrasterias apiculata var. *brachyptera* f. *rectangularis* Irénée-Marie & Hilliard 1963, Hydrobiologia 21(1/2): 109. Pl. 2, Fig. 17.
Micrasterias apiculata var. *simplex* Irénée-Marie 1952, Hydrobiologia 4(1/2): 142. Pl. 13, Fig. 2.
Micrasterias brachyptera var. *americana* Wolle 1883, Bull. Torrey Bot. Club 10: 18. Pl. 27, Fig. 19.

Cells medium-sized to large, 1 1/3 to 1 1/2 times as long as broad; sinus open, usually widely, from a sharp-angled interior; lateral lobes short and stout, usually twice divided, the ultimate lobules ending in 2 (rarely 3) stout, somewhat curved teeth; polar lobe exserted, broad (about 1/3 the width of the cell), typically swollen just above the base and tapering slightly toward the apex, but sometimes nearly rectangular or slightly dilated, with a slight apical notch bounded on each side by a low tumor which usually bears a short spine; lateral extensions of polar lobe short, diverging, ending in paired curved teeth; incisions between polar lobe and upper lateral lobes opening widely from a sharp-angled interior; membrane with superficial spines typically in single sparse rows along the isthmus and outer margin of polar lobe, but spines extremely variable, sometimes also on margins and scattered over the surface, sometimes nearly or completely absent; semicells in lateral view broadly rounded at base, in vertical view broadly fusiform, ending in a spine. L. (138)182–234 μm. W. 131–182 μm. W. at apex (42)54–80 μm. I. 24–43 μm. Th. to 72 μm. Zygospore unknown.

Characteristic of the species is the broad, exserted polar lobe, rather widely separated from the lateral lobes, but there is considerable variation in its shape. Variable also is the degree of dissection of the relatively short lateral lobes, and the presence and distribution of surface spines.

DISTRIBUTION: Alaska, Michigan, Minnesota, Montana, Utah, Wisconsin. New Brunswick, Newfoundland, Ontario, Québec. Great Britain, Europe, Russia, Arctic.

PLATE CXX, figs. 1-5.

9. Micrasterias conferta Lundell 1871, Nova Acta Reg. Soc. Sci. Upsal., III, 8: 14. Pl. 1, Fig. 5 var. conferta.

Micrasterias granulata Wood 1874, Smithson. Contrib. to Knowledge No. 241. 19: 146. Pl. 21, Fig. 16.

A small species, very little longer than broad, deeply constricted, the sinus sometimes closed by overlapping of basal lobes, other incisions narrow and usually closed, not deep; polar lobe subcuneate, apex convex but retuse in the middle, with 2 or 3 papillae or teeth on the outer part of the dorsal margin, and 1 on either side of the median hollow; lateral lobes equal, 2 to 3 times divided, the third division, if present, very shallow, ultimate lobules bidentate or retuse-emarginate; surface spines sometimes present; semicell in vertical view elliptic, slightly swollen in the middle and with a papilla at each end, in lateral view rectangular-oblong, slightly retuse in the middle, 3-papillate at the ends. L. 80-122 μm. W. 70-107 μm. W. at apex 36-44 μm. I. 11-19 μm. Th. c. 23 μm. Zygospore unknown.

DISTRIBUTION: Widely distributed in U.S. and Canada. Great Britain, Europe, Asia, Arctic.

PLATE CXXIII, figs. 1-4.

Key to the North American Varieties and Forms of *Micrasterias conferta*

1. Semicells without large supraisthmial protrusions.
  2. Lateral margins of polar lobe not excavate below angles.   9. var. *conferta*.
  2. Lateral margins of polar lobe excavate below angles, which are thus downwardly uncinate.
    3. Dorsal margin and angles of polar lobe bearing teeth or spines.
                                   9a. var. *hamata* f. *hamata*.
    3. Dorsal margin and angles of polar lobe without teeth or spines.
                                   9b. var. *hamata* f. *glabra*.
1. Semicells with large supraisthmial protrusions.   9c. var. *phymatophora*.

9a. Micrasterias conferta var. hamata Wolle 1883, Bull. Torrey Bot. Club 10: 19. Pl. 27, Fig. 1 f. hamata.

Micrasterias conferta var. *hamata* f. *spinosa* Prescott & Scott 1943, Pap. Michigan Acad. Sci., Arts & Lettr. 28: 71. Textfig. 2.

A variety characterized chiefly by the shape of its polar lobe, whose lateral margins are excavate below the angles, which are thus downwardly uncinate; cells usually larger, with deeper incisions, the lateral lobes more divided, their extremities more strongly toothed; surface spines sometimes present. L. 74-128 μm. W. 58-107 μm. W. at apex 34-54 μm. I. 11-20 μm.

This variety is more common than the typical in North America and is quite variable in its shape and surface markings. Dichotypical forms with the typical are known (West & West 1905a, p. 90. Pl. 4, Fig. 9).

DISTRIBUTION: Alaska, Florida, Louisiana, Maine, Massachusetts, Michigan, Minnesota, Mississippi, Wisconsin. British Columbia, Labrador, Ontario, Québec. Great Britain, Europe, South America.

PLATE CXXIV, figs. 1, 2; PLATE CXXV, figs. 1, 4.

9b. Micrasterias conferta var. hamata f. glabra (Presc. & Scott) Prescott comb. nov.

Micrasterias conferta var. *glabra* Prescott & Scott 1943, Pap. Michigan Acad. Sci., Arts & Lettr. 28: 70. Textfig. 1.

A form differing from the typical in its lack of spines on and within the dorsal margin of the polar lobe, and in having a mammillate protuberance at the base of each angle of the polar lobe; angles of polar lobe sometimes mammillate; extremities of lateral lobules without teeth or spines. L. 77–102 μm. W. 64–90 μm. I. 12–15 μm. Th. c. 19 μm.

A form from Louisiana, seen only once, seems to combine features of the typical variety and of var. *hamata* f. *glabra* (Pl. CXXIV, Fig. 3).

One form (semicell only) was seen in material from Florida with a large circular protuberance above the isthmus, and with a smaller one on either side. Although the semicell is smaller and its lobes less divided this seems to lead over into var. *phymatophora* (Pl. CXXIV, Fig. 5).

DISTRIBUTION: Florida, Louisiana, Mississippi.

PLATE CXXIV, figs. 3–7.

9c. Micrasterias conferta var. phymatophora (Presc. & Scott) Prescott comb. nov.

Micrasterias phymatophora Prescott & Scott 1952, Trans. Amer. Microsc. Soc. 71(3): 242. Pl. 4, Fig. 3.

Micrasterias phymatophora f. *circularis* Prescott & Scott 1952, *l.c.* 244. Pl. 4, Fig. 4.

A variety differing in the presence of a very large supraisthmial protrusion with an inwardly directed protrusion on either side; cells circular or somewhat elongate (1.02–1.3 times longer than broad); polar lobe bearing a small mammillate protuberance within each angle; lateral lobes 2 or 3 times divided, the ultimate divisions irregularly crenate; semicells in lateral view relatively thick, the base transversely oval, the apex vertically oblong-spatulate, in vertical view narrowly elliptic with a prominent protrusion on each side in the midregion, the wall undulate between the midregion and the bluntly pointed poles. L. 122–142 μm. W. 97–143 μm. I. 20–26 μm. Th. 42–45 μm.

DISTRIBUTION: Florida.

PLATE CXXIII, figs. 5, 6.

10. Micrasterias crux-melitensis (Ehrenb.) Ralfs 1848, Brit. Desm., p. 73. Pl. IX, Fig. 3 f. crux-melitensis.

Cells of medium-size, mostly a little longer than broad, very deeply constricted; sinus open with acute or linear interior; lateral lobes twice evenly divided, the incisions open, ends of lobules emarginate-bidentate; polar lobe with lower margins parallel, upper margins sharply diverging, apex widely retuse, angles produced into short diverging processes with emarginate-bidentate extremities; chloroplasts with relatively few pyrenoids. L. 85–164 μm. W. 78–153 μm. W. at apex 35–60 μm. I. 11–28 μm. Th. 20–35 μm. Zygospores very rare, so far only 1 immature spore known, found in NW Germany by Homfeld (1929, Pl. 4, Fig. 28), spherical with strong conical spines, its diameter not including spines 63 μm, including spines 92 μm.

This species is characterized by its short, stout, twice-divided lateral lobules, emarginate or bidentate at the extremities, by the polar lobe with lower margins parallel, and by its relatively wide-angled incisions. Twenty-five varieties and named forms have been recognized at one time or another. Krieger (1939, pp. 64–67) recognizes 8 varieties, but even among these there are many intermediate and dichotypical forms. We believe that all 6 of the varieties recorded for North America, noted below, might better be considered only as forms or variants of the typical plant. See Förster 1968 and 1970.

DISTRIBUTION: Widely distributed in U.S. Québec. Great Britain, Europe, Asia, Tropical Africa, South America, Celebes, Arctic.

PLATE CXIII, figs. 1–3.

Key to the North American Forms of *Micrasterias crux-melitensis*

1. Lateral lobes twice divided (to the second order) or in part 3 times divided (to the third order).
   2. Lateral lobes never more than twice divided; terminal teeth of lateral lobes not bent.
      3. Sinus opening from the interior; polar lobe longer than the body of the semicell below it.
         4. Without teeth at apex.   10. f. *crux-melitensis.*
         4. With a tooth on either side of the apical concavity.   10a. f. *evoluta.*
      3. Sinus closed throughout most of its length; polar lobe shorter than body of semicell below it.   10b. f. *janeira.*
   2. At least upper portion of lateral lobes again divided; terminal teeth of polar lobe often bent.   10f. f. *superflua.*
1. Lateral lobes mostly only once divided; terminal teeth of lobules irregular; sinus widely open.
   5. Lateral lobules at least in part bidentate; teeth on polar lobes not bent.
      6. Lateral lobes evenly and shallowly only once divided; polar lobe shorter than body of semicell below it.   10c. f. *rabenhorstii.*
      6. Lateral lobe irregularly divided and toothed; polar lobe longer than body of semicell below it.   10d. f. *simplex.*
   5. Lateral lobules ending in 1 tooth; teeth on polar lobes sometimes sharply bent.   10e. f. *spinosa.*

10a. Micrasterias crux-melitensis f. evoluta Turner 1892, Kongl. Svenska Vet.-Akad. Handl. 25(5): 92.

Micrasterias crux-melitensis in Woodhead & Tweed 1960, Hydrobiologia 15: 348. Pl. 2, Fig. 12.

A form differing in having a distinct tooth, up to 3 μm long, on each side of the apical concavity of the polar lobe. L. 94–128 μm. W. 84–121 μm.

Krieger (1939, p. 68, Pl. 116, Fig. 3), who differentiated *M. crux-melitensis* from *M. radians* Turn. on depth of incisions, transferred Turner's form to *M. radians.* Turner, however, (*l.c.,* p. 91) himself established *M. radians* without mentioning depth of incisions, differentiating it on its stelloid shape, with exserted polar lobe. On the next page he placed f. *evoluta* under *M. crux-melitensis.* We think that Turner's combination should be upheld.

DISTRIBUTION: Newfoundland. India, S. & E. Africa.

PLATE CXV, figs. 3, 4.

10b. Micrasterias crux-melitensis f. janeira (Racib.) Croasdale comb. nov.

Micrasterias janeira Raciborski 1885, Pamiet. Wydz. III, Akad. Umiej. w Krakowie 10: 97. Pl. 14, Fig. 4.
Micrasterias crux-melitensis var. *janeira* (Racib.) Grönblad 1920, Acta Soc. Fauna Flora Fennica 47: 35. Pl. 6, Figs. 17, 18.

Polar lobe shorter and broader with the apex more nearly straight; lateral lobes shorter and broader, the incisions shorter, the lobules abruptly truncate L. 78-114 μm. W. 69-103 μm. W. at apex 30-45 μm. I. 14-21 μm. Th. c. 30 μm.

Dick (1926, p. 447. Pl. 19, Figs. 4, 5) comments upon and illustrates transitional forms between this form and the typical plant. Nevertheless, he decided to maintain the variety, and later authors followed his example. We believe, however, that it is not sufficiently clearly distinct from the species to warrant being considered a variety as did Grönblad (*l.c.*).

Pl. CXIX, Fig. 1 shows an aberrant form from Florida.

DISTRIBUTION: Florida. Europe, South America.

PLATE CXIII, figs. 4-7; PLATE CXIX, fig. 1.

10c. Micrasterias crux-melitensis f. rabenhorstii (Kirchn.) Croasdale comb. nov.

Micrasterias rabenhorstii Kirchner 1878, Kryptogamen-Flora Schleisen 2(1): Algen, p. 163.
Micrasterias crux-melitensis var. rabenhorstii (Kirchn.) Krieger 1939, Rabenhorst's Kryptogamen-Flora 13(1:2): 66. Pl. 115, Fig. 4.

Lateral lobes only once shallowly divided, the lobules tapered and bidentate at the extremities; polar lobe short, its apical extremity forked into short, diverging processes; all incisions opening widely. L. 84-88 μm. W. 72-85 μm. I. 14-18 μm.

Krieger (*l.c.*) refers to the report of a dichotypical plant by Ducellier (1918a, p. 138. Pl. 4, Fig. 20) and suggests that this variety may be only a teratological form. Grönblad (1963, p. 22) states the same. We do not believe that the epithet *rabenhorstii* should be upheld as a variety of *M. crux-melitensis*.

DISTRIBUTION: Iowa, Minnesota, North Carolina. Europe.

PLATE CXIV, figs. 1, 2.

10d. Micrasterias crux-melitensis f. simplex (Borge) Croasdale stat. nov.

Micrasterias crux-melitensis var. simplex Borge 1894, Bih. Kongl. Svenska Vet.-Akad. Handl. 19, III(5): 35. Pl. 3, Fig. 40.

A form with the upper lobules of the lower lateral lobe and the lower lobules of the upper lateral lobes reduced to a single tooth; sinus opening widely from the interior; subpolar incision broad, rounded at base; apex of polar lobe only slightly concave, sometimes with a small incision in the middle. L. 97-130 μm. W. 70-120 μm. W. at apex c. 36-52 μm. I. 13-21 μm.

Although better defined than some of the other varieties of the species, this form is connected with the typical and other varieties by a series of forms (*e.g.* Grönblad 1921, Pl. 1, Fig. 16 and Gutwinski 1892, Pl. 3, Figs. 28, 29).

DISTRIBUTION: Alaska, Minnesota. Europe, Northern Russia, Japan, E. New Guinea.

PLATE CXIV, figs. 3, 4.

10e. Micrasterias crux-melitensis f. spinosa (Roll) Croasdale stat. nov.

Micrasterias crux-melitensis var. spinosa Roll 1925, Arch. Russ. Protistol. 4: 248, 252, Pl. 14, Fig. 3.

A form differing in having the lateral lobules terminating in only a single tooth; wall sometimes coarsely punctate; teeth on polar lobe sometimes curved. L. 100, 150 μm. W. 95, 115 μm. I. 25, 17.5 μm.

Most authors agree that this is only an accidentally reduced form of doubtful taxonomic value.

DISTRIBUTION: Québec. Europe, Russia.
PLATE CXIV, figs. 5, 6.

10f. Micrasterias crux-melitensis f. superflua (Turn.) Croasdale stat. nov.

Micrasterias crux-melitensis var. *superflua* Turner 1885, Jour. Roy. Microsc. Soc., II, 5(6): 936. Pl. 15, Fig. 11.

A form differing in having the extremities of the upper lateral lobules tridentate and the extensions of the polar lobe bearing curved teeth. L. 116–135 μm. W. 102–135 μm. I. 17–20 μm.

Wolle, in a ms. note on his copy of Turner's reprint (*l.c.*), questions the validity of this variety. Turner adds to his description: "Several specimens seen, having one or both segments as figured," which indicates that he saw dichotypical forms. And such a form is figured as the typical in Ralfs (1848, Pl. 9, Fig. 3b). West & West (1905a, p. 116) include it in the type. Krieger (1939, p. 67. Pl. 115, Fig. 7) describes and figures a rather different-appearing plant, its lateral lobes divided to the third order. It is probable that most records are of plants conforming to Krieger's rather than Turner's description.

DISTRIBUTION: Massachusetts, Michigan, Washington. Great Britain, Germany, Japan, Russia.
PLATE CXV, figs. 1, 2.

11. Micrasterias decemdentata (Näg.) Archer 1861, in Pritchard's Infus. p. 726.

Cells small, somewhat broader than long; polar lobe broadly conical, flattened at apex, lateral extensions ending in a rather long single tooth; lateral lobes relatively short, not much taller than polar lobe, horizontally extended, once or twice divided, ending in usually long teeth; subpolar incision wide-angled, resulting in horizontal upper margin of lateral lobe; sinus closed for inner half or third of its length, then opening rather widely; semicells in vertical view swollen, spindle-shaped; in lateral view elliptic with swollen base. L. 40–80 μm. W. 40–85 μm. W. at apex 24–42 μm. I. 7–13 μm. Th. 14–16 μm. Zygospore spherical with strong blunt spines, diameter 46 μm, including spines 60 μm.

Typical specimens of this species are easily recognized by the relatively short, horizontally extended lateral lobes, widely separated from the large polar lobe. But forms are seen approaching *M. truncata* var. *pusilla* G. S. West; compare also *M. abrupta* West & West and its var. *borgei* Krieg. *M. decemdentata* var. *expansa* (Turn.) Krieg. is, we believe, better placed as a form of *M. pinnatifida*, as Turner had it (Turner 1892, p. 89. Pl. 5, Fig. 3c) because of the relatively greater breadth at base and upward direction of the lateral lobes; also there are many connecting forms between this variety and forms of *M. pinnatifida*. Wolle's large form (Wolle, 1892, p. 126. Pl. 37, Figs. 5, 6) because of its large size, relatively greater length, and short polar lobe, seems much closer to *M. truncata* var. *neodamensis* (A. Br.) Dick.

DISTRIBUTION: Florida, Louisiana, Montana, North Carolina. British Columbia. Great Britain, Europe, Asia, Africa, Australia, South America.
PLATE XCIV, figs. 7–11.

12. Micrasterias denticulata Bréb. ex Ralfs 1848, Brit. Desm., p. 70, Pl. VII, Fig. 1. **var. denticulata.**

?Micrasterias cornuta Bennett 1886, Jour. Roy. Microsc. Soc., II, 6: 7. Pl. 1, Fig. 6. (a questionable form).

Cells large, 1 1/10 to 1 1/4 times longer than broad; sinus linear, closed, lateral lobes divided to the third or fourth order, their extremities very slightly retuse or crenate, typically with rounded angles; polar lobe narrowly cuneate, at the apex flattened or more often slightly retuse, with rather deep open notch and rounded angles; all incisions closed and moderately deep (the primary incision of the lateral lobes extending slightly more than halfway to the center of the semicell); surface without ornamentation; semicells in vertical view fusiform, with a central protuberance on each side flanked by a lesser one, in lateral view narrowly pyramidal with markedly swollen base. L. 180–350 μm. W. 165–300 μm. W. at apex 50–76 μm. I. 20–42 μm. Th. 55–64 μm. Zygospore commonly found, spherical, bearing forked processes (spines) 33–40 μm long; diameter of zygospore 71–116 μm, including processes 140–195 μm.

The processes of the zygospores vary considerably. Ralfs (*l.c.*) states that the degree of forking of the process increases with age (our Pl. CXXXV, Figs. 2–5); Homfeld (1929, 34. Pl. 4, Fig. 41) states that the forking can vary with the locality (our Pl. CXXXV, Figs. 8–11).

DISTRIBUTION: Widely distributed in U.S., Canada, and all continents. Arctic.

PLATE CXXXIV, figs. 7–9; PLATE CXXXV, figs. 1, 2–5, 8–11.

Key to the North American Varieties and Forms of *Micrasterias denticulata*

1. Extremities of lateral lobes retuse or emarginate with rounded angles.
    2. Without supraisthmial protuberances visible in face view.
        3. Wall not strongly granular.
            4. Polar lobe long (more than half as long as semicell), its lateral margins concave; in vertical view with 3 median swellings.   12. var. *denticulata*.
            4. Polar lobe short (less than half as long as semicell), its lateral margins straight; in vertical view with single large median swelling.   12a. var. *angulosa*.
        3. Wall strongly granular.   12c. var. *granulosa*.
    2. With supraisthmial protuberances visible in face view.
        5. Without prominent swellings on polar lobe.   12e. var. *taylorii* f. *taylorii*.
                                                                12b. var. *basiprotuberans*.
        5. With prominent swelling on polar lobe on either side of notch.
                                                                12f. var. *taylorii* f. *mamillata*.
1. Extremities of lateral lobes with sharp angles.   12d. var. *intermedia*.

12+. Micrasterias denticulata var. denticulata forma.

Two specimens of a form were observed from Louisiana which, in their proportions, could be taken as a hybrid between *M. denticulata* and *M. jenneri* Ralfs. They were 1 1/3 times longer than broad; the lateral lobes were divided barely to the third order, the polar lobe was broad with somewhat flattened apex, rounded angles, and deep notch; the surface was observed to be punctate in the smaller plant and granulate in the larger; the size was variable, but about midway between the 2 species. L. 125, 155 μm. W. 100, 120 μm. I. 22, 25 μm.

The identity might be determined if the plants were found again and a vertical view observed.

DISTRIBUTION: Louisiana.

PLATE CXXXV, fig. 6.

12a. Micrasterias denticulata var. angulosa (Hantzsch) West & West 1902, Trans. Roy. Irish Acad. 32B(1): 30.

Micrasterias angulosa Hantzsch in Rabenhorst 1862, Alg. Sachsens No. 1407.
Micrasterias denticulata var. *angustosinuata* Gay 1884a, Mongr. loc. Conjug. 52. Pl. 1, Fig. 4; 1884, Bull. Soc. Bot. France 31: 334.

A variety with shallower incisions, the polar lobe shorter, its lateral margins nearly straight; semicells in end view nearly rhomboid, with single large median swelling. L. 210–324 μm. W. 175–262 μm. W. at apex 42–84 μm. I. 26–58 μm. Th. 50–74 μm. Zygospore spherical with forked processes as in the typical variety, diameter 114–130 μm, not including processes, processes c. 32 μm.

DISTRIBUTION: Alaska, California, Colorado, Florida, Massachusetts, Michigan, Montana, New Hampshire, North Carolina, Oregon, Washington. British Columbia, Labrador, Newfoundland, Northwest Territories, Ontario, Québec. Great Britain, Europe, Japan, New Zealand, Africa, South America, Arctic.

PLATE CXXXV, figs. 7, 12–14.

12b. Micrasterias denticulata var. basiprotuberans Bullard 1914, in Collins, Holden & Setchell 1914, Phycotheca Boreali-Americana 40: No. 1724.

A variety with the face of the semicell bearing 3 more or less rounded supraisthmial projections, and sometimes a smaller swelling on the polar lobe on either side of the notch; isthmus broad. L. 227–230 μm. W. 195–205 μm. W. at apex 50–58 μm. I. 38–46 μm.

Bullard (l.c.) accompanied his material in the Phycotheca with only this brief description; "A variety with the base of the semicell furnished with a three-parted projection. This variety has been found at two stations and shows no tendency to pass into the typical form." Examination of his dried material has revealed several specimens in not very good condition, but recognizable as belonging to M. denticulata, and probably to the plant subsequently described as var. taylorii Krieger (1939, p. 108). If this could be proven, the varietal epithet basiprotuberans would, by priority, surplant the epithet taylorii, and would include the plant called f. mammillata Prescott & Scott (1952, 234. Pl. 4, Fig. 1). But we hesitate to make this change with such an incomplete description and figure. (The basal projections were particularly difficult to see in the dried material.)

DISTRIBUTION: Massachusetts.

PLATE CXXXVI, figs. 3, 4.

12c. Micrasterias denticulata var. granulosa Irénée-Marie 1951, Nat. Canadien 78(7/8): 183. Pl. 1, Figs. 5, 6.

A variety differing from the typical in its smaller size, relatively shorter polar lobe (approaching var. angulosa (Hantzsch) West & West), in its strongly granular wall, somewhat obscuring the incisions, and in its rather angular outline, due to the fact that the margins of the upper and lower lateral lobules are truncate, or slightly concave rather than convex (which makes it resemble somewhat M. jenneri Ralfs). L. 159.5–196 μm. W. 138.5–177 μm. W. at apex 54.7–62.8 μm. I. 29–32 μm.

Varieties named "granulosa" are often questionable taxa because the granular effect may be caused by the extrusion of pectic substances from the pores (Taft 1949, p. 214), but this variety has other distinguishing features which make it difficult to fit into the typical or one of the established varieties or forms. A vertical view would be most helpful.

DISTRIBUTION: Québec.

PLATE CXXXVI, fig. 5.

12d. Micrasterias denticulata var. intermedia Nordstedt 1880, in Wittrock & Nordstedt 1880, Alg. exsic. Bot. Notiser 1880 No. 370, p. 119 (also in Wittrock & Nordstedt 1889, Algae Aquae Dulcis Exsic. p. 35).

A variety approaching *M. thomasiana* Arch., with the basal and intermediate lobes of the semicell 3 times dichotomously divided, the extremities of the ultimate divisions excavate with pointed (but not dentate) angles; polar lobe with apex flattened on either side of notch. L. 210–220 μm. W. 174–180 μm. Th. 58–62 μm.

DISTRIBUTION: Alaska, Connecticut, Florida, Louisiana, North Carolina, Oregon, Tennessee, Washington. Europe, Australia, South America.

PLATE CXXXVI, figs. 6, 7.

12e. Micrasterias denticulata var. taylorii Krieger 1939, Rabenhorst's Krypto-gamen-Flora 13(1:2): 108. Pl. 137, Fig. 7 f. taylorii.

Micrasterias thomasiana Arch. var. Taylor 1935, Pap. Michigan Acad. Sci., Arts & Lettr. 20: 215. Pl. 45, Fig. 7.

A variety with the face of the semicell near the isthmus bearing 3 projections directed toward the isthmus, the lateral mammillate, the central 3-lobed with the middle lobe broadly truncate. I. 226–227 μm. W. 188–209 μm. W. at apex 65 μm. I. 25–26 μm.

This variety approaches *M. thomasiana* Arch. in the supraisthmial swellings, but differs in the rounded extremities of the ultimate lateral lobules and the angles of the polar lobe; also in the lack of other facial ornamentation.

DISTRIBUTION: Florida, Louisiana, Virginia. Newfoundland, Québec.

PLATE CXXXVI, fig. 1.

12f. Micrasterias denticulata var. taylorii f. mammillata Prescott & Scott 1952, Trans. Amer. Microsc. Soc. 71(3): 234. Pl. 4, Fig. 1.

A form differing from the typical by having 2 prominent swellings on the polar lobe, 1 on either side of the notch, and in having the central of the 3 supraisthmial protuberances very broadly elliptic, not 3-lobed; semicell in basal view narrowly elliptic with 3 prominent swellings on either side of the midregion, of which the center one is more prominent and broadly convex. L. 187–208 μm. W. 178–184 μm. I. 24–25 μm. Th. 52 μm.

DISTRIBUTION: Florida.

PLATE CXXXVI, fig. 2.

13. Micrasterias depauperata Nordstedt 1870, Vidensk. Medd. Naturh. Foren. Kjöbenhavn 1869: 222. Textfig. on p. 222 var. depauperata.

Cells medium-sized, about one-third longer than broad; polar lobe produced, widely separated from the lateral lobes by a broad, rounded sinus, apex sub-truncate and emarginate, angles laterally extended, bidentate at the extremities; lateral lobes once shallowly, evenly divided, ends bidentate, teeth diverging; basal lobule horizontal, upper one somewhat diverging; semicells in vertical view broadly fusiform; wall brownish, evidently punctate. L. 84–113 μm. W. 64–85 μm. W. at apex 48–59 μm. I. 15–16 μm. Th. 23–33 μm. Ratio of width to thickness 4.5 to 2. Zygospore unknown.

The figures seen of most plants recorded from North America seem to be of *M. depauperata* var. *wollei* Cushman, or of some other species. We illustrate Nordstedt's plant from Brazil for comparison with our more common American varieties.

DISTRIBUTION: Massachusetts, Pennsylvania, Wisconsin. ?British Columbia (no figure). Brazil.

PLATE CIII, fig. 1.

Key to the North American Varieties and Forms of *Micrasterias depauperata*

1. Lateral lobes and lateral extensions of polar lobe bidentate at extremities.
    2. Lateral lobules short, occupying less than 2/3 the height of the semicell.
        3. Cells less than 100 μm broad, lateral lobes evenly divided.   13. var. *depauperata.*
        3. Cells more than 110μm broad, upper lateral lobules shorter than lower.
                                                                          13c. var. *wollei* f. *wollei.*
    2. Lateral lobes tall, occupying 2/3 to 3/4 the height of the semicell.
        4. Dorsal margin of lateral lobe convex; extensions of polar lobe short, stout,
            diverging.   13a. var. *convexa.*
        4. Dorsal margin of lateral lobe strongly concave; extensions of polar lobe long and
            narrow, slightly converging.   13b. var. *kitchelii.*
1. Lateral extensions of polar lobe bearing a single stout tooth; lateral lobes undivided,
    bearing 1–3 teeth.   13d. *var. wollei* f. *apiculata.*

13a. Micrasterias depauperata var. **convexa** Prescott & Scott 1943, Pap. Michigan
Acad. Sci., Arts & Lettr. 28: 73. Pl. 5, Figs. 3, 4.

Cells with the dorsal margin of the lateral lobes strongly convex, sloping
from a shallower subpolar sinus, and bearing a stout tooth at about the middle,
and 2 stout teeth at the basal extremity; sinus linear and closed; semicells in
vertical view ovate-fusiform, the ratio of width to thickness three to two. L.
130–140 μm. W. 97–100 μm. L. 19.8–20.5 μm. Th. c. 66 μm.
DISTRIBUTION: Florida.
PLATE CIII, fig. 2.

13b. Micrasterias depauperata var. **kitchelii** (Wolle) West & West 1896a, Trans.
Linn. Soc. London, Bot., II, 5(5): 239.

Micrasterias kitchelii Wolle 1884a, Desm. U.S., p. 116. Pl. 37, Figs. 1–3; 1892, Desm. U.S.,
    p. 129. Pl. 41, Figs. 1–3 (not Wolle 1880, p. 45. Pl. 5, Fig. M).

Cells about as long as broad; polar lobe short, apex flatly convex and slightly
retuse, the angles horizontally extended, slightly converging, ending in stout
diverging teeth; subpolar sinus rounded, much less extensive than in the type;
lateral lobe tall, more than 2/3 the height of the semicell, once divided by an
incision about as broad as, but less deep than, the subpolar sinus, ends stoutly
bidentate; sinus opening widely from a sharp angle. L. 125 μm. W. 125 μm. W. at
apex 75 μm. I. c. 22 μm.
Wolle first described *M. kitchelii* in 1880, from Massachusetts. In 1884 he
figured a very different-appearing plant under this name, including under it the
Massachusetts record. In 1892 he figured the 1884 plant as typical of *M. kitchelii*
and showed the 1880 plant on another plate (Pl. 42, Fig. 2) as "*M. kitchelii* var."
West & West (1896a, p. 239), in changing the name *M. kitchelii* to *M. depauperata* var. *kitchelii*, refer to and partially describe the figure in Wolle 1884a, not
1880. Therefore we assume it is the plant figured as *M. kitchelii* by Wolle in
1884a (not 1880), which is the archetype for *M. depauperata* var. *kitchelii*. The
distribution is uncertain since in 1884a Wolle had not yet distinguished between
this plant and what he later called a variety, which plant Cushman in 1904a (p.
396, Fig. 3) named *M. depauperata* var. *wollei* Cushm.
DISTRIBUTION: Alaska, Florida, Massachusetts, New Jersey, North Carolina, Wisconsin, "New England." Labrador.
PLATE CIII, fig. 3.

13c. Micrasterias depauperata var. **wollei** Cushman 1904a, Bull. Torrey Bot. Club
31: 396. Textfig. 3 on p. 396 f. **wollei.**

Micrasterias kitchelii Wolle 1880, Bull. Torrey Bot. Club 7: 45. Pl. 5, Fig. M.
Micrasterias kitchelii Wolle "var." Wolle 1892, Desm. U.S., Pl. 42, Fig. 2.
Micrasterias depauperata "forma" West & West 1896a, Trans. Linn. Soc. London, Bot., II, 5(5): 238. Pl. 4, Fig. 1.
Micrasterias depauperata var. *kitchelii* (Wolle) West & West 1896a, in Smith 1924, Wisconsin Geol. Nat. Hist. Surv. Bull. 57(2): 42. Pl. 59, Fig. 4; in Krieger 1939, Rabenhorst's Kryptogamen-Flora 13(1:2): 39. Pl. 106, Figs. 3, 4; in Croasdale & Grönblad 1964, Trans. Amer. Microsc. Soc. 83: 172. Pl. 9, Fig. 1; in Prescott & Scott 1943, Pap. Michigan Acad. Sci., Arts & Lettr. 28: 73. Pl. 3, Fig. 5.

A variety differing in being one-half larger, the sinus more open, the apical lobe shorter and the upper portion of the lateral lobes considerably more compressed than the lower. L. 120–150 μm. W. 112–150 μm. W. at apex 80–96 μm. I. 19–29 μm. Th. 30 μm. Ratio of width to thickness two to one.

We think that Krieger (1939, p. 39) was wrong in absorbing Cushman's var. *wollei* into Wests' variety *kitchelii*. The latter differs strongly from the typical and from var. *wollei* in having a very short and narrowly extended polar lobe, in being relatively broader at the subpolar incision, and nearly quadrate, as opposed to the broadly pyramidal form of var. *wollei*. Nordstedt's typical plant has not been figured in North America since Wolle's original record. Most of our American forms quite definitely seem to belong to var. *wollei*.

DISTRIBUTION: Alaska, Florida, Georgia, Idaho, Louisiana, Massachusetts, Michigan, Mississippi, New Hampshire, North Carolina, Oregon, Utah, Washington, Wisconsin. British Columbia, Labrador, Ontario, Québec.

PLATE CIII, figs. 4–7; PLATE CIV, fig. 1.

13d. **Micrasterias depauperata** var. **wollei** f. **apiculata** Irénée-Marie 1952, Hydrobiologia 4: 146. Pl. 14, Figs. 2–5.

Micrasterias depauperata var. kitchelii (Wolle) West & West f. *apiculata* (Irénée-Marie) Graffius 1963, Dissert., p. 128. Pl. 7, Fig. 5.

A form differing from the typical plant in the abrupt curvature and greater convexity of the polar lobe, usually convex, rarely retuse at the summit, and by the angles, which usually are terminated by 1 straight, stout, acute, horizontal tooth; lateral lobes semielliptical, terminated by 1 to 3 short, acute, slightly diverging teeth; subpolar sinus broader and more open than the typical, but always deep; median sinus usually linear for a short distance from the interior, rounded within and broadened to the exterior; dimensions of the type.

DISTRIBUTION: Michigan. Québec.

PLATE CIV, figs. 2–6.

14. **Micrasterias floridensis** Salisbury 1936, Ohio Jour. Sci. 36: 58. Pl. 1, Fig. 2
var. **floridensis** f. **floridensis**.

Cells large, about as long as broad, subhexangular, deeply constricted, sinus linear, subpolar incisions and primary incisions of lateral lobes deep and linear, secondary incisions of lateral lobes V-shaped, extending only one-quarter way to center of cell; polar lobe long-cuneate, broadening strongly at apex, the apical margins slightly retuse, its lateral extensions ending in a single short tooth; vertical view of semicell fusiform. L. 198–228 μm. W. 190–232 μm. W. at apex 75–97 μm. I. 19.5–23 μm. Zygospore unknown.

DISTRIBUTION: Alaska, Florida, Georgia, North Carolina.

PLATE CVIII, fig. 1.

Key to the North American Varieties and Forms of *Micrasterias floridensis*

1. Cells without spines at the margins of the incisions.
  2. Cells with polar lobe smooth and only slightly retuse.
    3. Lateral lobules short, not tapering, the ends excavate.
      4. Cells more than 200 μm long.   14. var. *floridensis* f. *floridensis.*
      4. Cells less than 170 μm. long.   14b. var. *floridensis* f. *minor.*
    3. Lobules long and tapering, usually ending in 2 rather long prongs or teeth.
                                          14e. var. *subjohnsonii.*
  2. Cells with a tooth or small spine-tipped mammillate protuberance near the lateral
    extremities of the polar lobe, which is more deeply retuse.
                                 14a. var. *floridensis* f. *canadensis.*
1. Cells with spines on the margins of the major incisions.   14c. var. *spinosa.*

14a. Micrasterias floridensis var. **floridensis** f. **canadensis** (Irénée-Marie) Croasdale comb. nov.

Micrasterias johnsonii West & West var. *papillata* Taylor f. *canadense* Irénée-Marie 1957, Hydrobiologia 9: 75. Figs. 2, 3.

Cells circular or slightly longer than broad, major incisions linear and closed or slightly open, straight or slightly curved, so that lateral lobes and lobules are mostly broad and straight-walled, their ends excavate or with short teeth; polar lobe more deeply retuse than in typical plant, bearing a small, spine-tipped mammillate protuberance on each face near the lateral extremities, wall otherwise smooth. L. 165–225 μm. W. 155–202 μm. I. 16–25 μm.

This form, with *M. johnsonii* var. *bipapillata* Taylor (1935, p. 213, Pl. 48, Fig. 5), tends to connect *M. floridensis* to *M. johnsonii* West & West. Irénée-Marie (1952, Pl. 14, Figs. 7, 8) first published 2 figures: *M. johnsonii* var. *papillata* Taylor (he obviously meant var. *bipapillata*) and *M. Johnsonii* var. *papillata* forma. His Fig. 7 probably does belong to Taylor's variety; Fig. 8 differs in having less deep incisions, therefore relatively broader lobes and lobules, which end in shorter teeth. In 1957 (*l.c.*) he included the form represented by Fig. 8 in a new form he called *M. johnsonii* var. *papillata* Taylor f. *canadense*. The figures he shows in this paper represent forms with even broader straight-sided lobes and lobules, terminating in a deep excavation rather than paired teeth. This approaches *M. floridensis* more closely than it does *M. johnsonii*. His "forma," represented by Fig. 8 in 1952, probably belongs here also, as he states (1957, p. 75).

DISTRIBUTION: Alaska. Québec.
PLATE CVIII, figs. 2–4.

14b. Micrasterias floridensis var. **floridensis** f. **minor** Prescott & Scott 1952, Trans. Amer. Microsc. Soc. 71: 234. Pl. 2, Fig. 7.

A form smaller than the typical, broader than long, the base of the polar lobe relatively less stout; all incisions more open. L. 166–169 μm. W. 172–173 μm. I. 19–22.5 μm.

DISTRIBUTION: Florida.
PLATE CVIII, figs. 5, 6.

14c. Micrasterias floridensis var. **spinosa** (Whelden) Prescott & Scott 1952, Trans. Amer. Microsc. Soc. 71: 236. Pl. 2, Fig. 8.

Micrasterias floridensis f. *spinosa* Whelden 1941, Jour. Elisha Mitchell Sci. Soc. 57(2): 269. Pl. 6, Fig. 8.

A variety differing from the typical in the presence of 2 or more intramarginal spines located on the polar lobe and usually also on the lateral lobules about one-third or one-quarter way in from the ends, the arrangement forming a circle on the face of the semicells; sometimes with a small papilla or tooth on the dorsal margin of the polar lobe, near the extremity (Pl. CIX, Fig. 2). In one plant it was seen that although the margins of the lobules were sharp and smooth, the terminal teeth showed irregular lumps and swellings (Pl. CIX, Fig. 3). Semicells in vertical view narrowly oblong, with sharp extremities and constricted middle area. L. 172–227 μm. W. 162–230 μm. L. 18–24 μm.

DISTRIBUTION: Alaska, Florida, Georgia, Michigan, Mississippi, North Carolina.

PLATE CIX, figs. 1–4.

14d. **Micrasterias floridensis** var. **subjohnsonii** Prescott & Scott 1952, Trans. Amer. Microsc. Soc. 71: 236. Pl. 8, Figs. 1–3.

A variety differing from the typical in having the secondary incisions of the lateral lobes of the semicell deeper, the lobules longer, relatively narrower, and more strongly tapered; angles of the polar lobe sometimes bifid; subpolar incision deeper; sinus deep and narrow or closed. L. 199–221 μm. W. 207–234 μm. I. 22–24 μm.

Pl. CIX, Fig. 6 shows a specimen from Florida that seems to be dichotypical between var. *subjohnsonii* and var. *spinosa* (Whelden) Presc. & Scott.

DISTRIBUTION: Florida, North Carolina, Wisconsin.

PLATE CIX, figs. 5–7.

15. **Micrasterias foliacea** Bailey ex Ralfs 1848, Brit. Desm., p. 210. Pl. XXXV, Fig. 3 var. **foliacea.**

Micrasterias foliacea var. *granulifera* Cushman 1908, Rhodora 10: 111.

Cells small, united in filaments by the interlocking of polar lobes; cells rectangular, about as broad as long, or broader; sinus sublinear; lateral lobes divided to the second or third order, dissimilar, the lower lobule horizontal, the upper diverging, its upper part reduced to a conical projection, other extremities bidentate or emarginate; polar lobe very distinctive: basal fourth narrow with erect parallel sides, upper three-quarters of the lobe greatly expanded, anvil-shaped, with a very broad, deep subrectangular excavation in the middle of the apex, angles stout, ending in 2 widely divergent teeth; depressed portion of apex bearing 2 stout spines on each side, unequal in size; semicell in vertical view narrowly fusiform, the central part obliquely rhomboidal, showing the unequal apical spines; in lateral view swollen at base and expanded at apex. L. 58–99 μm. W. 56–92 μm. W. at apex 30–58 μm. I. 12–22 μm. Th. 16–23 μm.

This species is readily recognized by its peculiar apex, which interlocks the cells into filaments up to 182 cells long. Cushman's var. *granulifera (l.c.)* was described without a figure as "like the type but with the surface covered with large irregularly disposed granules." We agree with Krieger (1939, p. 76) that this should probably be included with the typical.

DISTRIBUTION: Alabama, Florida, Georgia, Louisiana, Massachusetts, Michigan, Mississippi, New Hampshire, New Jersey, New York, North Carolina, Rhode Island, South Carolina, Virginia, Wisconsin. Newfoundland, Nova Scotia, Ontario, Québec. Europe, Asia, Africa, Australia, South America, Panama, Philippines, Celebes.

PLATE CXXXIX, figs. 3–8.

Key to the North American Varieties of *Micrasterias foliacea*

1. Without intramarginal spines along major incisions.   15. var. *foliacea*.
1. With 1-3 stout intramarginal spines on lateral lobes at isthmus and subpolar incision.
                                                                          15a. var. *ornata*.

15a. **Micrasterias foliacea** var. **ornata** Nordstedt 1870, Vidensk. Medd. Naturh. Foren. Kjöbenhavn 1869(14/15): 221. Pl. 2, Fig. 16.

A variety with 1 to 3 small spines on the upper margin of the upper lateral lobe near its base and the lower margin of the lower lateral lobes near the isthmus; incision between polar lobe and lateral lobes usually more widely open. L. 66-83 μm. W. 67-83 μm. W. at apex 30-37 μm. I. 12-16 μm. Zygospore reported from Sunda (Krieger 1932, 218. Pl. 24, Fig. 2), spherical with relatively few large sharp spines, diam. 35 μm, including spines 70 μm.

Salisbury (1936, p. 58) questions the validity of Krieger's zygospore since it differs markedly from that of the typical, seen by him (*l.c.*) and Turner (1892, 94, Pl. 6, Fig. 15). However, Hirano (1967, 63. Pl. 12, Fig. 2) figures one for the typical from SE Asia which, except for more numerous spines, is quite similar to Krieger's for var. *ornata*.

DISTRIBUTION: Florida, Louisiana. Southern Asia, Australia, Africa, South America.

PLATE CXL, figs. 1, 2.

16. **Micrasterias horrida** Taft 1949, Trans. Amer. Microsc. Soc. 68(3): 215. Pl. 2, Fig. 4.

Cells large, slightly longer than broad, deeply constricted; sinus and all major incisions narrowly linear; lateral lobes uneven, the upper portion larger, divided to the fourth order, the lower portion mostly to the third order, the extremities shallowly emarginate with rounded angles; polar lobe cuneate with concave lateral margins and retuse apex, with deep open notch and rounded angles; cell wall coarsely punctate, with 3 oval rows of scrobiculations just above the isthmus, and with long, pointed hairs located near the apex and base of the polar lobe, near the lateral incisions, and just above the isthmus; semicell in vertical view fusiform with poles acute and with 3 scrobiculate undulations at the middle of each side. L. 218-238 μm. W. 181-208 μm. I. 16-23 μm.

This species should be compared to *M. denticulata* Bréb. ex Ralfs and its varieties, from which it differs in the supraisthmial scrobiculations and the long spinelike hairs.

DISTRIBUTION: New Jersey (Pine Barrens).

PLATE CXXXVII, fig. 1.

17. **Micrasterias jenneri** Ralfs 1848, Brit. Desm., p. 76. Pl. XI, Fig. 1 var. **jenneri** f. **jenneri**.

Cells of medium size, 1 1/3 to 1 1/2 or more times longer than broad; sinus and major incisions linear and often closed, incisions between lateral lobules of second order very shallow and open; lateral lobules of third order often only slightly indicated and more or less irregular; polar lobe short, less than half the height of the semicell, very widely cuneate with rounded angles, apex broadly convex but retuse in middle; semicells in lateral view oblong-elliptic, in vertical view fusiform with rounded ends, but varying as shown by different authors; wall usually rough with what have frequently been described as granules, but are more likely extrusions from pores. L. 127–188 μm. W. 80–130 μm. W. at apex 50–81 μm. I. 18.5–34 μm. Th. c. 49 μm. Zygospore unknown.

The species is characterized by its relative length, its few shallow incisions, and the rough wall. Ralfs (*l.c.*) comments that this species seems to unite *Micrasterias* with *Euastrum*, but "a transverse view shows the absence of the inflated protuberances always found in true species of *Euastrum*." In central Europe this is an outstanding Sphagnophile.

DISTRIBUTION: Florida, Kentucky, Louisiana, Maine, Massachusetts, Michigan, Mississippi, Montana, North Carolina, Pennsylvania, South Carolina. Nova Scotia, Québec. Great Britain, Europe, Swedish Lappland, New Zealand, Indonesia.

PLATE CVI, figs. 1–7.

Key to the North American Varieties and Forms of *Micrasterias jenneri*

1. Cells more than 100 μm long.
  2. Lateral lobules of third order always indicated by at least a notch.
                                                           17. var. *jenneri* f. *jenneri*.
  2. Lateral lobules of third order often absent or barely indicated.   17b. var. *simplex*.
1. Cells less than 100 μm long.   17a. var. *jenneri* f. *minor*.

17a. **Micrasterias jenneri** var. **jenneri** f. **minor** Croasdale f. nov.

Cellulae magnitudine c. dimidium plantae typicae et relative latiores, 1.2 ad 1.32 plo longiores quam latae; lobus polaris relative longior, plus quam dimidium altitudinis cellulae; membrana levior, tribus granulis, autem, (vel extrusionibus maioribus a poris?) in quoque margine lobi polaris exteriore interdum visis. Cellulae 65–79 μm. long., 53–60 μm lat. (1.2–1.34 x), 13–15 μm. lat. ad isthmum.

ORIGO: E stagno liliis pleno distante 2 milia passuum (3.2 km) ad meridiem a loco Kiln, Hancock Co., Mississippi dicto.

HOLOTYPUS: A. M. Scott Coll. "Miss. 23."

ICONOTYPUS: Plate CVI, figs. 8, 9.

Cells about half as large as typical and relatively broader, 1.2 to 1.32 times longer than broad; polar lobe relatively longer, more than half the height of the cell; wall less strongly roughened with pore extrusions, but with three granules (or larger pore extrusions?) sometimes present on each outer margin of the polar lobe). L. 65–79 μm. W. 52–60 μm. (1.2–1.32 x), I. 12–15 μm.

This form was very abundant in a lily pond in Mississippi.

DISTRIBUTION: Louisiana, Mississippi.

PLATE CVI, figs. 8. 9.

17b. **Micrasterias jenneri** var. **simplex** West 1890, Jour. Roy. Microsc. Soc. 1890: 287. Pl. 6, Fig. 34.

A variety with simpler outline; lateral lobes of second order absent or indicated only by short, acute-angled incision; wall punctate, in one instance described as scrobiculate with shallow pits. L. 127–200 (205) μm. W. 91–132 μm. W. at apex 53–72 μm. Th. 43–51 μm.

DISTRIBUTION: Florida, Louisiana, Massachusetts, Mississippi, North Carolina. Newfoundland. Great Britain, Europe, Asia, Africa, South America, Indonesia.

PLATE CVI, figs. 10–13.

18. **Micrasterias johnsonii** West & West 1898a, Linn. Soc. Jour. Bot. 33: 297. Pl. 16, Fig. 5 var. **johnsonii** f. **johnsonii**.

Cells large, a little broader than long, sinus deep and narrow with sigmoid-sinuose sides; lateral lobes divided to the second order (twice), the lobes strongly swollen at the base and ending in 2 long, diverging teeth; polar lobe narrow, its lower part cylindrical, or nearly so, above broadening into 2 conical projections, each ending in a long tooth; polar lobe and lateral lobules bearing a single row of small surface spines near the incisions. L. 200–299 μm. W. 200–275 μm. W. at apex 26–70 μm. I. 20–30 μm. Zygospore unknown.

Bourrelly (1966, p. 96. Pl. 14, Fig. 2) describes from Québec a small form (L. 200–210 μm, W. 200 μm) with a larger polar lobe (reaching 70 μm lat.), otherwise similar to the type (Pl. CX, Fig. 2).

DISTRIBUTION: Florida, Louisiana, Massachusetts, Mississippi, New Hampshire, North Carolina. Ontario, Québec.

PLATE CX, figs, 1–3.

Key to the North American Varieties and Forms of *Micrasterias johnsonii*.

1. Cells with a single row of short spines on the surface along each major incision.
  2. Cells less than 300 μm long.
    3. Lateral extremities of polar lobe ending in 1 tooth.   18. var. *johnsonii* f. *johnsonii*.
    3. Lateral extremities of polar lobe ending in 2 teeth.   18a. var. *johnsonii* f. *bispinata*.
  2. Cells 300 μm or more long.   18b. var. *johnsonii* f. *maior*.
1. Cells without surface spines along the incisions.
  4. Polar lobe separated from lateral lobes by a very deep open incision, narrowed at base and narrowed again below its very broadly horizontally expanded apex, which bears a small projection on either side of the shallow median incision.
                                                18c. var. *bipapillata*.
  4. Polar lobe with moderately deep subpolar incision, not narrowed at base, and widening to its moderately expanded apex, which lacks small projections.
    5. Lateral extremities of polar lobe usually ending in 1 tooth, all terminal teeth smooth; sinus usually open.   18d. var. *ranoides* f. *ranoides*.
    5. Lateral extremities of polar lobe ending in 2 teeth; sometimes teeth and adjacent portion of lobules gently and irregularly undulate; sinus nearly closed.
                                                18e. var. *ranoides* f. *novae-angliae*.

18a. **Micrasterias johnsonii** var. **johnsonii** f. **bispinata** Prescott & Scott 1952, Trans. Amer. Microsc. Soc. 71: 238. Pl. 3, Fig. 2.

A smaller, more compact form with stouter polar lobe and lateral lobules, and with the sinus closed in all but its outer part; the lateral extremities of the polar lobe bidentate. L. 212–228 μm. W. 210–224 μm. W. at apex 83 μm. I. 24–27 μm.

In general outline this closely resembles var. *ranoides*, but it has the surface spines of the typical.

DISTRIBUTION: Florida, Mississippi.

PLATE CX, fig. 4.

18b. **Micrasterias johnsonii** var. **johnsonii** f. **maior** Prescott f. nov.

Micrasterias johnsonii West & West var. *johnsonii* forma Prescott & Scott 1952, Trans. Amer. Microsc. Soc. 71(3): 236. Pl. 3, Fig. 1.

Forma a varietate typica differens quod aliquantulo maior et quod ordinatum spinarum diversam habet, ordinem spinarum a lobulis lateralibus inferioribus trans superficiem semicellulae extendentem media in parte praebens; membrana speciminis unius in Florida reperti (Pl. CX, Fig. 5) in lobulis lateralibus infra dentes terminales incrassata per c. dimidium longitudinis lobuli. Cellulae 304–320 μm., 266–287 μm lat., 18–27 μm. lat. ad isthmum.

ORIGO: E stagno distante 6 milia passuum (9.6 km.) ad meridiem a loco Masaryk, Hernando Co., Florida dicto.

HOLOTYPUS: A. M. Scott Coll. 769,771.

ICONOTYPUS: Plate CX, figs. 5, 6.

A form differing from the typical by being somewhat larger and by having a different pattern of spine arrangement, there being in the middle of the lower lateral lobules a row of spines that continues across the face of the semicell in the midregion; in one plant, from Florida (Pl. CX, Fig. 5) the wall was thickened on the lateral lobules below the terminal teeth for about half the length of the lobule. L. 304–320 μm. W. 266–287 μm. I. 18–27 μm.

DISTRIBUTION: Florida, Louisiana.

PLATE CX, figs. 5, 6.

18c. **Micrasterias johnsonii** var. **bipapillata** Taylor 1935, Michigan Acad. Sci., Arts & Lettr. 20: 213. Pl. 48, Fig. 5.

A slightly smaller variety with shorter terminal teeth on the lateral lobes, without surface spines, and bearing 2 mammillate-subaculeate projections on each side of the terminal lobe near the top; wall smooth. L. 165–250 μm. W. 154–240 μm. I. 16–22.5 μm.

DISTRIBUTION: Newfoundland, Québec.

PLATE CX, fig. 7.

18d. **Micrasterias johnsonii** var. **ranoides** (Salisb.) Krieger 1939, Rabenhorst's Kryptogamen-Flora 13(1:2): 73. Pl. 119, Fig. 3 f. **ranoides**.

Micrasterias ranoides Salisbury 1936, Ohio Jour. Sci. 36: 59. Pl. 1, Fig. 1.

Cells more compact with shorter lobes, cells as long as broad or a little longer; polar lobe stouter, evenly expanded from base; surface spines lacking; cells in lateral view fusiform with polar lobes about half as thick as the cell. L. 208–302 μm. W. 202–290 μm. W. at apex 93–110 μm. Max. W. polar lobe 74–79 μm. I. 21–27 μm. Th. 45–50 μm. Teeth 20–25 μm.

DISTRIBUTION: Florida, Georgia, Mississippi, New Hampshire, North Carolina. Ontario.

PLATE CXI, figs. 1, 2.

18e. **Micrasterias johnsonii** var. **ranoides** f. **novae-angliae** (Wheld.) Croasdale comb. nov.

Micrasterias johnsonii var. *novae-angliae* Whelden 1942, Rhodora 44: 178. Fig. 5.

Cells large, somewhat longer than broad, without surface spines; terminal teeth of lateral lobes stout and blunt; lateral extremities of polar lobe sometimes bearing an extra tooth; in some specimens the surface of the teeth and even of the adjacent part of the lobule is gently and irregularly undulate (Pl. CXI, Fig. 3). L. 280–358 μm. W. 270–323 μm. Max. W. polar lobe 93–116 μm. I. 19–30 μm.

DISTRIBUTION: New Hampshire.

PLATE CXI, figs. 3, 4.

19. **Micrasterias laticeps** Nordstedt 1870, Vidensk. Medd. Naturh. Foren. Kjöbenhavn 1869: 220. Pl. 2, Fig. 14 var. **laticeps** f. **laticeps**.

Micrasterias laticeps f. *major* Nordstedt 1870, Vidensk. Medd. Natur. Foren. Kjöbenhavn 1869: 220. Pl. 2, Fig. 14.

Cells medium-sized, 1 1/10 to 1 1/3 times broader than long; sinus narrow within, then opening widely; lateral lobe entire, horizontally extended, long-conical, often somewhat constricted before the ends, terminating in 2 short teeth; subpolar incision acute-angled; lateral extensions of polar lobe very similar to the lateral lobes and nearly as broadly extended, their ends bent slightly downward and terminating in 1 short tooth; cell in vertical view fusiform, in lateral view short-elliptic. L. 112–221 μm. W. 122–265 μm. I. 15–37 μm. Th. 37–64 μm. Zygospore unknown.

Nordstedt (1870, p. 220) described his new species under "f. *major*" and "f. *minor*." He illustrated only f. *major* and this has been taken as the type. For cases of polymorphism in this species, see a recent paper by Bicudo & Sormus (1972).

DISTRIBUTION: Widely distributed in U.S. and Canada. Asia, South America, Panama, Philippine Islands.

PLATE LXXXVIII, figs. 1–4.

Key to the North American Varieties and Forms of *Micrasterias laticeps*

1. Cells broader than long; sinus open throughout most of its length.
  2. Lateral lobes at base about as broad as lateral extensions of polar lobe.
    3. Cells more than 110 μm long.   19. var. *laticeps* f. *laticeps*.
    3. Cells less than 90 μm long.   19a. var. *laticeps* f. *minima*.
  2. Lateral lobes at base 1/2 to 2 times as broad as lateral extensions of polar lobe.
                                       19d. var. *inflata*.
1. Cells longer than broad; sinus closed throughout at least 1/2 of its length.
  4. Sinus closed for less than 2/3 of its length; dorsal margin of lateral lobes not strongly convex.   19b. var. *crassa* f. *crassa*.
  4. Sinus closed throughout most of its length; dorsal margin of lateral lobes strongly convex.   19c. var. *crassa* f. *robustior*.

19+. Micrasterias laticeps var. **laticeps** forma ad var. **crassum** Prescott accedens.

A form approaching var. *crassa* in its greater breadth of lateral lobes at their bases and lateral extensions of polar lobes, and in its relative length, which, equaling the breadth, is greater than that of var. *laticeps*, although less great than that of var. *crassa*. L. 133 μm. W. 133 μm. W. at apex 128 μm. I. 21 μm.

DISTRIBUTION: Florida.

PLATE LXXXVIII, fig. 6.

19a. Micrasterias laticeps var. **laticeps** f. **minima** Prescott f. nov.

Cellulae multo minores, 1 1/5 plo latiores quam longae; pars apicalis aeque lata ac lobi laterales. Cellula 79 μm. long., 94 μm lat. 14 μm lat. ad isthmum.

ORIGO: Lacus Durant Lake, Hamilton Co., N.Y. dictus.

HOLOTYPUS: R. Pyle Coll. NY–5,

ICONOTYPUS: Plate LXXXVIII, fig. 5.

Cells very much smaller, 1 1/5 times broader than long; breadth of apical part as great as that of lateral lobes. L. 79 μm. W. 94 μm. I. 14 μm.

This plant has the small size of var. *acuminata*, but differs from it in the typical 2-toothed lateral lobe, thicker apical extensions, and less retuse apex.

DISTRIBUTION: Adirondacks, New York.

PLATE LXXXVIII, fig. 5.

19b. **Micrasterias laticeps** var. **crassa** Prescott, in Prescott & Magnotta 1935, Pap. Michigan Acad. Sci., Arts & Lettr. 20: 166. Pl. 25, Fig. 7 f. **crassa.**

Cells somewhat longer than broad, polar lobe strongly convex at apex, its lateral extensions relatively stout, terminating in downward-pointing teeth; lateral lobes strongly convex on dorsal surface, resulting in a sinus closed for more than half its length; the cells presenting a much more compact and stouter appearance than those of the typical plant. L. 108–125 μm. W. 111–118 μm. W. at apex c. 105 μm. I. 19 μm.

Krieger (1939, p. 15) includes here, with a query, a form illustrated by Wolle (1884a, p. 115. Pl. 37, Fig. 4), called by him *M. laticeps*. Because of its smaller polar lobe with a strongly retuse apex, and its more open sinus, Wolle's form seems closer to var. *inflata* Krieger.

DISTRIBUTION: Alaska, Florida, Louisiana, Massachusetts, Michigan, Mississippi, Montana.

PLATE LXXXIX, fig. 1.

19c. **Micrasterias laticeps** var. **crassa** f. **robustior** Förster 1964, Hydrobiologia 23(3/4): 378. Pl. 14, Fig. 6: Pl. 45, Photos. 3, 4.

Micrasterias laticeps var. *crassa* forma Prescott & Scott 1943, Pap. Michigan Acad. Sci., Arts & Lettr. 28: 75. Pl. 2, Fig. 6.

Cells definitely longer than broad, sinus linear, closed in interior, soon open, lobes broadly rounded; wall colorless, scrobiculate. L. 127–134 μm. W. 103–130 μm. W. polar lobe 96–120 μm. I. 20–22 μm.

DISTRIBUTION: Louisiana, Mississippi.

PLATE LXXXIX, figs. 2, 3.

19d. **Micrasterias laticeps** var. **inflata** Krieger 1939, Rabenhorst's Kryptogamen-Flora 13(1:2): 15. Pl. 99, Fig. 1.

Lateral lobes very broad at base and somewhat swollen; apical lobe smaller, with strongly retuse apex. L. 138–143 μm. W. 143–151 μm. W. at apex 118–122 μm. I. 25–29 μm.

DISTRIBUTION: U.S., Michigan. South America.

PLATE LXXXIX, figs. 4, 5.

20. **Micrasterias mahabuleshwarensis** Hobson 1863, Quart. Jour. Microsc. Sci. 1863: 169. fig. on p. 168 var. **mahabuleshwarensis** f. **mahabuleshwarensis.**

Micrasterias mahabuleshwarensis var. *americana* Wolle 1881, Bull. Torrey Bot. Club 8(1): 1. Pl. 6, Fig. 1.
Micrasterias mahabuleshwarensis var. *compacta* Nordst. f. *americana* Nordstedt 1888a, Kongl. Svenska Vet.-Akad. Handl. 22(8): 31.
Micrasterias mahabuleshwarensis f. *dichotoma* Smith 1921, p.p., Trans. Wisconsin Acad. Sci., Arts & Lettr. 20: 345. Pl. 9, Fig. 12.

Cells of medium size, slightly longer than broad; sinus sharp-angled, opening widely; polar lobe in lower half quadrate, above widely expanding into narrow diverging processes at the angles, with more nearly erect accessory processes at their base, arising asymmetrically, one at the front and one at the back of the apex, apex between processes nearly straight, often with a pair of intramarginal granules in middle; lateral lobes once divided into narrow tapering processes; subpolar sinus very broad with rounded base; all processes with serrate or denticulate margins, the teeth often stouter at the base of the lateral lobules on their inner side, ends 3- or 4-denticulate; semicell with a row of acute intramarginal granules along the inner part of the major incisions, and with a small granulate or denticulate central protuberance above the isthmus; semicells in lateral view evenly tapered from a rounded basal swelling, the apex showing the asymmetrical arrangement of the 4 processes; in vertical view rhomboid-fusiform, the small central protuberance and asymmetrically disposed apical processes showing clearly. L. 100–220 μm. W. 85–190 μm. W. at apex 55–105 μm. I. 15–31 μm. Th. c. 40 μm. Zygospore reported from S. India (Ramanathan 1962, p. 42, Fig. 19), spherical with elongate spines which are twice branched and end in short spines. Diameter 56–68 μm, including spines 133–147 μm. Spines 40–45 μm long.

DISTRIBUTION: Widely distributed in U.S. British Columbia, Québec. Great Britain, Europe, Asia, Afric, Australia, New Zealand, South America.

PLATE CXLIII, figs. 1, 2, 4, 5.

Key to the North American Varieties and Forms of *Micrasterias mahabuleshwarensis*.

1. Apex with accessory processes.
  2. Lateral lobes once divided.
    3. Lateral lobules slender, at least 4 times as long as broad.
      4. Without long cylindrical central protuberance, and without strong intramarginal spines along all incisions.   20. var. *mahabuleshwarensis* f. *mahabuleshwarensis*.
      4. With long cylindrical central protuberance, and with strong intramarginal spines along all incisions.   20e. var. *surculifera*.
    3. Lateral lobes short and swollen, no more than 2 times as long as broad.
                                20b. var. *ampullacea*.
  2. One or both of lateral lobules again divided.
    5. Both lobules divided, the inner part of each shorter, as in *M. americana*.
                      20a. var. *mahabuleshwarensis* f. *dichotoma*.
    5. Only upper lobule divided, the outer part sometimes shorter.   20f. var. *wallichii*.
1. Apex without accessory processes.
  6. Median area above isthmus with circle of granules on small protuberance.
                         20c. var. *ringens* f. *ringens*.
  6. Median area with large smooth protuberance.   20d. var. *ringens* f. *glabra*.

20a. Micrasterias mahabuleshwarensis var. **mahabuleshwarensis** f. **dichotoma** Smith 1921, Wisconsin Acad. Sci., Arts & Lettr. 20: 345. Pl. 9, Figs. 13, 14.

A form in which the lateral lobules are again divided, but often quite irregularly and often varying in the semicells of the same plant, the inner member always smaller, effecting a resemblance to *M. americana* (Ehrenb.) Ralfs. L. 130–165 μm. W. 110–165 μm. W. at apex 72–89 μm. I. 20–36 μm.

DISTRIBUTION: Florida, Michigan, Minnesota, Nevada, New Hampshire, New York, North Carolina. Nova Scotia, Ontario, Québec. Asia, South America.

PLATE CXLIV, figs. 1, 2.

20b. Micrasterias mahabuleshwarensis var. **ampullacea** (Mask.) Nordstedt 1888a, Kongl. Svenska Vet.-Akad. Handl. 22(8): 30. Pl. 2, Fig. 16.

Micrasterias americana (Ehrenb.) Ralfs f. *ampullacea* Maskell 1888, Jour. Roy. Microsc. Soc. 1888: 9. Pl. 2, Fig. 13.

A variety with the lateral lobules and lateral extensions of the polar lobe short and thick, tapering evenly to rounded, toothed extremities; upper lateral lobe sometimes with a short process; all margins serrate; semicells in end view rhombic-fusiform, slightly swollen in the middle. L. 91–205 μm. W. 67–170 μm. W. at apex c. 75 μm. I. 20–40 μm. Th. c. 45 μm.

The American forms are mostly smaller with even shorter lobules.

DISTRIBUTION: Florida. Asia, Australia, South America, East Indies, Java. PLATE CXLIII, figs. 3, 6.

20c. **Micrasterias mahabuleshwarensis** var. **ringens** (Bail.) Krieger 1939, Rabenhorst's Kryptogamen-Flora 13(1:2): 52. Pl. 110, Fig. 9 f. **ringens.**

Micrasterias ringens Bailey 1851, Smithson. Contrib. Knowledge 2(Art.8): 37. Pl. 1, Fig. 11.
Micrasterias ringens var. *serrulata* Wolle 1885a, Bull. Torrey Bot. Club 12(12): 128. Pl. 51, Fig. 15.
?Micrasterias baileyi Ralfs 1848, Brit. Desm., p. 211. Pl. XXXV. Fig. 4.
Micrasterias mahabuleshwarensis var. *serrulata* (Wolle) Smith 1924a, Roosevelt Wildlife Bull. 2(2): 143. Pl. 16, Fig. 2; Irénée-Marie 1938, Flore Desmid. Montréal, p. 231. Pl. 40, Fig. 1.

A variety differing in the absence of the accessory processes on the polar lobe; semicell in lateral and vertical views slender; with a circle of granules in the middle above the isthmus. L. 119–150 μm. W. 94–150 μm. W. at apex 60–70 μm. I. 12–22 μm.

DISTRIBUTION: Florida, Georgia, Massachusetts, Michigan, Minnesota, New Hampshire, New Jersey, New York, North Carolina, Pennsylvania, Rhode Island, South Carolina. British Columbia, Nova Scotia, Ontario, Québec. South America. PLATE CXLIV, figs. 3, 4.

20d. **Micrasterias mahabuleshwarensis** var. **ringens** f. **glabra** Prescott & Scott 1952, Trans. Amer. Microsc. Soc. 71(3): 240. Pl. 6, Fig. 4.

A form differing from the typical by having a smooth, rounded protuberance just above the isthmus (without a circle of granules), and with two prominent granules on either side of the apices of the median sinus; the intramarginal granules along the primary divisions of the cell more prominent than in the typical; wall densely punctate, punctations coarser in the area of the central protuberance L. 128–138 μm. W. 109–115 μm. W. at apex 58 μm. I. 20–21 μm. Th. c. 41 μm.

DISTRIBUTION: Florida. PLATE CXLIV, figs. 5, 6.

20e. **Micrasterias mahabuleshwarensis** var. **surculifera** Lagerheim 1888, Bih. Kongl. Svenska Vet.-Akad. Handl. 13, III (9): 5. Pl. 1, Fig. 1.

A variety in which the facial protuberance of the semicell is extended to form an elongate process arising at right angles from the face of the semicell, the apex of the process encircled by toothlike granules; incision between lateral lobules deep and open, lobules relatively long with spiny margins and intramarginal spines along major incisons; with a row of more pronounced granules below the apex. L. 142–180 μm. W. 132–158 μm. W. at apex 74–92 μm. Th. not including protuberance 31–33 μm, including protuberance 59–74 μm.

DISTRIBUTION: Florida. Southeast Asia, Africa, South America. PLATE CXLIII, fig. 7.

20f. **Micrasterias mahabuleshwarensis** var. **wallichii** (Grun.) West & West 1905a, Monogr. II: 122. Pl. 54, Figs. 7, 8; Pl. 55, Figs. 1–3.

Micrasterias americana f. *hermanniana* (Reinsch) Maskell 1888, Jour. Roy. Microsc. Soc. 1888: 9. Pl. 2, Fig. 11.
Micrasterias americana var. *hermanniana* Reinsch in Wolle 1892, Desm. U.S., p. 124. Pl. 36, Fig. 5.

A variety with the upper lateral lobule again divided, the upper member either equal to or smaller than the lower; central protuberance small, truncate, with a circle of granules at the apex; sometimes with 1 or 2 strong granules or spines above the isthmus. L. 132–223 μm. W. 120–208 μm. W. at apex 67–115 μm. I. 21–33 μm. Th. 48.5–51 μm.

Wolle (1892, *l.c.*) shows a plant which probably belongs here, with the lower lobes likewise divided.

DISTRIBUTION: Florida, Massachusetts. Québec. Great Britain, Europe, Asia, Australia, New Zealand, Philippines, Celebes, Samoa, Arctic.

PLATE CXLV, figs. 1, 2.

21. **Micrasterias muricata** (Bail.) Ralfs 1848, Brit. Desm., p. 210 var. **muricata** f. **muricata.**

Euastrum muricatum Bailey 1846, Sillim. Amer. Jour. Sci. & Arts n.s., 1: 126, Fig. 1, 2.

Cells of medium size, about 1.5 times longer than broad, isthmus opening widely from a narrow interior; polar lobe broad and rectangular in basal half, widely diverging at each end into a long, narrow horizontal process, with a similar accessory process in the same plane, arising asymmetrically, one at the front and the other at the back of the apex, which is slightly convex with a shallow median depression; lateral lobes once divided in the vertical plane, the lobules remote with widely rounded sinus (incision) between them (this situation is often interpreted as being separate lateral lobes, 2 on each side); lobules long and narrow, similar to the apical processes; upper lateral lobule nearly always once divided in the horizontal plane, the lower lobule divided into 3 members in the horizontal plane; the lateral lobules and apical processes making a series of 3 whorls on an elongated semicell body, all processes and lobules little-tapering and 2- or 3-dentate at the extremities; surface of semicell without ornament; semicell in vertical view elliptic, with slightly diverging processes and a straight middle one at the poles; in lateral view oblong, the base slightly swollen. L. 150–224 μm. W. (83) 100–160 (208) μm. W. at apex 78–140 μm. I. 18–32 μm. Zygospore unknown.

A triradiate form has been observed (Pl. CXLVII, Fig. 3).

DISTRIBUTION: Widely distributed in U.S. and Canada.

PLATE CXLV, figs. 3, 6.

Key to the North American Varieties and Forms of *Micrasterias muricata.*

1. Lateral lobules and apical processes horizontally divided.
   2. Lateral lobules not vertically divided.
      3. Area above isthmus smooth.
         4. Cells more than 125 μm long.
            5. Cells elliptic in vertical view.   21. var. *muricata* f. *muricata.*
            5. Cells much swollen in vertical view.   21a. var. *tumida.*
         4. Cells less than 125 μm long.   21b. var. *muricata* f. *minor.*
         3. Area above isthmus with 1 (−3) tubercles or verrucae.
                                                              21a. var. *muricata* f. *basituberculata.*
   2. Lower lateral lobule once divided vertically as well as horizontally.   21c. var. *laevigata.*
1. Lateral lobes and apical processes undivided.   21d. var. *simplex.*

21a. **Micrasterias muricata** var. **muricata** f. **basituberculata** Scott ex Croasdale f. nov.

Micrasterias muricata in Grönblad 1956, Soc. Sci. Fennica, Comm. Biol. 15(12): 26. Pl. 5, Figs. 38, 39; Bourrelly 1966, Int. Rev. Ges. Hydrobiol. 51(1): 96. Pl. 13, Fig. 3; Förster 1972, Nova Hedwigia 23: 543. Pl. 11, Fig. 1.

A form having a large distinct granule, rounded or geminate (sometimes 2 or 3) above the isthmus; semicell in vertical view somewhat tumid, with granule visible, in lateral view rectangular, tapering very slightly from the apex, granule visible. L. 140–220 μm. W. 95–140 μm. I. 23–30 μm. Th. c. 36 μm.

This form is perhaps commoner than the typical. It is possible that the granule has been overlooked in some specimens illustrated without it, but Smith and Scott, both careful observers, have illustrated both conditions.

DISTRIBUTION: Florida, Louisiana, Massachusetts, Mississippi. Ontario, Québec.

PLATE CXLVI, figs. 1–3.

21b. **Micrasterias muricata** var. **muricata** f. **minor** Cushman 1905, Rhodora 7: 254.

A form very much smaller but otherwise quite similar to the typical. L. 110–123 μm. W. 69–84 μm. W. at apex 52–84 μm. I. 15, 16 μm.

DISTRIBUTION: New Hampshire, Wisconsin.

PLATE CXLVII, fig. 2.

21c. **Micrasterias muricata** var. **laevigata** Irénée-Marie 1952, Hydrobiologia 4: 148. Pl. 14, Fig. 9.

Micrasterias muricata "abnormal semicells" in West & West 1896a, Trans. Linn. Soc. London, Bot. II, 5: 239. Pl. 14, Figs. 5, 6.
Micrasterias muricata var. *furcata* Förster 1972, Nova Hedwigia 23: 543. Pl. 11, Fig. 2.

A variety differing in the form of the basal lobe, which is horizontally tripartite as in the typical but whose middle member is broadly rectangular in face view and divided vertically to form 2 processes, equal in length. L. 132–172 μm. W. 90–104 μm. W. at apex 69–80 μm. I. 13–25 μm.

West & West's "abnormal semicells," from the U.S., are at the large end of the size range. Förster's variety differs from Irénée-Marie's form only in having the cells more compact, with the sinuses between the arms sharp-angled, and the extremities furnished with 3 or 4 teeth.

DISTRIBUTION: "U.S.," Mississippi. Québec.

PLATE CXLVI, figs. 4, 5.

21d. **Micrasterias muricata** var. **simplex** Grönblad ex Krieger 1939, Rabenhorst's Kryptogamen-Flora 13(1:2): 74. Pl. 120, Figs. 1, 2.

?*Micrasterias pseudofurcata* Wolle 1892, Desm. U.S., p. 122. Pl. 39, Fig. 4 (not Wolle 1881, Bull. Torrey Bot. Club 8(1): 1. Pl. 6, Fig. 3).
?*Micrasterias pseudofurcata* var. *minor* Wolle 1892, Desm. U.S., p. 122. Pl. 41, Fig. 11.
?*Micrasterias baileyi* Ralfs in Wolle 1892, Desm. U.S., p. 130. Pl. 41, Fig. 6.

A variety in which the accessory apical processes are lacking and the lateral lobules are simple, without division in the horizontal plane. L. 152 μm. W. 125 μm. W. at apex 103 μm. I. 23 μm. Th. 32 μm.

The two Wolle taxa listed above as synonyms are doubtful, so their dimensions and distribution are not included.

Teiling (1957, p. 55), discussing this variety, states: "The differences between this desmid and the related *muricata* and *Nordstedtiana* are in all respects so comprehensive that it must be considered a separate species which may perhaps have its correspondence in *Micr. Baileyi* as it is depicted by Wolle (1884a, Tab. 37:6)."

DISTRIBUTION: Labrador.

PLATE CXLVI, fig. 6.

21e. Micrasterias muricata var. **tumida** West & West 1896a, Trans. Linn. Soc. London, Bot., II, 5: 240. Pl. 14, Fig. 7.

A variety differing by having a protuberance at the base of the semicells, seen best in vertical view, where it is also noticeable that the isthmus is circular, not elliptic, and the outer members of the lower lateral lobule are shorter and bent slightly toward the median line. L. 165–195 μm. W. 115–122 μm. I. 16.5–18 μm.

Irénée-Marie (1938, 233. Pl. 41, Fig. 5), who gives us what seems to be the only face view of this variety, shows without comment an additional distinguishing feature: the markedly cylindrical structure of the lower part of the polar lobe.

DISTRIBUTION: California, "U.S." Québec.

PLATE CXLV, figs. 4, 5.

22. Micrasterias nordstedtiana Wolle 1884, Bull. Torrey Bot. Club 11: 15: 1884a, Desm. U.S., p. 113. Pl. 52, Figs. 3–5 f. **nordstedtiana.**

Cells large, about as broad as long, deeply constricted with widely open sinus; polar lobe elongate with erect, subparallel sides that bear on each side of the median portion a short mammillate or conical projection; upper portion of polar lobe concave or shallowly indented, with angles continued into long, thin, very slightly diverging processes, and at the base of each a similar but shorter horizontal process, asymmetrically disposed, the process of one side projecting toward the front, that of the other toward the back; lateral lobes about halfway divided with diverging lobules; all lobules and processes ending in typically 3, but in some cases 2 or 4, short teeth; semicells in lateral view elongate with sides parallel above a swollen base; the apex showing 3 very short processes; in vertical view spindle-shaped with sharply swollen middle, and pointed processes. L. 130–220 μm. W. 115–195 μm. W. at apex 108–150 μm. I. 14–28 μm. Zygospore unknown.

DISTRIBUTION: Michigan, New Hampshire, New Jersey, Pennsylvania, Washington, Wisconsin. Nova Scotia, Ontario, Québec.

PLATE CXLVII, figs. 1, 4–6.

Key to the North American Forms of *Micrasterias nordstedtiana*

1. Lateral margins of polar lobe with a single process on each side.   22. f. *nordstedtiana.*
1. Lateral margins of polar lobe with a pair of processes on each side.   22a. f. *gemina.*

22a. Micrasterias nordstedtiana f. **gemina** Prescott & Scott 1952, Trans. Amer. Microsc. Soc. 71(3): 242. Pl. 8, Fig. 6.

A form differing from the typical by having a pair of short bifid processes, rather than 1, on the lateral margins of the polar lobe. L. 126–133 μm. W. 124–244 μm. I. 21–24 μm.

DISTRIBUTION: Florida.

PLATE CXLVII, fig. 7.

23. Micrasterias novae-terrae (Cushm.) Krieger 1939, Rabenhorst's Kryptogamen-Flora 13(1:2): 91. Pl. 127, Figs. 6, 7 var. **novae-terrae.**

Micrasterias conferta Lund. var. *novae-terrae* Cushman 1904, Bull. Torrey Bot. Club 31(1): 583. Pl. 26, Fig. 8; Taylor 1935, Pap. Michigan Acad. Sci., Arts & Lettr. 20: 212. Pl. 43, Fig. 3; Pl. 48, Fig. 1.
Micrasterias speciosa Wolle f. in Taylor 1935, (*l.c.*) 215. Pl. 48, Fig. 2.

Cells medium-sized or less, up to 1/3 times longer than broad, but sometimes nearly circular; sinus and all incisions fairly deep and open; polar lobe narrow, lower portion with parallel margins, upper third diverging, its apex showing a shallow notch flanked by paired spines, the angles stout, ending in 2 stout teeth; lateral lobes unequal, the upper one twice divided, the lower one narrower, once divided, all ultimate lobules emarginate or ending in 2 stout teeth; wall densely punctate. L. 93.6–150(200) μm. W. 86.4–116(165) μm. W. at apex 34–57 μm. I. 10–20 μm. "Zygospore spherical with large, slightly curved spines" (unconfirmed, and dimensions not given).

DISTRIBUTION: Alaska, Florida, Massachusetts, Michigan, Minnesota, Wisconsin. Newfoundland, Québec.

PLATE CXXV, figs. 2, 3, 5, 6.

Key to the North American Varieties of *Micrasterias novae-terrae*.

1. Intramarginal spines lacking. 23. var. *novae-terrae*.
1. Intramarginal spines present along principal incisions. 23a. var. *speciosa*

23a. **Micrasterias novae-terrae** var. **speciosa** (Wolle) Krieger & Bourrelly 1956, Ergeb. deutsch. limnol. Venezuela-Expedition 1952, 1: 152. Pl. 6, Figs. 56–69.

Micrasterias speciosa Wolle 1885, Bull. Torrey Bot. Club 12: 4; 1887, Freshwater Algae. U.S., p. 38. Pl. 56, Figs. 1, 2.
Micrasterias speciosa Wolle forma West & West 1896a, Trans. Linn. Soc. London, Bot., II, 5: 240. Pl. 14, Fig. 10 (not 11).
Micrasterias papillifera Bréb. var. *speciosa* (Wolle) Krieger 1939, Rabenhorst's Kryptogamen-Flora 13(1:2): 90. Pl. 130, Fig. 3.
Micrasterias novae-terrae var. *spinosa* Prescott & Scott 1952, Trans. Amer. Microsc. Soc. 71(3): 242. Pl. 2, Figs. 1, 2.

A variety differing from the typical by having stout intramarginal spines along the sinus and major incisions, as well as a few, sometimes, on the surface; the angles of the polar lobe may be more horizontally extended and are tipped with 2 (rarely 3) stout teeth; semicells in apical view spindle-shaped with a slight median swelling, pointed poles, and marginal spines. L. 90–195 μm. W. 84–172 μm. W. at apex 41.4–58 μm. I. 10–35 μm. Th. c. 30 μm. Zygospore (from Venezuela) spherical with stout processes trifurcate at their tips; diameter of zygospore 80–85 μm, including processes 140–150 μm.

DISTRIBUTION: Connecticut, Florida, Massachusetts, Montana, Nebraska, New Jersey, North Carolina, Utah, Wisconsin. Newfoundland, Québec. Europe, Russia, South America, Swedish Lappland, Arctic.

PLATE CXXV, fig. 7; PLATE CXXVI, figs. 1, 2, 4.

24. **Micrasterias oscitans** Ralfs 1848, Brit. Desm., p. 76. Pl. X, Fig. 2 var. **oscitans.**

Micrasterias mucronata (Dixon) Rabenh. f. *intermedia* Nordstedt 1873, Acta Univ. Lund. 9: 6.

Cells of medium size, usually slightly longer than broad; semicells 3-lobed, polar lobe wide, fusiform, with convex apex, tapering strongly to ends bearing single or paired spines; lateral lobes very broad at the base, horizontal, conical, with dorsal margin somewhat more convex, tapering strongly to bidentate ends; subpolar incision shallow and open; sinus linear within, then opening broadly; chloroplast with numerous irregular longitudinal lamellae and relatively few pyrenoids; semicells in lateral view elliptic-pyramidal with rounded ends, in vertical view elliptic-fusiform with acuminate (or rarely bifid) ends; ratio of width to thickness about 3 to 1. L. 125–163 μm. W. 100–148 μm. W. at apex 75–95 μm. I. 23–30 μm. Th. 44–52 μm. Zygospore unknown.

DISTRIBUTION: Florida, Maine, Massachusetts, Michigan, Mississippi, New Jersey, North Carolina, Rhode Island, Utah. British Columbia, Newfoundland, Ontario. Great Britain, Europe, Asia, Africa, East Indies.

PLATE XCIII, figs. 3–5, 8.

Key to the North American Varieties of *Micrasterias oscitans*

1. Lateral lobes conical, evenly tapered; apex convex; semicell with ratio of width to thickness c. 3–1.   24. var. *oscitans*.
1. Lateral lobes more convex on upper surface; margin often angled; apex flattened; semicell with ratio of width to thickness c. 2–1.   24a. var. *mucronata*.

24a. **Micrasterias oscitans** var. **mucronata** (Dixon) Wille 1880, Christiana Vid.-Selsk. Forh. 1880(11): 21. Pl. 1, Fig. 3.

Micrasterias mucronata (Dixon) Rabenhorst 1868, Flora Europ. Algar. 3: 187. Textfig. 68b on p. 109.

Cells proportionally longer and thicker than typical; polar lobe flattened, or barely convex, the terminal teeth usually directed downwards; lateral lobe larger, its dorsal margin much more convex, often angled and bearing a tooth on the angle; subpolar incision shallower; ratio of width to thickness of semicell about 2 to 1. L. 125–160 μm. W. 102–140 μm. W. at apex 74–87 μm. I. 23–33 μm. Th. 58–64 μm.

DISTRIBUTION: Florida, Louisiana, Michigan, New York, Rhode Island. Great Britain, Europe, Asia, Faeroes, East Indies.

PLATE XCIII, figs. 6, 7; PLATE XCVII, fig. 6.

25. **Micrasterias papillifera** Bréb. ex Ralfs 1848, Brit. Desm., p. 72. Pl. IX, Fig. 1 var. **papillifera** f. **papillifera**.

Micrasterias papillifera f. *major* West & West 1905a, Monogr. II: p. 92. Pl. 44, Fig. 3.
Micrasterias papillifera var. *varvicensis* Turner 1893, Naturalist 18: 345. Fig. 13.
Micrasterias papillifera var. *verrucosa* Schmidle 1896, Österr. Bot. Zeit. 46: 23. Textfig. on p. 23.
Micrasterias sol (Ehrenb.) Kütz. var. *ornata* Nordst., in Taft 1934, Trans. Amer. Microsc. Soc. 53(2): 96. Pl. 6, Fig. 24.

Cells of medium size, usually a little longer than broad, moderately deeply constricted; sinus and major incisions closed except at ends; polar lobe cuneate with slightly concave sides, apex concave with slight median notch; lateral lobes of first order equal, each again twice divided, the ultimate lobules bidentate or emarginate; surface with an intramarginal row of spines along sinus and major incisions; semicells in lateral view slightly swollen at base and pointed at pole, in vertical view fusiform with a slight median swelling and pointed ends. L. 95–170 (200) μm. W. 95–170 μm. W. at apex 33–51 μm. I. 13–30 μm. Th. 19–39 μm. Zygospore compressed, globose in front view, broadly oval in lateral view, with numerous strong spines, simple or slightly furcate at apex, 76–95 μm x 57–75 μm.

DISTRIBUTION: Widely distributed in U.S. and Canada. Great Britain, Europe, Asia, New Zealand, Africa, South America, Arctic.

PLATE CXXVI, figs. 3, 5, 6.

Key to the North American Varieties and Forms of *Micrasterias papillifera*

1. Intramarginal spines present along the principal incisions.

2. Extremities of lobules shallowly bidentate or excavate.
   3. Dorsal margin of polar lobe bearing a simple spine on either side of the apical
      incision.   25. var. *papillifera* f. *papillifera*.
   3. Dorsal margin of polar lobe bearing a short, sharp-pointed process on either side of
      the apical incision.   25b. var. *papillifera* f. *novae-scotiae*.
2. Extremities of lobules with 2 long, stout, somewhat incurved spines.
                                                 25a. var. *papillifera* f. *mucronata*.
1. Intramarginal spines lacking.
   4. Polar lobe cuneate with curved margins as in the type.   25c. var. *glabra*.
   4. Polar lobe more rectangular and less tapered at base, with nearly straight margins; all
      incisions less deep.   25d. var. *rousseauiana*.

25a. Micrasterias papillifera var. **papillifera** f. **mucronata** Dick 1930, Mitt. Pfälz. ver. Naturkde., Pollichia 3: 33. Pl. 6, Fig. 2.

Incisions deeper than in the typical; lateral lobes of the last order with two long, stout, mostly somewhat incurved spines. L. 150 μm. W. 144 μm. W. at apex 38 μm. I. 15 μm.
     DISTRIBUTION: Massachusetts. Germany.
     PLATE CXXVII, fig. 1.

25b. Micrasterias papillifera var. **papillifera** f. **novae-scotiae** (Turn.) Croasdale stat. nov.

Micrasterias papillifera var. *novae-scotiae* Turner 1885, Jour. Roy. Microsc. Soc., II, 5(6): 937.
   Pl. 15, Fig. 16.

A form in which the spines or teeth on the polar lobe, on either side of the median notch, are replaced by short, sharp-pointed processes. L. 152 μm. W. 133 μm.
     We agree with Krieger (1939, p. 90) that this is possibly only a teratological form.
     DISTRIBUTION: Nova Scotia (only one record, by Turner).
     PLATE CXXVII, fig. 2.

25c. Micrasterias papillifera var. **glabra** Nordst., in Wittrock & Nordstedt 1882, Alg. exs., Nr. 466, Bot. Notiser 1882: 53 et Wittrock & Nordstedt 1889, Fasc. 21, p. 35.

A smooth-walled variety without the intramarginal rows of spines along the sinus and major incisions. L. 84–160 μm. W. 84–135 μm. W. at apex c. 41 μm. I. 15–23 μm.
     DISTRIBUTION: Connecticut, Iowa, Massachusetts, Michigan, Oklahoma, South Carolina, Washington. British Columbia, Newfoundland, Québec. Great Britain, Europe, Asia, Africa, South America, Arctic.
     PLATE CXXVII, figs. 3, 4.

25d. Micrasterias papillifera var. **rousseauiana** Irénée-Marie 1949a, Nat. Canadien 76(8/10): 290. Pl. 6, Fig. 12.

A variety with broadly oval cells, 1 1/4 times longer than broad, differing from the typical in having the polar lobes more rectangular, less tapered at the base, less retuse at the apex, rarely nearly straight; polar lobe with two very small spines near the outer extremities; wall smooth; sinus typical; all incisions relatively shallow and linear. L. 147 μm. W. 122 μm. I. 20 μm.
     DISTRIBUTION: Québec.
     PLATE CXXVII, fig. 5.

26. **Micrasterias pinnatifida** (Kütz.) Ralfs 1848, Brit. Desm., p. 77. Pl. X, Fig. 3 var. **pinnatifida** f. **pinnatifida**.

Micrasterias oscitans Ralfs var. *pinnatifida* (Kütz.) Rabenhorst 1868, Flor. Europ. Algar. 3: 189.

Cells small, mostly somewhat broader than long, deeply constricted, sinus opening widely from a sharp interior; lateral lobes single, undivided, conical, horizontally extended, somewhat constricted just before the bifid extremity; polar lobe widely horizontally spreading, its apex flat or slightly convex or retuse; its extensions smaller than the lateral lobes, bifid at the extremities; subpolar incision deep and semicircular within; cells in vertical view fusiform with acuminate poles, in lateral view narrowly pyramidal, constricted below the rounded apex; chloroplast with erect curved lamellae and few pyrenoids. L. 40–80 μm. W. 37–84 μm. W. at apex 31–59 μm. I. 8–20 μm. Th. 15–20 μm. Zygospore globose, furnished with long, stout, acute spines, each arising from a blunt protuberance; diameter 44 μm., including spines 67 μm.

Plants of this species are exceedingly variable; rarely is a specimen seen with all 4 lateral lobes alike; the varieties intergrade with each other, with many dichotypical forms. It seems better to reduce most of the named varieties to forms, and possibly these are merely variants of the type. (For cases of polymorphism in this species, see the recent paper by Sormus & Bicudo 1974).

DISTRIBUTION: Widely distributed in U.S. and Canada. Great Britain, Europe, Asia, Australia, New Zealand, Africa, South America, Faeroes, Cuba.

PLATE LXXXIX, figs. 6–8; PLATE XC, figs. 1–8.

Key to the North American Varieties and Forms of *Micrasterias pinnatifida*

1. Semicells compressed in vertical view.
  2. Lateral lobes undivided, ending in 1, 2, or 3 teeth.
    3. Lateral lobes ending in 2 teeth.
      4. Lateral lobes not much stouter than extensions of polar lobe.
        5. Subpolar incision more or less semicircular; extensions of polar lobe stiffly horizontal, their terminal teeth not directed upwards.
                                        26. var. *pinnatifida* f. *pinnatifida*.
        5. Subpolar incision quadrate; extensions of polar lobe sinuously horizontal or diverging.
          6. Extensions of polar lobe straight, directed somewhat upwards.
                                        26a. var. *pinnatifida* f. *angusta*.
          6. Extensions of polar lobe sinuous, horizontal, their teeth directed upwards.
                                        26d. var. *pinnatifida* f. *quadrata*.
      4. Lateral lobes much stouter than extensions of polar lobe.
        7. Extremities of lateral lobes not or barely narrowed, broader than extensions of polar lobe.   26c. var. *pinnatifida* f. *inflata* p.p.
        7. Extremities of lateral lobes narrowed, similar to extensions of polar lobe.
                                        26g. var. *pseudoscitans*.
    3. Lateral lobes ending in 3 teeth.
      8. Lateral lobes much thicker than subpolar incision.   26c. var. *pinnatifida* f. *inflata* p.p.
      8. Lateral lobes not thicker than subpolar incision.   26e. var. *pinnatifida* f. *tridentata*.
  2. Lateral lobes forked at extremities, each division bidentate.
                                        26b. var. *pinnatifida* f. *furcata*.
1. Semicells triangular in vertical view.   26f. var. *pinnatifida* f. *trigona*.

26a. **Micrasterias pinnatifida** var. **pinnatifida** f. **angusta** Prescott & Scott 1952, Trans. Amer. Microsc. Soc. 71(3): 246. Pl. 7, Fig. 7.

A form differing from the typical in having the sinus linear and narrowly closed inwardly, then opening abruptly and forming the lower margins of the lateral lobes which are upwardly directed and slightly sigmoid, bifid at their extremities; the upper margins of the lateral lobes nearly straight and horizontal; the extensions of the polar lobe straight and somewhat upwardly directed, bifid at their extremities. L. 69 μm. W. 67 μm. I. 10 μm.

DISTRIBUTION: Florida, Louisiana, Mississippi.

PLATE XCI, fig. 1.

26b. Micrasterias pinnatifida var. pinnatifida f. furcata (Krieg.) Croasdale stat. nov.

Micrasterias pinnatifida var. *furcata* Krieger 1939, Rabenhorst's Kryptogamen-Flora 13(1:2): 18. Pl. 99, Figs. 11, 12.

Micrasterias pinnatifida var. *divisa* West 1891, Jour. Bot. 29: 354. Pl. 315, Fig. 8; Irénée-Marie 1949, Nat. Canadien 76(1/2): 25. Pl. 2, Figs. 5, 6 (not Pl. 1, Figs. 7, 8).

A form differing from the typical in that the lateral lobes are broader at the base and more or less deeply forked, the ends of the forks being two-toothed. L. 65–74 μm. W. 66–80 μm. W. at apex 38–51 μm. I. 14–15.5 μm.

No illustration of this form shows all four lateral lobes equally bifurcate. West's original figure on which the variety is based (West 1891, Pl. 315, Fig. 8) is less regular than Krieger has redrawn it (Krieger 1939, Pl. 99, Fig. 11). American figures show great diversity in degree of bifurcation. The only reasonably dependable records are from Maine (West 1891) and Québec (Irénée-Marie, 1949).

DISTRIBUTION: Maine. Québec.

PLATE XCI, figs. 2–5.

26c. Micrasterias pinnatifida var. pinnatifida f. inflata (Wolle) Croasdale stat. nov.

Micrasterias oscitans Ralfs var. *inflata* Wolle 1876, Bull. Torrey Bot. Club 6: 122.

Micrasterias pinnatifida var. *inflata* Wolle 1881, Bull. Torrey Bot. Club 8: 1. Pl. 6, Fig. 5.

Micrasterias pinnatifida var. *inflata* f. *ornata* Irénée-Marie 1938, Flore Desm. Montréal, p. 220. Pl. 33, Fig. 14.

Lateral lobes almost rectangular, with widely separated terminal teeth; extensions of polar lobe rather short and stout. L. 55–69 μm. W. 53–71 μm. W. at apex 40–52 μm. I. 8–12 μm.

This is an exceedingly variable form, and seems to grade by dichotypical forms into the typical and into f. *tridentata*. There may be an extra spine on 1 or more of the lateral lobes; the subpolar incision may be widely open or nearly closed. Because many of the specimens might as well be considered as belonging to the type or to f. *tridentata*, it is difficult to be definite about the distribution.

DISTRIBUTION: Louisiana, Maine, Michigan, New Hampshire, Oregon, Pennsylvania, Vermont. British Columbia, Québec. Europe, Asia, Africa.

PLATE XCII, figs. 1–5.

26d. Micrasterias pinnatifida var. pinnatifida f. quadrata (Bail.) Croasdale comb. nov.

Micrasterias quadrata Bailey 1851, Smithson. Contrib. Knowledge 2 (Art. 8): 37. Pl. 1, Fig. 5.

A large form with the lateral extensions of the polar lobe slender and extending horizontally to nearly the breadth of the lateral lobes, the upper terminal tooth directed upward, the apex somewhat depressed; lateral lobes greatly extended horizontally and sometimes rather abruptly tapered from a broad base. L. 81–86 μm. W. 100–102 μm. W. at apex 71–80 μm. I. 13–15 μm.

Krieger (1939, p. 20), dismissing Bailey's drawing as "incomplete," takes as type a figure of Borge (1918, p. 66) described by Borge as a "forma" and differing from Bailey's in its generally stouter aspect. We illustrate both.

DISTRIBUTION: Florida, Georgia, Minnesota, New York, South Carolina. Europe, Brazil.

PLATE XCI, figs, 6, 7.

26e. **Micrasterias pinnatifida** var. **pinnatifida** f. **tridentata** (Krieg.) Croasdale stat. nov.

Micrasterias pinnatifida var. *tridentata* Krieger 1939, Rabenhorst's Kryptogamen-Flora 13(1:2): 20. Pl. 100, Fig. 8, 9.

A form whose lateral lobes bear a supplementary third tooth. L. 45–65 μm. W. 43–70 μm. I. 14 μm.

As Krieger (1939, p. 21) himself suggests, this is probably only a teratological condition. The presence of the third tooth seems to be an inadequate character on which to establish a variety, since it is encountered in cells of various shapes, and is generally not constant on the four lateral lobes of any one specimen. West's form, however, (West 1893, Pl. 20, Fig. 9), on which the variety is based, is unlike any other form we have seen of *M. pinnatifida*, and is consistent in this feature on all 4 lateral lobes.

DISTRIBUTION: Florida, Idaho, Oregon, Washington, Wisconsin. British Columbia, Great Britain, Europe.

PLATE XCII, fig. 6.

26f. **Micrasterias pinnatifida** var. **pinnatifida** f. **trigona** (West) Croasdale stat. nov.

Micrasterias pinnatifida var. *trigona* West 1889a, Jour. Bot. 27: 206. Pl. 291, Fig. 15.

A triradiate form, 3-angled in vertical view. Reported only once in North America. L. c. 60 μm. W. c. 75 μm. W. at apex c. 51 μm. I. c. 27 μm.

It seems more appropriate to consider a trigonal expression as a form rather than a variety.

DISTRIBUTION: Maine.

PLATE XCII, figs. 8–11.

26g. **Micrasterias pinnatifida** var. **pseudoscitans** Grönblad 1920, Acta Soc. Fauna Flora Fennica 47(4): 36. Pl. 6, Figs. 7, 8.

?Micrasterias pinnatifida f. *rhomboidea* Brunel 1938, Nat. Canadien 65: 71. Fig. 1b.

A variety with the lateral lobes biconvex, somewhat inflated; semicells broader in apical view; cells in outline somewhat similar to *M. oscitans* Ralfs but much smaller and with less convex apical lobe. L. 54–70 μm. W. 53–80 μm. W. at apex 42–55 μm. I. 10–16.5 μm.

Brunel (Litt. 1961) changed the name of his plant from f. *rhomboidea* to var. *pseudoscitans* Grönblad. Nevertheless, Brunel's plant is quite different from Grönblad's in its nearly closed sinus and more compressed polar lobe. Perhaps his name f. *rhomboidea* should be retained.

DISTRIBUTION: Alaska, Colorado, Massachusetts, Minnesota, Montana, Oklahoma, Oregon, Wisconsin. Québec. Europe, Asia, South America, Arctic, Indonesia.

PLATE XCIII, figs. 1, 2.

27. **Micrasterias piquata** Salisbury 1936, Ohio Jour. Sci. 36: 59. Pl. 1, Figs. 3–5 var. **piquata** f. **piquata**.

Cells medium-sized, 1 1/4 to 1 1/2 times as long as broad, subrectangular with broadly truncate polar lobe and rectangular undivided lateral lobes; all angles terminating in 2 (rarely 1) short teeth (which may be reduced to mere swellings on the margins, or even be absent); subpolar incision small with rounded base, sometimes partially closed by teeth at the exterior; semicell in vertical view broadly oval, in lateral view rounded-pyramidal. L. 100–143 μm. W. including teeth 83–124 μm. I. 17–30 μm. Th. c. 37 μm. Zygospore unknown.

Scott found this plant frequently in the southeastern United States, and from his unpublished figures we have expanded Salisbury's original description.

DISTRIBUTION: Florida, Georgia, Louisiana, Mississippi.

PLATE CV, figs. 5–7.

Key to the North American Varieties and Forms of *Micrasterias piquata*

1. Lateral lobes subrectangular, not significantly broader at base.
  2. Subpolar incision with walls parallel.   27. var. *piquata* f. *piquata.*
  2. Subpolar incision open widely, its lower margin horizontal.
                                                                27a. var. *piquata* f. *picayunensis.*
1. Later lobes trapeziform, broadest at base.   27b., var. *lata.*

27a. **Micrasterias piquata** var. **piquata** f. **picayunensis** Prescott & Scott 1943, Pap. Michigan Acad. Sci., Arts & Lettr. 28: 76. Pl. 1, Fig. 3.

A somewhat larger form, the polar lobe strongly convex or slightly flattened, not at all emarginate, higher than in the typical, separated from the lateral lobes by a wider incision, the dorsal margins of the lateral lobes being almost horizontal; lateral lobe undivided, its angles not or barely extended, bidentate; sinus usually opening sooner and more widely. L. 128–151 μm. W. 105–123 μm. I. 25–30 μm.

One of Scott's specimens (Pl. CV, Fig. 7) seems to be intermediate between this form and the typical.

DISTRIBUTION: Louisiana, Mississippi.

PLATE CV, fig. 8.

27b. **Micrasterias piquata** var. **lata** Prescott & Scott 1952, Trans. Amer. Microsc. Soc. 71: 246. Pl. 5, Fig. 9.

A variety differing by having the lower angles of the lateral lobes extended so that the semicell is broader at the base; dorsal margin of polar lobe broadly convex; lateral lobes undivided, their upper angles more extended than in the typical; subpolar incision deeper and narrow but enlarged internally; the median sinus closed for the inner half, then abruptly opening. L. 112–116 μm. W. 92–108 μm. W. at apex 72–75 μm. I. 22–24 μm.

DISTRIBUTION: Florida.

PLATE CV, fig. 9.

28. **Micrasterias radians** Turner 1892, Kongl. Svenska Vet.-Akad. Handl. 25(5): 91. Pl. 5, Fig. 6.

Cells of medium size, usually slightly longer than broad; lateral lobes evenly twice divided, lobules sometimes slightly swollen at base, terminating in paired teeth; polar lobe slightly exserted, the lower margins parallel, the upper part diverging, the apex deeply concave between the 2 short bidentate processes; all incisions rather widely open. L. 100–184 μm. W. 82–144 μm. W. at apex 32–67 μm. I. 8–21 μm. Th. c. 25 μm. Zygospore unknown.

This species is very similar to *M. crux-melitensis* (Ehrenb.) Ralfs, distinguished by the exserted polar lobe (Turner, *l.c.*) and deeper incisions (Krieger 1939, p. 67), as well as by having the apex more deeply concave and the apical teeth sometimes curved. We show a connecting form from Florida (Pl. CXIX, Fig. 2). Grönblad, Prowse & Scott (1958, p. 19) and Thomasson (1960, p. 22) both call attention to the close relationship between the 2 species, and it would probably be advisable to reduce *M. radians* to a variety of *M. crux-melitensis*, a variety that might lead over into *M. radiata* Hass.

DISTRIBUTION: Florida, Montana, Virginia. SE Asia, Africa, Madagascar, South America.

PLATE CXV, figs. 5-7.

29. **Micrasterias radiata** Hassall 1845, Hist. British Freshw. Algae 1, p. 386, Pl. 90, Fig. 2 var. radiata f. radiata.

Micrasterias radiata f. *deflexa* Irénée-Marie 1951, Nat. Canadien 78: 188. Pl. 2, Figs. 2-4.
Micrasterias radiata var. *simplex* (Wolle) Smith 1921, Trans. Wisconsin Acad. Sci., Arts & Lettr. 20: 344. Pl. 9, Figs. 1-3.
Micrasterias furcata Ag. ex Ralfs 1848, Brit. Desm. p. 73. Pl. IX, Fig. 2.
Micrasterias furcata var. *decurta* Turner 1885, Jour. Roy. Microsc. Soc., II, 5(6): 936. Pl. 16, Fig. 10.
Micrasterias furcata var. *simplex* Wolle 1885a, Bull. Torrey Bot. Club 12: 128. Pl. 51, Figs. 6, 7.
Micrasterias pseudofurcata Wolle 1881, Bull. Torrey Bot. Club 8: 1. Pl. 6, Fig. 3.

Cells of medium size, usually slightly longer than broad, sinus deep and open; lateral lobes sometimes entire but usually once or twice divided, the degree of division often varying within the same cell or semicell; lateral lobes or lobules usually slender, sometimes swollen at base, slightly tapering, deeply bidentate at tip, the teeth usually spreading; polar lobe normally exserted, with slender, parallel-walled erect part and widely divergent slender extensions, strongly bidentate at tip; all incisions widely open; cells in vertical and lateral views fusiform with pointed ends and swollen basal region; chloroplasts with irregularly disposed lamellae; pyrenoids chiefly in the midregion. L. 126-250 μm. W. 102-208 μm. W. at apex 48-94 μm, Max. W. polar lobe 60-98 μm. I. 12-30 μm. Th. 27-40 μm. Zygospore unknown.

This is a common species in North America, especially in the warmer regions. It is exceedingly variable, especially in the degree of division of the lateral lobes, and many forms and varieties have been described and named. For the sake of distribution records we are keeping separate the 10 varieties and forms currently accepted for North America, but believe they could probably all be reduced to 2 forms, a compact form, such as var. *pseudocrux* Grönbl. and an extenuated form such as var. *gracillima* Smith.

DISTRIBUTION: Widely distributed in the U.S. and Canada. On all continents.

PLATE CXVI, figs. 3-8.

Key to the North American Varieties and Forms of *Micrasterias radiata*

1. Lateral lobes at least on 1 semicell once or twice divided (when only once divided the lateral lobes considerably longer than the extensions of the polar lobe).
  2. All lobes and lobules slender; erect portion of polar lobe longer than broad; lateral lobules longer than the breadth of lateral lobe at base.
   3. Lateral lobes not tapered sharply from swollen base.
    4. Lobules and polar lobe extensions not exceptionally long and slender (length of latter less than 9 times its breadth).

5. Extensions of polar lobe diverging.
   6. Upper margin of lateral lobe inflated at base, forming sharp angle with polar
      lobe.   29b. var. *alata.*
   6. Upper margin of lateral lobe not strongly inflated at base.
                                                         29. var. *radiata* f. *radiata.*
5. Extensions of polar lobe nearly horizontal, paralleling lateral lobules.
                                                   29a. var. *radiata* f. *parallela* p.p.
4. Lobules and extrusions of polar lobe very long and slender.
   7. Lateral lobes mostly twice divided.   29c. var. *dichotoma.*
   7. Lateral lobes once divided (var. *gracillima*).
      8. Cells circular or longer than broad; upper lateral lobules usually straight.
         9. Undivided portion of lateral lobes more than 1/2 total length; margins of
            erect portion of polar lobe not converging.   29d. var. *gracillima* f. *gracillima.*
         9. Undivided portion of lateral lobe less than 1/2 total length; margins of
            erect portion of polar lobe converging.   29e. var. *gracillima* f. *brevis.*
      8. Cells broader than long; upper lateral lobules usually curving upward.
                                                         29f. var. *gracillima* f. *laxata.*
   3. Lateral lobes strongly swollen at base, and thereafter tapering sharply.
                                                         29g. var. *lagerheimii.*
2. Lobes and lobules short; erect portion of polar lobe shorter than broad; lateral
   lobules shorter than breadth of lateral lobe at base.   29h. var. *pseudocrux.*
1. Lateral lobes completely undivided, barely exceeding in length the extensions of the
   polar lobe.
   10. Lateral lobules and lateral extensions of polar lobe horizontal.
                                                   29a. var. *radiata* f. *parallela* p.p.
   10. Lateral lobes diverging.   29i. var. *unica.*

29a. **Micrasterias radiata** var. **radiata** f. **parallela** Prescott & Scott 1952, Trans.
Amer. Microsc. Soc. 71: 248. Pl. 7, Figs. 1–3.

A form in which the lateral lobes and the extensions of the polar lobe are
extended horizontally, or more nearly so than in the typical; lateral lobes variable,
either simple or once divided, L. 88–118 μm. W. 122–142 μm. I. 14–15 μm.

The polar lobe is formed and its extensions directed in such a way as to
suggest *M. arcuata* var. *gracilis* West & West. In the latter, however, the extensions
are relatively longer and more slender, curve downward, and are not bifid at the
tips. Whelden (1941, Pl. 6, Fig. 7) figures forms of *M. pinnatifida* (Kütz.) Ralfs
from Florida which are quite similar to this form although the size is smaller than
ours.
DISTRIBUTION: Florida.
PLATE CXVII, figs. 1, 2.

29b. **Micrasterias radiata** var. **alata** Prescott & Scott 1952, Trans. Amer. Microsc.
Soc. 71: 247. Pl. 3, Fig. 4.

A variety differing from the typical variety in its relatively longer lateral
lobules and polar lobe extensions, all bidentate with long teeth; differing also in
the upper lateral lobules, which are swollen on their upper basal margins somewhat in the manner of *M. alata* Wall. L. 184–248 μm. W. 184–246 μm. Max. W.
polar lobe 119–154 μm. I. 21–23 μm.
DISTRIBUTION: Florida, Virginia.
PLATE CXVII, fig. 3.

29c. **Micrasterias radiata** var. **dichotoma** (Wolle) Cushman 1908, Rhodora 10:
108.

Micrasterias dichotoma Wolle 1884a, Desm. U.S., p. 111, Pl. 52, Fig. 2.

A variety with the lateral lobes twice divided; all parts more slender; erect part of polar lobe with parallel margins. L. 160–250 μm. W. 142–211 μm. W. polar lobe 85–125 μm. I. 13–23 μm.

Intermediate and dichotypical forms connect this variety with the typical (e.g. Pl. CXVIII, Fig. 4) and with var. *lagerheimii* Krieg. (Pl. CXVIII, Fig. 6).

DISTRIBUTION: Florida, Louisiana, Massachusetts, Minnesota, Mississippi, New Jersey, North Carolina, Oregon, Washington. Newfoundland, Nova Scotia, Ontario, Québec. Europe, Asia, Africa, South America, Central America, Arctic.

PLATE CXVII, figs. 4, 5.

29d. **Micrasterias radiata** var. **gracillima** Smith 1921, Trans. Wisconsin Acad. Sci. 20: 344. Pl. 9, Figs. 6–11 f. **gracillima**.

Micrasterias radiata Hass., in Brown 1930, Trans. Amer. Microsc. Soc. 49: 114. Pl. 11, Fig. 12.

A slender variety with the lateral lobes only once, but deeply divided; erect portion of polar lobe long and slender, with parallel margins, its length nearly twice the breadth of the lateral lobes; extensions of polar lobe slender and widely divergent, angle between polar and lateral lobes nearly 90°. L. 135–190 μm. W. 130–170 μm. W. at apex 65–118 μm. I. 15–18 μm.

DISTRIBUTION: Alaska, Florida, Georgia, Louisiana, Massachusetts, Michigan, New York, North Carolina, Virginia, Wisconsin. Labrador, Newfoundland, Nova Scotia, Ontario, Québec. South America.

PLATE CXVII, fig. 6; PLATE CXVIII, fig. 1.

29e. **Micrasterias radiata** var. **gracillima** f. **brevis** Prescott & Scott 1952, Trans. Amer. Microsc. Soc. 71: 247. Pl. 3, Fig. 3.

A variety differing from the typical by having the undivided part of the basal lobes much shorter, the lobules correspondingly longer, and symmetrically tapering; the margins of the erect portion of the polar lobe slightly converging rather than subparallel. L. 181 μm. W. 179 μm. I. 19 μm.

DISTRIBUTION: Florida.

PLATE CXVIII, fig. 5.

29f. **Micrasterias radiata** var. **gracillima** f. **laxata** Prescott & Scott 1952, Trans. Amer. Microsc. Soc. 71: 248. Pl. 1, Fig. 2.

A form differing from the typical variety by having the lobules of the lateral lobes and the extensions of the polar lobe much more extended, the upper lateral lobules curving toward the poles; the cells decidedly broader than long. L. 189–210 μm. W. 226–239 μm. W. at apex 166–181 μm. I. 16–18 μm.

DISTRIBUTION: Florida.

PLATE CXVIII, fig. 2.

29g. **Micrasterias radiata** var. **lagerheimii** Krieger 1939, Rabenhorst's Kryptogamen-Flora 13(1:2): 71. Pl. 118, Fig. 2.

Micrasterias dichotoma Wolle, in Lagerheim 1885, Öfv. Kongl. Vet.-Akad. Förhandl. 42(7): 231. Pl. 27, Fig. 4.

A variety with the lateral lobes swollen at the base, then twice divided into very slender lobules; erect portion of polar lobe slender with parallel margins, its upper angles extended into long, very narrow, diverging processes; polar and lateral lobes forming a right angle. L. 220 μm. W. 165 μm. W. at apex 111 μm. I. 18 μm.

This variety is very close to varieties *dichotoma* and *gracillima*; from the former it differs in its wider angle between the lateral and polar lobes, from the latter in the additional division of the lateral lobes, and from both in the swollen basal part of the lateral lobes. All the characters are variable, and these three varieties probably should be united.

DISTRIBUTION: Florida, Massachusetts (as *M. dichotoma* Wolle), Virginia.
PLATE CXVIII, fig. 3.

29h. Micrasterias radiata var. pseudocrux Grönblad 1920, Acta Soc. Fauna, Flora Fennica 47(4): 37. Pl. 6, Figs. 12–14.

Cells smaller, less deeply incised; polar lobe excluding processes scarcely less broad than long; cells thicker in vertical view. L. 137–142 μm. W. 124–125 μm. W. at apex 57–61 μm. I. 21 μm. Th. 42 μm.

In its shorter, stouter processes this variety forms a link with *M. crux-melitensis* (Ehrenb.) Hass., but differs in the longer, narrower extensions of the polar lobe, the rather abrupt narrowing of the lateral lobules, and the thicker vertical view with more tapered poles.

DISTRIBUTION: Michigan (forma), Montana (forma). Finland.
PLATE CXIX, figs. 3, 4.

29i. Micrasterias radiata var. unica Prescott & Scott 1952, Trans. Amer. Microsc. Soc. 71: 248. Pl. 1, Fig. 1.

A variety in which all the basal lobes of the semicell are extended into long, simple, undivided processes, and the extensions of the polar lobe are almost as long and nearly horizontal. L. 128–216 μm. W. 151–242 μm. W. at apex 122–193 μm. I. 13–16 μm.

The extreme expressions of this variety are very striking, but there are less definite connecting forms (Pl. CXIX, Fig. 5).

DISTRIBUTION: Florida.
PLATE CXIX, figs. 5, 6.

30. Micrasterias radiosa Ralfs 1848, Brit. Desm., p. 72. Pl. 8, Fig. 3 var. **radiosa** f. **radiosa**.

Micrasterias sol (Ehrenb.) Kütz. 1849, Spec. Alg. p. 171; West & West 1905a, Monogr. II, p. 95. Pl. 46, Figs. 1, 2.
Micrasterias radiosa var. *aculeata* Krieg. f. Prescott & Scott 1943, Pap. Michigan Acad. Sci., Arts & Lettr. 28: 77 Pl. 6, Fig. 7.
Micrasterias radiosa var. *taylorii* Irénée-Marie 1952, Hydrobiologia 4(1/2): 154. Pl. 15, Fig. 8.

Cells medium-sized, mostly circular; sinus linear, narrow; upper lateral lobes broader than lower, usually divided to the fourth order; lower lateral lobes usually divided to the third order, all incisions narrow and very deep, with margins of lobules often slightly undulate; lobules strap-shaped with furcate-dentate extremities; polar lobe narrow with subparallel sides, expanding at apex, which is sometimes exserted, apical margin retuse-emarginate, usually with 1 or 2 spines on a small swelling on either side of the central indentation; angles of polar lobe typically bearing 2 stout spines. Surface spines lacking; semicell in lateral view linear-oblong, very gradually tapered to a tridentate apex, in vertical view linear-fusiform, with acute poles and a very slightly flattened median protuberance. L. 121–320 μm. W. 120–314 μm. W. at apex 24–50 μm. I. 12–32 μm. Zygospore of type unknown.

There is considerable disagreement in literature as to whether this taxon should be called *M. radiosa* or *M. sol*. We follow Smith (1924, p. 46), who states:

"W. & G. S. West have discussed the nomenclature of this alga and shown that what Ralfs calls *M. radiosa* is not *M. radiosa* (Lyngb.) Agardh but *Micrasterias Sol* (Ehrenb.) Kützing (*Euastrum Sol* Ehrenberg). In spite of the fact that the name *M. Sol* has priority, the ruling that Ralfs' British Desmidieae is to be taken as the starting point for the nomenclature of the Desmids requires the retention of the specific name *radiosa*."

The species is distinguished from other deeply incised circular forms by the deeper incisions, greater subdivision, and undulate margins of the lateral lobes.

DISTRIBUTION: Widely distributed in U.S. and Canada. Great Britain, Europe, Brazil, Arctic.

PLATE CXXVIII, fig. 6; PLATE CXXIX, figs. 2-4.

Key to the North American Varieties and Forms of *Micrasterias radiosa*

1. Basal lobules not exserted beyond other lobules; polar lobe typically not much broader at apex than at base.
  2. Lobules narrow, incisions to fourth (rarely fifth) order.
    3. Incisions moderately deep, nearly closed.
      4. Intramarginal spines lacking along the incisions.
        5. Polar lobe narrow, bearing less than 10 spines at the apex.
          6. Wall not strikingly punctate.   30. var. *radiosa* f. *radiosa*.
          6. Wall strikingly punctate.
            7. Puncta not in concentric rings; cells circular.   30i. var. *punctata* f. *punctata*.
            7. Puncta in concentric rings; cells somewhat angular.
                                      30j. var. *punctata* f. *concentrica*.
        5. Polar lobe broad, bearing 10 spines at the apex.   30a. var. *radiosa* f. *mistassiniensis*.
      4. With intramarginal spines along the incisions.
        8. Intramarginal spines small and sparse, no more than 12 along any incision.
          9. Cells less than 250 μm long.   30f. var. *ornata* f. *ornata*.
          9. Cells more than 250 μm long.   30h. var. *ornata* f. *laurentiana*.
        8. Intramarginal spines large, conspicuous and crowded, more than 12 along major incisions.   30g. var. *ornata* f. *aculeata*.
    3. Incisions very deep, often open.   30b. var. *elegantior*.
  2. Lobules broad; incisions only to third order.
    10. Strong intramarginal spines along major incisions.
      11. Incisions narrower than the lobules.   30c. var. *murrayi* f. *murrayi*.
      11. Incisions mostly broader than lobules.   30d. var. *murrayi* f. *elegantior*.
    10. Wall without intramarginal spines.   30e. var. *murrayi* f. *glabra*.
1. Basal lobules exserted beyond other lobules; polar lobe usually broader at apex than at base.   30k. var. *swaine*.

30a. Micrasterias radiosa var. **radiosa** f. **mistassiniensis** Irénée-Marie 1949a, Nat. Canadien 76: 293. Pl. 4, Fig. 34.

A form with the polar lobe broader, the apical depression deeper; polar lobe bearing 10 spines; three on each outer margin and 2 on either side of the apical depression; surface spines lacking. L. 251-284 μm. W. 228-261 μm. W. at apex 51.5-54 μm. I. 24.5-32 μm.

DISTRIBUTION: Québec.

PLATE CXXIX, fig. 5.

30b. Micrasterias radiosa var. **elegantior** (G. S. West) Croasdale comb. nov.

Micrasterias sol (Ehrenb.) Kütz. var. *ornata* Nordst. f. *elegantior* G. S. West 1914, Mém. Soc. Neuchatel. Sci. Nat. 5: 1035. Pl. 22, Fig. 44.
Micrasterias sol (Ehrenb.) Kütz. var. *elegantior* G. S. West f. *glabra* Bourrelly 1966, Int. Rev. ges. Hydrobiol. 51(II): 98. Pl. 14, Fig. 4.
Micrasterias radiosa f. West & West 1896a, Trans. Linn. Soc. London, Bot., II, 5(2): 240. Pl. 13, Fig. 30.

Micrasterias radiosa var. *ornata* f. *elegantior* (G. S. West) Smith 1924, Wisconsin Geol. & Nat. Hist. Surv., Bull. 57(II): 47. Pl. 60, Fig. 4.

A variety with all the incisions very deep, so that the central part of the semicell is much reduced, and the length of the polar lobe is more than three-quarters of the total length of the semicell; polar lobe and lateral lobules thin, incisions usually open; surface spines usually present along major incisions; spines on either side of the apical notch present or lacking. L. 143–248 µm. W. 136–258 µm. W. at apex 32–39 µm. I. 12–28 µm.

This deeply incised form is very striking, but the presence of intramarginal spines along the major incisions is variable. It seems better, therefore, not to consider this a form of var. *ornata* Nordst.

DISTRIBUTION: Georgia, Louisiana, Michigan, Montana, North Carolina, South Carolina, Virginia, Wisconsin. Newfoundland, Ontario, Québec. Europe, Australia, Africa, South America.

PLATE CXXXII, figs. 2–4.

30c. **Micrasterias radiosa** var. **murrayi** (West & West) Allorge & Allorge 1931, Rev. Algol. 5(3/4): 352. Pl. 7, Fig. 6 f. **murrayi**.

Micrasterias murrayi West & West 1903, Linn. Soc. Jour. Bot. 35: 538. Pl. 15, Figs. 1, 2.
Micrasterias papillifera Bréb. ex Ralfs f. *murrayi* (West & West) Kossinskaja 1960, Flora Plant. Cryptog. URSS 5. Conjugatae 2, Desmidiales 1: 463. Pl. 72, Figs. 8–10.

A variety with the lateral lobes broad, divided usually only to the third order; all incisions more open than the typical; sinus very widely open; rather evident intramarginal spines along the major incisions. L. 116–158 µm. W. 104–157 µm. W. at apex 38–44 µm. I. 12–20 µm. Zygospore spherical with numerous long, thin spines, simple or once-forked at the ends, the hooklets small and short; diam. 58–64 µm, including spines 100–110 µm.

DISTRIBUTION: Florida, Massachusetts, Michigan. British Columbia, Québec. Great Britain, Europe, Africa, Arctic.

PLATE CXXXI, figs. 4, 5; PLATE CXXXII, fig. 1.

30d. **Micrasterias radiosa** var. **murrayi** f. **elegantior** Prescott & Scott 1952, Trans. Amer. Microsc. Soc. 71(3): 249. Pl. 2, Fig. 4.

A form differing from the typical variety by having the lobules more slender and elongated, and the median incision narrow, the sides more nearly parallel. L. 124 µm. W.125 µm. I. 12.5 µm.

DISTRIBUTION: Florida.

PLATE CXXXI, fig. 6.

30e. **Micrasterias radiosa** var. **murrayi** f. **glabra** (Irénée-Marie) Croasdale comb. nov.

Micrasterias murrayi West & West var. *glabra* Irénée-Marie 1949a, Nat. Canadien 76: 289. Pl. 3, Fig. 32.

A variety differing in having the median sinuses less open, the polar lobe slightly exserted, and the wall altogether without intramarginal surface spines. L. 142–161 µm. W. 122–148 µm. W. at apex 35.4–40 µm. I. 17.7–26 µm.

The wall is described by Irénée-Marie (*l.c.*) as "lightly granular."

DISTRIBUTION: Québec.

PLATE CXXXI, fig. 7.

30f. Micrasterias radiosa var. ornata Nordst. 1870, Vidensk. Medd. Naturh. Foren. Kjöbenhavn 1869: 223. Pl. 2, Fig. 11 f. **ornata.**

Micrasterias radiosa Ralfs p.p. in Wolle 1892, Desm. U.S., p. 119. Pl. 35, Fig. 3.
Micrasterias radiosa var. *wollei* Cushman 1904a, Bull. Torrey Bot. Club 31: 394.
Micrasterias radiosa f. *papillifera* Irénée-Marie 1951, Nat. Canadien 78(7/8): 190. Pl. 2, Fig. 8.

A variety with a row of small intramarginal spines along some or all of the major incisions. L. 112–248 μm. W. 110–250 μm. W. at apex 27–38 μm. I. 12–28 μm. Th. c. 22 μm. Zygospore reported from Florida by Salisbury (1936, 59. Pl. 1, Fig. 10). Diam. not including spines 83–86 μm, spines 38–45 μm long., 9 μm thick at base.

Forms of *M. radiosa* with intramarginal spines are commoner than forms without. There is so much variation in the number and position of spines, some barely discernible, that this variety should probably be joined to the typical.

DISTRIBUTION: Widely distributed in U.S. and Canada. Great Britain, Europe, Asia, South America, Arctic.

PLATE CXXX, figs. 3–8.

30g. Micrasterias radiosa var. ornata f. aculeata (Krieg.) Croasdale comb. nov.

Micrasterias sol (Ehrenb.) Kütz. var. *aculeata* Krieger 1939, Rabenhorst's Kryptogamen-Flora 13(1:2): 94. Pl. 131, Fig. 2.

A form differing in its intramarginal spines, which are longer, stouter, and more numerous, and occur along the primary and secondary incisions. L. 186–226 μm. W. 182–221 μm. W. at apex 36–50 μm. I. 14–24 μm.

Krieger (*l.c.*) bases his variety on *M. sol* var. *ornata* forma Borge (1925, 26. Pl. 5, Figs. 18, 19), but illustrates it with only one of Borge's figures (Fig. 18), which has more numerous and stouter apical spines on the polar lobe. Borge (*l.c.*) does not mention the polar lobe and shows in his figure 19 (our Pl. CXXXI, Fig. 2) an apex with fewer, smaller spines.

The North American records are not supported by illustrations and are therefore open to doubt.

DISTRIBUTION: Florida, Kansas, Mississippi, Minnesota, North Carolina. South America.

PLATE CXXXI, figs. 1–3.

30h. Micrasterias radiosa var. ornata f. laurentiana Brunel 1938, Nat. Canadien 65: 73. Fig. 2b.

Micrasterias radiosa f. *laurentiana* Brun., in Irénée-Marie 1951, Nat. Canadien 78:189. Pl. 2, Fig. 5.

A large form differing in the polar lobe, which is usually more dilated above, sometimes swollen at base, and which bears a pair of teeth rather than a single tooth on either side of the median incision, and 3, sometimes even 4, teeth, but never only 2, at the extremities; lobules of the last 2 orders shorter, their extremities only slightly incised; with small spines along the major incisions. L. 240–290 μm. W. 203–260 μm. W. at apex 20–65 μm. I. 24–32 μm.

DISTRIBUTION: Florida, Mississippi. Québec.

PLATE CXXIX, fig. 8; PLATE CXXX, figs. 1, 2.

30i. Micrasterias radiosa var. punctata West 1889, Jour. Roy. Microsc. Soc. 1889: 20. Pl. 2, Fig. 12 f. **punctata.**

A variety differing in having the wall distinctly punctate and the divisions of the lateral lobes rather like those of *M. papillifera* Bréb. ex Ralfs especially the ultimate ones; general outline also more angular. L. c. 165 μm. W. c. 145 μm. W. at apex 45 μm. I. 25 μm.

Krieger (1939, p. 89) puts this variety as a synonym of *M. papillifera* Bréb. var. *glabra* Nordst., but we think West's plant is better under *M. radiosa*, to which he assigned it, because of the deeper incisions, with division to the fourth (not third) order, and because of the larger size; it differs also in its distinctively large punctations.

DISTRIBUTION: Massachusetts. Québec.

PLATE CXXIX, fig. 7.

30j. Micrasterias radiosa var. punctata f. concentrica Croasdale f. nov.

Micrasterias radiosa var. *punctata* West, in Irénée-Marie 1951, Nat. Canadien 78(7/8): 192. Pl. 3, Fig. 1.

Forma differens quod plus quam sesqui maior quam varietas typica, et quod puncta anulis concentricis distribuuntur, anuli magnitudine numeroque variare possunt, et puncta, ut videtur, densitate differunt. Cellulae 251-285 μm long., 241-263.5 μm lat., 51.5-60 μm lat. ad apicem, 29-36 μm lat. ad isthmum.

ORIGO: Locus La Pointe du Lac, Quebec dictus.

ICONOTYPUS: Irénée-Marie 1951, Nat. Canadien 78(7/8): Pl. 3, Fig. 1.

A form differing in being more than 1 1/2 times as large as the typical variety and in having the punctations distributed in concentric rings, which may vary in size and number, and in which the punctations are apparently of different density. L. 251-285 μm. W. 241-263.5 μm. W. at apex 51.5-60 μm. I. 29-36 μm.

DISTRIBUTION: Québec.

PLATE CXXIX, fig. 6.

30k. Micrasterias radiosa var. swainei (Hast.) West & West 1896a, Trans. Linn. Soc. London, Bot., II, 5(2): 240.

Micrasterias swainei Hastings in Wolle 1892, Desm. U.S., p. 119. Pl. 42, Fig. 1.

Micrasterias radiosa var. *extensa* Prescott & Scott 1943, Pap. Michigan Acad. Sci., Arts & Lettr. 28: 77. Pl. 4, Figs. 7, 8, 10; Prescott & Scott 1952, Trans. Amer. Microsc. Soc. 71(3): 249. Pl. 1, Fig. 5 (as "forma").

A variety with the lowermost lobules of the lateral lobes (and rarely the lobules adjacent to the polar lobe) usually extended beyond the other lobules into a stout spine; the polar lobe usually broadening markedly at the apex, its angles ending in 1 or 2 teeth and often bearing an extra tooth on the dorsal surface. Incisions rather deep and usually narrow, the margins of the lobules sometimes undulate, with a row of small surface spines sometimes seen along the margins of the principal incisions. L. 132-223 μm. W. 135-272 μm. W. at apex 24-48 μm. I. 12-21 μm.

This is a very controversial variety. Borge (1893, p. 397), West & West (1896a, p. 240), Cushman (1905, p. 254; 1908, p. 105), and Krieger (1939, p. 97) believe it should be considered a variety of *M. radiosa*. But Wolle (1892, p. 119), Irénée-Marie (1952, p. 157), and Mix (1965, p. 174) believe it should stand as a separate species. Irénée-Marie (1952, *l.c.*) would keep *M. radiosa* var. Prescott & Scott (1943, *l.c.*) separate, and Mix (1965, *l.c.*), from a study of American material in culture, would reinstate *M. swainei* and include in it *M. radiosa* var. *extensa*. Mix's photomicrographs of the collected material and the forms cultured from it make a convincing argument for combining these 2 taxa, but we agree with Borge, Wests, and others that they cannot be wholly separated from *M. radiosa*.

DISTRIBUTION: Florida, Louisiana, Massachusetts, Mississippi, New Hampshire. Québec.
PLATE CXXXII, figs. 5–7.

31. **Micrasterias rotata** (Grev.) Ralfs 1848, Brit. Desm., p. 71, Pl. VIII, Fig. 1a var. **rotata f. rotata.**

Micrasterias rotata f. *inermis* Irénée-Marie 1951, Nat. Canadien 78(7/8): 193. Pl. 3, Fig. 5.
Micrasterias rotata f. *nuda* (Wolle) Irénée-Marie 1938, Flore Desm. Montréal, p. 230. Pl. 37, Fig. 1.
Micrasterias fimbriata Ralfs var. *nuda* Wolle 1884a, Desm. U.S., p. 110. Pl. 36, Fig. 4.

Cells large, 1 1/10 to 1 1/7 times as long as broad, incisions moderately deep, sinus linear, closed; lateral lobules usually unequally divided, the upper larger, divided to the fourth order, only incisions of the third and fourth order slightly open, ultimate lobules deeply excavate or shortly bidentate at the extremities; polar lobe gradually broadening to apex, which is slightly exserted with bidentate angles; apex retuse, usually with median notch, sometimes showing larger punctations at margin; subapical and surface spines lacking; semicell in lateral view ovate-lanceolate with broadly truncate apex and swollen base, in vertical view narrowly elliptic-rhomboid, with acute poles and a small inflation in the middle of each side. L. 190–366 μm. W. 165–305 μm. W. at apex 43–80 μm. I. 26–43 μm. Th. c. 50 μm. Zygospores frequently observed: spherical with numerous, stout, sometimes slightly curved spines. Diam. not including spines 108–136 μm. Spines 13–35 μm long.

The two forms of Irénée-Marie listed as synonyms above, varying in degree of exsertion of polar lobe and strength of dentation of ultimate lobules, seem to fall within the normal range of the typical. We show Irénée-Marie's forms on our Pl. CXXXIII, figs. 7, 8.

DISTRIBUTION: Widely distributed in U.S. and Canada. Cosmopolitan. Arctic.
PLATE CXXXIII, figs. 1–8.

Key to the North American Varieties and Forms of *Micrasterias rotata*

1. Surface of semicell wholly without spines.
  2. Extremities of lobules and angles of polar lobe bidentate or excavate.
    3. Polar lobe broader at apex, which is sharply depressed with notch, angles extended and bidentate.   31. var. *rotata* f. *rotata.*
    3. Polar lobe rectangular, not broader at apex, which is shallowly depressed without notch; angles barely extended, excavate.   31a. var. *rotata* f. *clausa.*
  2. Most or all of extremities rounded-truncate.   31c. var. *japonica.*
1. Semicells with spine or tooth at apex of polar lobe on either side of apical notch.
                                                                    31b. var. *rotata* f. *evoluta.*

31a. **Micrasterias rotata** var. **rotata** f. **clausa** Irénée-Marie 1951, Nat. Canadien 78(7/8): 193. Pl. 3, Fig. 4.

A form with the polar lobe exserted, rectangular throughout its length, at the apex shallowly retuse without apical notch, its angles barely extended and only shallowly excavate; all incisions closed except uppermost part of that between the polar and lateral lobes. L. 300–320 μm. W. 230–235 μm. W. at apex 65–70 μm. I. 65–70 μm.

The dimensions that Irénée-Marie gives for the polar lobe and isthmus are inconsistent with the shape of the cell. He gives 88–90 μm for the breadth of the polar lobe and 45–47 μm for the isthmus, but a glance at the figure shows that these 2 dimensions are about equal, and a comparison with the length and breadth of the cell brings them to 65–70 μm.

DISTRIBUTION: Québec.

PLATE CXXXIII, fig. 9.

31b. Micrasterias rotata var. **rotata** f. **evoluta** Turner 1892, Kongl. Svenska Vet.-Akad. Handl. 25(5): 167. Pl. 23, Fig. 1.

A form having a small spine on a low protuberance on either side of the apical notch; angles of polar lobes and extremities of lateral lobules often more strongly dentate. L. 210–306 μm. W. 169–265 μm. I. c. 40 μm.

In other taxa of *Micrasterias* the presence of a spine on either side of the apical notch is a variable feature, and we are not sure that this is a good form. Plants of this form should be compared with plants of *M. apiculata* var. *fimbriata* (Ralfs) Nordst., from which they differ only in less open incisions and shorter terminal teeth. These 2 taxa, if indeed they are distinct, make a close link between the 2 species.

DISTRIBUTION: California, Connecticut, Michigan, Montana, Wisconsin, Wyoming. New Brunswick, Québec. Great Britain, Europe, India, South Africa, South America, West Indies.

PLATE CXXXIV, figs. 1–4.

31c. Micrasterias rotata var. **japonica** Fujisawa 1936, Nat. Hist. Mag. 34(58): 14, Pl. 2, Figs. 6a, b.

Micrasterias denticulata Bréb. f. *mucronata* Irénée-Marie 1951, Nat. Canadien 78(7/8): 182. Pl. 1, Fig. 4.

Micrasterias denticulata Bréb. var. *minnesotensis* Turner 1885, Jour. Roy. Microsc. Soc., II, 5(6): 937. Pl. 16, Fig. 14.

Cells large, slightly longer than broad; lateral lobes divided to the third or fourth order, the extremities slightly retuse with rounded angles; polar lobe only slightly broader at apex, slightly exserted, its apex concave with shallow notch, its angles ending in a single mucro or very slightly excavate. L. 215–266 μm. W. 174–252 μm. W. at apex 30–63 μm. I. 29–42 μm.

As Krieger (1939, p. 103) suggests, this may be merely a reduced form.

Irénée-Marie's *M. denticulata* f. *mucronata* (*l.c.*) seems better placed in *M. rotata* var. *japonica* because of its polar lobe, which lacks an apical notch and is slightly exserted. His figure closely resembles our plant from Florida shown in Pl. CXXXIV, Fig. 5. Turner's *M. denticulata* var. *minnesotensis* (*l.c.*), although very incompletely illustrated by only the polar lobe and the adjacent lateral lobules of the last order, may possibly also belong here for the same reason.

DISTRIBUTION: California, Florida, Michigan, Minnesota, New Hampshire. Québec. Japan.

PLATE CXXXIV, figs. 5, 6.

32. Micrasterias sexspinata (Irénée-Marie & Hilliard) Croasdale & Prescott comb. nov.

Euastrum sexspinatum Irénée-Marie & Hilliard 1963, Hydrobiologia 21(1/2): 107. Pl. 2, Fig. 14.

Plant of a medium size, a little broader than long, with truncated ends on each side of the medium sinus, and ending with short spines, and with an upper lobe rounded at the apex which has a very shallow depression in the center; this upper lobe ends with a small spine at each corner where the concave sides meet with the convex apex: apical view rhomboidal, ending with a small spine and with a rounded protuberance in the middle of the longer arcs; lateral view almost circular, a little rounded at the base towards the isthmus, and with a protuberance in the middle of each side. L. 80–82.5 μm. W. 83–84.5 μm. I. 16–16.5 μm. Spines 0.33 μm long.

Irénée-Marie and Hilliard (*l.c.*) comment: "We know of no other species approaching this one, though the lower lobe of each semicell reminds one of *Euastrum pinnatifida*, but the apical lobe is absolutely different. This species is certainly a Desmid, because of the perfect symmetry of the two halves, but it is difficult to state whether it is a *Euastrum,* a *Cosmarium,* or a *Staurastrum.* Since there is at least a touch of an apical incision, we are inclined to refer this plant to the genus *Euastrum.*" The present authors believe that because of the nature of its apical lobe it more closely resembles a *Micrasterias*, possibly in the extremely variable *M. pinnatifida-M. oscitans* group.

DISTRIBUTION: Alaska.

PLATE XCII, fig. 7.

33. **Micrasterias suboblonga** Nordstedt 1887, Bot. Notiser 1887: 164; 1888a, Kongl. Svenska Vet.-Akad. Handl. 22(8): 78. Pl. 2, Fig. 18 var. suboblonga.

Cells medium-sized, nearly 1.5 times as long as broad; polar lobe depressed, semicircular, slightly retuse, about one-half the height of the semicell; lateral lobes undivided, retuse, with rounded and slightly projecting extremities; subpolar incision open, with rounded apex, often narrower at extremity; median sinus narrow, sinuous; cell wall strongly punctate; semicells in vertical view elliptic-fusiform and in lateral view oblong with slightly compressed sides. L. 140–158 μm. W. 90–114 μm. W. at apex 90–102 μm. W. at polar incision 50–56 μm. I. 24 μm. Th. 54–60 μm. Zygospore unknown.

Nordstedt (*l.c.*) comments that this plant seems to unite *Micrasterias* with *Euastrum* and states that he classifies it with *Micrasterias* because it completely lacks facial protuberances. He describes the wall as "closely scrobiculate-punctate," Scott & Prescott (1960, p. 326) write: "All species of *Micrasterias* have porose walls, but we have never seen, and do not know of any in which the wall can be called scrobiculate."

DISTRIBUTION: Australia (not New Zealand, as in Krieger 1939, p. 41). Not yet found in North America.

PLATE CV, fig. 1.

Key to the North American Varieties and Forms of *Micrasterias suboblonga*

1. Polar lobe nearly as high as lateral lobe; subpolar incision extending halfway to center
    line of semicell; lateral lobes mostly retuse.   33. var. *suboblonga.*
1. Polar lobe much less high than lateral lobe; subpolar incision extending only one-third
    way to center line of semicell; lateral lobes mostly rounded.
  2. Cells less than 100 μm long.   33a. var. *australis* f. *australis.*
  2. Cells more than 100 μm long.   33b. var. *australis* f. *maxima.*

33+. **Micrasterias suboblonga** var. **suboblonga** forma.

A small, less elongate form with the subpolar incision narrower and more inclined; all angles tipped with a single stout tooth; only 1 semicell seen. L. 115 μm. W. 94 μm. I. 35.5 μm.

This plant superficially resembles a form figured by Borge (1903, p. 118. Pl. 5, Fig. 18) as *M. decemdentata* (Näg.) Arch., but it is more than twice as large and less angular than that species.

DISTRIBUTION: Idaho (Selkirk Mts., Bonner Co.).

PLATE CV, fig. 2.

33a. **Micrasterias suboblonga** var. **australis** Krieger 1939, Rabenhorst's Krypto-gamen-Flora 13(1:2): 41. Pl. 105, Fig. 13 f. **australis.**

Cells only 1/3 as large as the type, oblong-elliptic, with closed sinus; lower angles of polar lobe and lateral lobes acute; polar lobe shallower than lateral lobe, its apex slightly retuse; margins of lateral lobes rounded. L. 49–50 µm. W. 30–32.5 µm. I. 5 µm.

The varietal epithet is poorly chosen because the species was originally described from Australia (not from New Zealand as Krieger (*l.c.*) states). The typical variety has not yet been reported from North America.

DISTRIBUTION: Australia.

33b. **Micrasterias suboblonga** var. **australis** f. **maxima** (Presc. & Scott) Prescott comb. nov.

Micrasterias suboblonga var. *maxima* Prescott & Scott 1943, Pap. Michigan Acad. Sci., Arts & Lettr. 28: 78. Pl. 1, Fig. 7.

A very large form, nearly 3 times as large as the typical; the polar lobe rounded, the lower angles of the lateral lobes tipped with a blunt, spinelike protuberance. L. 122–143 µm. W. 91–102 µm. I. c. 28–33 µm.

DISTRIBUTION: Florida, Maryland, Mississippi.

PLATE CV, figs. 3, 4.

34. **Micrasterias tetraptera** West & West 1898, Linn. Soc. Jour. Bot. 33: 296. Pl. 16, Fig. 6 var. **tetraptera** f. **tetraptera.**

Cells less than medium-sized, 1 1/6 times longer than broad; sinus linear; polar lobe cuneate, with lateral margins strongly concave, rounded at apex with median depression, smooth, angles evenly tapered to blunt point or papilla; lateral lobes unevenly divided, the upper portion broader and usually divided to the fourth order, the lower portion only to the third order, the extremities usually bidentate or emarginate; subpolar incisions rather widely open, other incisions open at interior, usually closed in outer portion; sinus and subpolar incision showing a row of intramarginal spines. L. 109–116 µm. W. 90–100 µm. W. at apex c. 56 µm. I. 16 µm. Zygospore unknown.

DISTRIBUTION: Florida, Georgia, Louisiana. Québec.

PLATE CXXVIII, figs. 1, 2.

Key to the North American Varieties and Forms of *Micrasterias tetraptera*

1. Polar lobe very broad, more than half the cell diameter; lateral lobes divided mostly to the third order.
    2. Cell circular; lateral lobes divided to the third, rarely fourth order.
        3. Extremities of apex curved downward, resulting in subpolar incision open in middle.   34. var. *tetraptera* f. *tetraptera.*
        3. Extremities of apex not curved downward, subpolar incision narrow.
                                                                    34a. var. *tetraptera* f. *taylorii.*
    2. Cells hexagonal; lateral lobes divided to the second, rarely third order.
        4. All incisions narrow, without spines.   34b. var. *angulosa* f. *angulosa.*
        4. Subpolar incision open in middle; intramarginal spines at subpolar incision and sinus.   34c. var. *angulosa* f. *alobulifera.*
1. Polar lobe narrow, about 1/3 the cell diameter; lateral lobes divided mostly to fourth order.
    5. Apex smooth.   34d. var. *longesinuata* f. *longesinuata.*
    5. Apex bearing a pair of short spines or teeth toward extremities.

6. Extremities not narrowed into short processes; semicell without basal protuberance.

34f. var. *longesinuata* f. *spinosa.*

6. Extremities narrowed into short processes; semicell with prominent basal protuberance.   34e. var. *longesinuata* f. *protuberans.*

34a. **Micrasterias tetraptera** var. **tetraptera** f. **taylorii** (Krieg.) Croasdale stat. nov.

Micrasterias tetraptera var. *taylorii* Krieger 1939, Rabenhorst's Kryptogamen-Flora 13(1:2): 93.
   Pl. 127, Fig. 3.
Micrasterias conferta Lund, in Taylor 1935, Pap. Michigan Acad. Sci., Arts & Lettr. 20: 212. Pl.
   46, Fig. 3.
Micrasterias jenneri Ralfs var. *lata* Woodhead & Tweed 1960, Hydrobiologia 15(4): 327. Pl. 2,
   Fig. 16.

A form broader than the typical, with more swollen apex, which is less deeply
retuse in the middle; extremities of lateral lobules bluntly emarginate; surface
spines wholly lacking. L. (87):107-150(187) μm. W. (74)96-115(148) μm. I.
13.6-27.3(40) μm.

This form does not seem sufficiently different from the typical to deserve to be
treated as a variety. In fact, Krieger & Bourrelly (1956, 153. Pl. 6, Figs. 52, 53),
listing it from the Andes, state: "Personally we think that it is only a variation
without scientific value." But the plant they show, larger and less deeply incised,
seems to form a link with the even larger and even less deeply incised plant called
M. jenneri var. *lata* Woodhead & Tweed (*l.c.*), which, although large, seems more
closely allied to *M. tetraptera* than it does to *M. jenneri.*
DISTRIBUTION: New Hampshire, Oregon, Newfoundland, Venezuela.
PLATE CXXVII, figs. 6. 7.

34b. **Micrasterias tetraptera** var. **angulosa** Irénée-Marie 1952, Hydrobiologia 4(1/2): 158. Pl. 15, Figs. 12, 13 f. **angulosa.**

A small variety differing from the typical by its hexagonal form, by the median
and subpolar sinuses, which are always linear, by the secondary and tertiary
incisions, which are less deep and less open, and by the almost complete absence
of intramarginal spines which ornament the principal incisions of the typical
plant; wall densely granular (from pore extrusions?) or smooth. L. 90-110 μm. W.
89.3-98 μm. I. 15.8-16.5 μm.
DISTRIBUTION: Québec.
PLATE CXXVIII, fig. 3.

34c. **Micrasterias tetraptera** var. **angulosa** f. **alobulifera** (Presc. & Scott) Prescott comb. nov.

Micrasterias tetraptera var. *alobulifera* Prescott & Scott 1952, Trans. Amer. Microsc. Soc. 71(3):
   249. Pl. 5, Fig. 10.

A form showing the same hexagonal shape and reduced size, but differing in
having all incisions open as in the typical and in having the typical intramarginal
spines along the major incisions. L. 90-101 μm. W. 75-92 μm. I. 12-17.5 μm.

Because of its hexagonal shape and small size this seems to be very closely
related to var. *angulosa*, which was described in the same year. Var. *angulosa* was
published in January and var. *alobulifera* was published in July. The name var.
angulosa therefore has priority, and this spinose, open-incision plant from the
South best might be considered a form of it.
DISTRIBUTION: Florida, Louisiana.
PLATE CXXVIII, fig. 4.

34d. Micrasterias tetraptera var. **longesinuata** Krieger 1939, Rabenhorst's Krypto-gamen-Flora 13(1:2): 93. Pl. 127, Fig. 2 f. **longesinuata.**

Micrasterias papillifera Bréb. ex Ralfs f. Taylor 1935. Pap. Michigan Acad. Sci., Arts & Lettr. 20: 214. Pl. 46, Fig. 2.

A large variety with the incisions somewhat deeper and mostly linear; lower lateral lobe sometimes equaling upper, sometimes narrower and less divided, divided mostly to the fourth order; polar lobe narrower, less convex at apex; intramarginal spines reduced and mostly restricted to the subpolar incision. L. 150 μm. W. 135 μm. I. 18 μm.

Krieger (*l.c.*) failed to portray the intramarginal spines on Taylor's figure (*l.c.*), on which he based his variety. Thomasson (1963, p. 105) suggests that this variety "probably has nothing to do with *M. tetraptera*."

DISTRIBUTION: Newfoundland. South America.

PLATE CXXVIII, fig. 5.

34e. Micrasterias tetraptera var. **longesinuata** f. **protuberans** (Presc. & Scott) Prescott comb. nov.

Micrasterias tetraptera var. *protuberans* Prescott & Scott 1943, Pap. Michigan Acad. Sci., Arts & Lettr. 28: 79. Pl. 5, Figs. 5, 6.

A form with the extremities of the polar lobe slightly extended into short processes and bearing apically a pair of short teeth or spines; with three intra-marginal spines within the upper margin of the polar lobe and the adjacent margin of the upper lateral lobules; semicell with a prominent rounded basal protuberance. L. 138-158 μm. W. 135-154 μm. I. 18.5-20 μm.

DISTRIBUTION: Florida.

PLATE CXXVII, fig. 8.

34f. Micrasterias tetraptera var. **longesinuata** f. **spinosa** Prescott & Scott 1952, Trans. Amer. Microsc. Soc. 71(3): 249. Pl. 2, Fig. 5.

A form differing in having a pair of teeth or spines on the dorsal margin of the extremities of the polar lobe, and 2 to 4 intramarginal spines within the upper margin of the polar lobe and the adjacent margins of the upper lateral lobules. L. 115-172 μm. W. 109-151 μm. I. c. 15 μm.

DISTRIBUTION: Florida.

PLATE CXXIX, fig. 1.

35. Micrasterias thomasiana Archer 1862, Quart. Jour. Microsc. Sci., II, 2: 239. Pl. 12, Figs. 1-5 var. **thomasiana.**

Cells large, as broad as long, or shorter, or up to 1 1/3 times longer, sinus and major incisions closed and rather deep; upper and lower portions of the lateral lobes about equal, divided mostly to the fourth order, extremities bidentate; polar lobe narrowly cuneate with slightly concave margins, a depressed apex with a rather deep notch having a broad-based tooth on either side of it, and with emarginate angles; wall sculpture variable, mostly with a round or pointed swelling over the isthmus, not always easily visible, and on either side of it a larger, outwardly curved, hollow process irregularly 2-3-toothed at the extremity; with a conical tooth at the base of the lateral lobules of the first, second, and sometimes third orders, and with 1 or 2 teeth, one above the other, within the base of the polar lobe, and sometimes a smaller tooth within the extremities; semicell in lateral view narrowly pyramidal with a rounded basal inflation, 2 acute projections on each side, and a truncate apex; in vertical view fusiform, the middle projection small, the lateral ones large and curved. L. 170–275 (303) μm. W. 139–241.5 (270) μm. W. at apex 36–53 μm. I. 30–36 (42.5) μm. Th. max. 57 μm. Zygospore unknown.

DISTRIBUTION: Alaska, Virginia, Wisconsin. New Brunswick, Nova Scotia, Québec. Great Britain, Europe, Asia, New Zealand, Australia, Arctic.

PLATE CXXXVII, figs. 2, 3.

Key to the North American Varieties of *Micrasterias thomasiana*

1. Surface spines present; supraisthmial processes obvious.
   2. Surface spines short and simple; lateral supraisthmial processes slightly bent and simple.   35. var. *thomasiana*.
   2. Surface spines at base of second-order lobules 2-toothed; lateral supraisthmial processes strongly bent and 2-parted.   35b. var. *pulcherrima*.
1. Surface spines absent; supraisthmial processes reduced or absent.   35a. var. *notata*.

35a. **Micrasterias thomasiana** var. **notata** (Nordst.) Grönblad 1920, Acta Soc. Fauna Flora Fennica 47(4): 38.

Micrasterias denticulata Bréb. var. *notata* Nordstedt 1888a, Kongl. Svenska Vet.-Akad. Handl. 22(8): 29. Pl. 2, Fig. 13.

A variety differing in the absence of the processes and wall spines; only 1 to 3 more or less developed swellings above the isthmus. L. 215–287 μm. W. 170–236 μm. W. at apex c. 42 μm. I. 24–30 μm. Th. 50–74 μm.

Grönblad (*l.c.*) points out that this variety seems to be intermediate between *M. thomasiana* and *M. denticulata*, but more closely resembles the former.

DISTRIBUTION: United States (throughout). British Columbia, Nova Scotia. Great Britain (as *M. denticulata* var. *notata*), Europe, Asia, Australia, New Zealand, Africa, South America, West Indies, Arctic.

PLATE CXXXVII, figs. 4–6.

35b. **Micrasterias thomasiana** var. **pulcherrima** G. S. West 1909, Linn. Soc. Jour. Bot. 39: 58. Pl. 4, Fig. 1.

A more ornamented variety, with a short bidentate process curving outward at the base of each lateral lobule of the second order, with a large conical tooth at the base of the upper lateral lobules of the first order; the basal processes on either side of the middle swelling larger, outwardly curved, and possessing a large dorsal tooth; upper lateral lobules divided to the fifth order. L. 219–250 μm. W. 197–219 μm. L. c. 30 μm. Th. c. 65 μm.

DISTRIBUTION: Montana. Finland, Asia, Australia (Yan Yean Reservoir).

PLATE CXXXVIII, fig. 1.

36. **Micrasterias torreyi** Bail. ex Ralfs 1848, Brit. Desm., p. 210. Pl. XXXV, Fig. 5 var. **torreyi**.

Micrasterias torreyi f. *punctata* Irénée-Marie 1951, Nat. Canadien 78: 194. Pl. 4, Figs. 2, 3.
Micrasterias torreyi var. *nordstedtiana* (Hieron.) Schmidle 1898a, Engler's Bot. Jahrb. 26: 48; forma in Prescott & Scott 1943, Pap. Michigan Acad. Sci., Arts & Lettr. 28: 79. Pl. 3, Fig. 2.

Cells large, about as broad as long or broader; sinus at first narrow, then broadly opening; upper lateral lobules broader than the lower, usually divided twice, lower lobules narrower and only once divided; all lobules cone-shaped, tapering to a single point or with a bidentate extremity; polar lobe with its lower margins horizontal, upper margins diverging to form 2 conical lateral extensions which end in single or paired teeth L. 250–410 μm. W. 250–450 μm. W. at apex 74–110 μm. I. 27–44 μm. Zygospore unknown.

Wolle (1892, p. 34) shows on one plate 8 rather strongly varying forms, noting that one plant may have semicells of widely differing shape. Most of these forms have been refound in our southern states, as well as an additional form with the apical part of the polar lobe rather deeply cleft and tapering into 2 somewhat erect and narrow lobules. This feature is seen in var. *nordstedtii* (Hieron.) Schmidle, but our forms do not show the overlapping lateral lobules of that variety. Grönblad (1956, p. 26) reports from New Hampshire, without a figure, a small form (L. 236 μm. W. 228 μm) with all lobules bidentate, polar lobe typical.

DISTRIBUTION: Alabama, Alaska, Florida, Georgia, Louisiana, Massachusetts, Michigan, Mississippi, New Hampshire, New Jersey, North Carolina, Pennsylvania, Vermont, Virginia, Wisconsin, Wyoming. Québec (as f. *punctata*). South America.

PLATE CXI, figs. 5–7; PLATE CXII, fig. 1.

Key to the North American Varieties and Forms of *Micrasterias torreyi*

1. Upper lateral lobe larger and more divided than the lower; cells 250 μm long or more.
                                                    36. var. *torreyi*.

1. Both lateral lobes of same size and degree of division; cells less than 225 μm long.
                                         36a. var. *crameri* f. *minor*.

36a. Micrasterias torreyi var. **crameri** (Bern.) Krieger 1939, Rabenhorst's Kryptogamen-Flora 13(1:2): 98. Pl. 134, Fig. 2 f. **crameri**.

Micrasterias lux Josh. var. *crameri* Bernard 1909, Dept. Agric. Indes-Néerl. 1909, p. 64. Pl. 5, Fig. 115.

Cells with upper and lower lateral lobes about equal in size and degree of division (to the third order); all lobules equally tapered and ending in 2 teeth. L. 350–360 μm. W. 315–320 μm. I. 28–30 μm.

The typical form of this variety has not yet been reported from North America. Probably this variety would be better placed under *M. hieronymusii* Schmidle (1898a, p. 49) as recommended by Thomasson (1960, p. 24).

36b. Micrasterias torreyi var. **crameri** f. **minor** Irénée-Marie 1959, Hydrobiologia 13: 334. Pl. 2, Fig. 8.

A very small form, less than two-thirds the size of the typical; all lobules equal and bidentate. L. 210 μm. W. 170 μm. I. 20 μm.

DISTRIBUTION: Québec.

PLATE CXII, fig. 2.

37. Micrasterias triangularis Wolle 1881, Bull. Torrey Bot. Club 8: 1. Pl. 6, Fig. 4.

Micrasterias pseudotorreyi Wolle 1883, Bull. Torrey Bot. Club 10(2): 19. Pl. 27, Fig. 2.

Cells large, 1 to 1.3 times longer than broad; sinus narrow, often closed except at the outside; subpolar incision usually angularly open from a closed or narrow interior; lateral lobes evenly divided to the third order, the ends emarginate to bidentate; polar lobe small, cuneate, its lateral extremities with one or two teeth, in the middle slightly retuse; semicell in lateral view elliptical with broadened base, in vertical view broadly fusiform with sharp ends. L. 201–272 μm. W. 108–224 μm. W. at apex c. 13 μm. I. 27–37 μm. Zygospore unknown.

Krieger (1939, p. 100) suggests that *M. pseudotorreyi* Wolle is probably a reduced form of *M. triangularis*. This seems reasonable, because of its smaller size, broader polar lobes, and more evenly divided lateral lobes (Pl. CVII, Fig. 7).

DISTRIBUTION: Florida, Georgia, Louisiana, Massachusetts, Mississippi, New Hampshire, New Jersey, North Carolina, Pennsylvania, Rhode Island, South Carolina. Newfoundland. Great Britain.

PLATE CVII, figs. 1-4, 7.

37+. **Micrasterias triangularis** forma Prescott & Scott 1952, Trans. Amer. Microsc. Soc. 71: 250. Pl. 8, Fig. 5.

A form with low protuberances on either side of the slight median invagination of the polar lobe. In one semicell of one specimen each protuberance was furnished with a short spine.

Prescott & Scott (*l.c.*), in describing this form from Florida, stated, "The Florida plants possibly could be described as a new variety, although the differentiating characters do not seem to be present consistently." L. 201-203 μm. W. 161-171 μm. I. 27-30 μm.

DISTRIBUTION: Florida.

PLATE CVII, figs. 5, 6.

38. **Micrasterias tropica** Nordstedt 1870, Vidensk. Medd. Naturh. Foren. Kjöbenhavn 1869: 219. Pl. 2, Fig. 15.

Cells medium-sized, a little longer than broad; sinus opening widely from a sharp-angled interior; polar lobe with erect, parallel-sided lower part and a diverging upper part, the apex flat between the long, spreading processes; lateral lobe undivided, horizontally extended, strongly tapered; with small stout marginal spines toward the extremities of lateral lobes and polar processes, and similar spines within the margins of all but the extremities; spines stouter or paired at the depressed apex and above the isthmus; semicells in vertical view fusiform with a blunt-angled median inflation; marginal and intramarginal spines present. L. 87-125 μm. W. 80-101 μm. W. at apex 38-52 μm. I. 16-18 μm. Th. 30-32 μm. Zygospore unknown.

DISTRIBUTION: Virginia (without figure, therefore to be confirmed, since this is essentially a tropical species). Russia and Poland (very rare), Southern Asia, Australia, Africa, South America.

PLATE CXL, figs. 3, 4.

39. **Micrasterias truncata** (Corda) Bréb. ex Ralfs 1848, Brit. Desm., p. 75. Pl. VIII, Fig. 4; Pl. X, Fig. 5 var. **truncata** f. **truncata**.

Micrasterias truncata f. *granulata* Raciborski 1889, Pamiet. Wydz. Mat.-Przyr. Krakow 17: 105. Pl. 7, Fig. 1.
Micrasterias truncata var. *quebecensis* Irénée-Marie 1951, Nat. Canadien 78: 196. Pl. 2, Figs. 6, 7.

Cells medium-sized or smaller, circular or very slightly longer than broad; polar lobe broad, usually broadly rounded, but often flattened or retuse in middle, bearing one or two short teeth at the angles; lateral lobes once or twice shallowly divided, the ends emarginate or with 2 teeth; subpolar incision usually narrow and sharp-angled; sinus narrow, closed for the inner half or all of its length; semicells in vertical view broad, the breadth about twice the thickness, fusiform; in lateral view elliptical with slightly tapered ends; chloroplast with raised, mostly forked longitudinal ridges and relatively few pyrenoids. L. 75-144 μm. W. 74-135 μm. W. at apex 55-99 μm. I. 10-30 μm. Th. 36-52 μm.

Zygospore globose with long, at times somewhat curved, spines. Diameter 75–80 μm. including spines 118–130 μm.

This is one of the commonest species of *Micrasterias*, and, as Krieger & Bourrelly (1956, p. 152) expressed it, "has been pulverized to the limit." In 1939 Krieger listed 17 varieties, and at least 6 more have been added to the literature since them. Grönblad (1921, p. 22) stated: "The whole group of forms called *M. truncata* includes so many different forms, always with a continuous series of intermediate forms, that it is not possible to make up a system of them." Förster (1970, p. 305. Pls. 14–16) reports only the typical form from Sud-Holstein but illustrates 42 "morphae" for it, in which he submerges several named varieties.

For North America 18 varities have been reported. Most of them are connected by janus- or intermediate forms, but it seems possible for this work to isolate 8 which might be considered as legitimate varities, the others being reduced to named forms and some unnamed variants. We present below an artificial key to these named varieties and forms, realizing that further "lumping" is probably advisable. See Thomaszewicz (1973, p. 592).

Forma *granulata* Racib. appears to be a cell with an eroded wall, the granules being described as unequal, "wall between granules often punctate." Var. *quebecensis* Irénée-Marie differs only in its polar lobe with straight or nearly straight dorsal margin, and deeper, more open principal lateral incisions. As illustrated by its author (our Pl. XCV, Fig. 4) this plant comes within the normal variation of the type.

DISTRIBUTION: Widely distributed in North America and the other continents, principally of the North Temperate Zone. Arctic.

PLATE XCV, figs. 1–5.

Key to the North American Varieties and Forms of *Micrasterias truncata*

1. Cells small, less than 67 μm long.
  2. Lateral lobes deeply divided; sinus soon open.
    3. Cells subcircular, not or very little broader than long; breadth of polar lobe 0.7 or less of the maximum breadth; teeth of polar lobe not extended.
          39m. var. *pusilla* f. *pusilla.*
    3. Cells broader than long; breadth of polar lobe 0.75 or more of the maximum breadth; teeth of polar lobe rather extended.   39n. var. *pusilla* f. *cuneata.*
  2. Lateral lobes barely divided; sinus mostly closed.
    4. With facial protuberance.   39l. var. *mississippiensis.*
    4. Without facial protuberance.
      5. Lateral extensions of polar lobe with convex lower margin and 2 or 3 teeth at end of upper margin.   39a. var. *truncata* f. *convexa.*
      5. Lower margin of polar lobe extensions not convex, no teeth above.
          39b. var. *truncata* f. *neodamensis* p.p.
1. Cells medium-sized, between 70 and 200 μm long.
  6. Cells in chains; apex flat.   39i. var. *concatenata.*
  6. Cells not in chains; apex convex or slightly retuse.
    7. Polar lobe outstanding, with open subpolar incision.
      8. Base of incision acute; lateral extensions of polar lobe bearing a single tooth.
          39h. var. *bahusiensis.*
      8. Base of incision broadly rounded; lateral extensions of polar lobe rounded-truncate or with 2 teeth.
        9. All lobes rounded.   39d. var. *truncata* f. *reducta.*
        9. All lobes with small, widely spaced teeth; extremities of polar lobe somewhat truncate.   39g. var. *truncata* f. *uralensis.*
    7. Polar lobe not outstanding, subpolar incision narrow or absent.
      10. Subpolar incision narrow; lateral lobes divided.
        11. Lateral lobes twice divided.
          12. Lateral lobules angularly rounded, without teeth.   39j. var. *crenata* f. *crenata.*

12. Lateral lobules sharply emarginate or with teeth.
  13. Lobules emarginate or with 2 teeth.
    14. Lobules emarginate or with short teeth.   39. var. *truncata* f. *truncata*.
    14. Lobules ending in long teeth.   39e. var. *truncata* f. *semiradiata*.
  13. Lobules ending in 3 teeth.   39f. var. *truncata* f. *tridentata*.
11. Lateral lobes once divided.
  15. Lobules emarginate at extremities.
    16. Cells distinctly longer than broad.   39k. var. *crenata* f. *carteri*.
    16. Cells about circular.   39b. var. *truncata* f. *neodamensis* p.p.
  15. Lobules toothed at extremities.
    17. Lobules with row of 3–7 small teeth at extremities.
                                    39c. var. *truncata* f. *quadrata*.
    17. Lobules with 2 stout teeth; semicells turgid, rhomboidal in vertical view.
                                  39o. var. *turgida* f. *turgida*.
10. Subpolar incision absent; lateral lobes undivided.
  18. Lateral lobes with teeth.   39q. var. *turgida* f. *elobata*.
  18. Lateral lobes without teeth.   39p. var. *turgida* f. *elevata*.

39a. Micrasterias truncata var. **truncata** f. **convexa** (Presc. & Scott) Croasdale stat. nov.

Micrasterias truncata var. *convexa* Prescott & Scott 1943, Pap. Michigan Acad. Sci., Arts & Lettr. 28: 80. Pl. 2, Fig. 3.

Cells small, suborbicular, the apex broadly convex or somewhat flattened or retuse; sinus narrowly linear, closed throughout all or nearly all of its length; semicells 3-lobed, the subpolar incision narrow and curved, the lateral extensions of the polar lobe convex on ventral as well as dorsal surface, the latter bearing 3 blunt teeth near its extremity; lateral lobes once divided, the incision rounded and very shallow, each lobule ending in 3 (rarely 2) short teeth. L. 49–66 µm. W. 49–61 µm. I. 12–15 µm.

This form differs chiefly in its polar lobe, convex on the lower surface and with 3 teeth. The lateral lobules resemble sometimes f. *neodamensis*, sometimes f. *tridentata*. Forma *neodamensis* and f. *convexa*, var. *crenata* and its forma *carteri* make up an association of forms that differ from the typical in their minimal incisions and reduced dentation. Prescott shows a form from Wisconsin (Pl. XCVIII, Fig. 2) in which one apex, the sinus, and the subpolar incisions are somewhat undulate.

DISTRIBUTION: Florida, Louisiana, Mississippi, Wisconsin.

PLATE XCVIII, figs. 1–3.

39b. Micrasterias truncata var. **truncata** f. **neodamensis** (A.Br.) Dick 1926, Krypt. Forsch. Bayer. Bot. Ges. 7: 448. Pl. 20 Fig. 8.

Micrasterias truncata var. *neodamensis* (A. Br.) Dick, in Krieger 1939, Rabenhorst's Kryptogamenfl. 13(1:2): 30.
Micrasterias truncata var. *rectangularis* Irénée-Marie 1957, Hydrobiologia 9: 86. Figs. 8, 9.
Micrasterias decemdentata (Näg.) Arch., in Miller 1936, Univ. Wyoming Publ. 2: 82. Pl. 5, Fig. 3; Wolle 1884a, Desm. U.S., p. 113. Pl. 33, Figs. 5, 6.
Micrasterias crenata Bréb., in Wailes 1931, Vancouver Museum & Art Notes 6(2): 75. Pl. 2, Fig. 19.

Cells nearly circular, usually slightly longer than broad; polar lobe broadly arched but usually somewhat retuse in middle, bearing typically 1 tooth at the lateral extremity; lateral lobes only once shallowly divided, the lobules short and truncate with short teeth at the ends; subpolar incision narrow or closed; sinus mostly closed. L. 57–136 µm. W. 55–119 µm. I. 12–30 µm.

Dick (*l.c.*) illustrates his form with a dichotypical figure, one of the semicells being like the typical, and he agrees with Grönblad that it is probably only an undeveloped cell of the typical plant. It seems doubtful that Dick meant to name it as a new form, but Krieger (1939, p. 30) made it into a variety attributed to Dick. We include it, reduced to a form, because of the many records in North America, but doubt its validity as a taxon.

DISTRIBUTION: California, Colorado, Florida, Idaho, Minnesota, Montana, Oregon, Wisconsin. British Columbia, Québec. Europe, South America.

PLATE XCVI, figs. 1–7.

39c. Micrasterias truncata var. truncata f. quadrata (Bulnh.) Croasdale stat. nov.

Micrasterias truncata var. *quadrata* Bulnheim 1861 (not 1859), Hedwigia 2: 52. Pl. 9, Fig. 5.

Cells nearly circular, polar lobe convex, but sometimes flat or retuse in the middle, ending in 1 or 2 teeth; lateral lobes once divided, each lobule truncate, bearing a vertical row of 3 to 7, typically 5 to 6, short, sharp teeth; subpolar incision shallow and narrow or closed, sinus closed for nearly all or inner half of its length. L. 85–131 μm. W. 76–130 μm. I. 19–25 μm. Th. c. 49 μm.

Bulnheim first described this plant in 1859 (p. 21, "Pl. 2, Fig. 3"). He obviously meant Fig. 2, not 3, but neither is a good illustration of his variety *quadrata*, which is well represented by him under this name in 1861 (*l.c.*). Grönblad (1921, p. 21, Pl. 1, Figs. 9, 10) defends this as a good variety, citing among other characters the thickness of the cell when seen in vertical view and the "highly arched polar lobe." However, the vertical view that he shows is in its proportions exactly that of the typical (the width twice the thickness), and our plant illustrated in Pl. XCVII, fig. 2 was named var. *quadrata* by Grönblad (litt.) but shows a flattened polar lobe. The condition of the lateral lobules ending in a vertical row of 5 to 6 teeth seems merely to reflect a tridentate form in which the secondary division of the lateral lobes is reduced or absent. It is never consistent throughout all the lobules. There seem to be no grounds for considering this a good variety.

DISTRIBUTION: Florida, Maine, Massachusetts, Michigan, Mississippi, Minnesota, New Hampshire, Washington, Wisconsin. Ontario. Europe, Russia, Arctic.

PLATE XCVII, figs. 1–3.

39d. Micrasterias truncata var. truncata f. reducta Prescott f. nov.

Forma circa 1.22 plo longior quam lata, omni angulo rotundato, interdum 1 2 dentibus brevibus obtusisque ad extremitates laterales lobi polaris praedita; lobi laterales non aut vix divisi; incisio subpolaris lata et non profunda; sinus intus inapertus, deinde latius apertus. Cellulae 78–84 μm long., 64–72 μm lat., 16–21 μm lat. ad isthmum.

ORIGO: E loco Helen Lake, Vilas Co., Wisconsin dicto.

HOLOTYPUS: G. W. Prescott Coll. 2W295.

ICONOTYPUS: Plate XCVIII, figs. 4, 5.

A form about 1.22 times as long as broad, with all the angles rounded, sometimes with 1 or 2 short, blunt teeth at the lateral extremities of the polar lobe; lateral lobes not or barely divided; subpolar incision broad and shallow; sinus closed interiorly, then opening rather widely. L. 78–84 μm. W. 64–72 μm. I. 16–21 μm.

This reduced form seems closest to f. *uralensis* because of the relative length of the cell, and the broad, shallow subpolar incision. But see also one of the variable or teratological forms of *M. truncata* var. *turgida* Taylor (1935, Pl. 47, Fig. 3).

DISTRIBUTION: Montana, Wisconsin.

PLATE XCVIII, figs. 4, 5.

39e. Micrasterias truncata var. truncata f. semiradiata (Näg.) Cleve 1864, Öfv. Kongl. Vet.-Akad. Förh. 20(10): 487.

Micrasterias truncata var. *minor* Wolle 1884a, Desm. U.S., p. 114, Pl. 38, Fig. 9.

Cells subcircular, mostly somewhat broader than long; apex broadly rounded, sometimes flattened, undulate or slightly retuse in the middle, with stout single or paired teeth at the ends; lateral lobes shallowly twice divided, all ends with long, stout, and sharp teeth; subpolar incision narrow, open, rounded in interior; sinus closed for inner half, then opening. L. 64-105 μm. W. 60-98.5 μm. W. at apex 55-76 μm. I. 13-21 μm.

A form differing in the longer, stronger teeth at the lateral extensions of the polar lobe and the ends of the lateral lobules, in the deeper, more open incisions and sinus, and in the depressed cells. A whole series of intermediate forms connects this with the typical. It seems not to be a very clear-cut form. Forma *tridentata*, f. *quadrata*, and f. *semiradiata* are a group of forms that vary from the typical, particularly in the nature of their dentation, which is a relatively unimportant character for diagnosis.

DISTRIBUTION: Alaska, Indiana, Louisiana, Massachusetts, Michigan, Montana, New Hampshire, New York, Ohio, Pennsylvania, Vermont, Virginia, Wisconsin. Labrador, Québec. Great Britain, Northern and Central Europe, Russia, Africa.

PLATE XCVII, figs. 4, 5.

39f. Micrasterias truncata var. truncata f. tridentata (Benn.) Croasdale stat. nov.

Micrasterias truncata var. *tridentata* Bennett 1890, Jour. Roy. Microsc. Soc. 1890: 7. Pl. 1, Fig. 10.

A form with the ends of the lateral lobules tridentate instead of bidentate. (Dimensions not given.) Krieger (1939, p. 33) accepts this variety, but West & West (1905a, p. 83) include it under the typical. We agree with West and West, since the tridentate condition is rarely expressed on every lobule. For convenience it is considered a form, possibly leading over into f. *quadrata*. See Forster 1972, p. 545. L. 82-130 μm. W. 85.5-130 μm. W. at apex 61-78 μm. I. 19-26 μm.

DISTRIBUTION: Florida, Massachusetts, Minnesota, Mississippi, New Hampshire, Vermont, Wisconsin. Québec. Great Britain, Poland.

PLATE XCVIII, fig. 8; PLATE XCIX, figs. 1, 2.

39g. Micrasterias truncata var. truncata f. uralensis (Krieg.) Croasdale stat. nov.

Micrasterias truncata var. *uralensis* Krieger 1939, Rabenhorst's Kryptogamen-Flora 13(1:2): 34. Pl. 104, Fig. 5.

Cells small, 1.14 to 1.26 times longer than broad; the polar lobe broadly convex, truncate at its lateral extremities; subpolar incision shallow and open, with rounded interior; lateral lobes very shallowly once divided, all angles armed with short teeth; sinus opening outward. L. 80-88 μm. W. 70-79 μm. I. 16-25 μm.

Until more specimens are seen it might be better to consider this only a form. Its first illustrator, Woronichin (1930, p. 43, Pl. 1, Fig. 3) called it merely

M. truncata, and showed only 1 semicell, the only illustration up to now. We show a specimen from Montana (Pl. XCVIII, Fig. 7) with 1 semicell, quite like Woronichin's, the other with the polar lobe less truncate at the ends and more like a reduced form of the typical. The kind of variation in this form is shown in a more extreme expression in *M. truncata* var. *truncata* f. *reducta* Prescott.

DISTRIBUTION: Montana. Russia (Polar-Urals).

PLATE XCVIII, figs. 6, 7.

39h. Micrasterias truncata var. bahusiensis Wittrock 1869, Nova Acta Reg. Soc. Upsala, III, 7: 9. Pl. 1, Fig. 2.

Micrasterias truncata var. *mauritiana* Irénée-Marie 1949, Nat. Canadien 76: 29. Pl. 3, Figs. 5, 6.
Micrasterias truncata var. *mauritiana* f. *triangularis* Irénée-Marie 1949, Nat. Canadien 76: 30. Pl. 3, Figs. 7, 8. (Spelling corrected from "*mauricianum*" by Irénée-Marie (1957, p. 86).

Polar lobe outstanding, separated from the horizontally extended lateral lobes by a wide, usually sharp-angled incision; sinus usually opening outwardly; polar lobe convex, sometimes slightly retuse or flattened, its lateral extensions ending in a single tooth; lateral lobes once or twice divided, the ends bearing short paired teeth. L. 75–132 μm. W. 74–110 μm. W. at apex 60.5–85 μm. I. 17.8–25 μm.

This variety is distinguished by having its polar lobe quite widely separated from the broad, horizontally directed lateral lobes. Apparently closely related forms are f. *uralensis* Krieger and var. *pusilla* G. S. West.

DISTRIBUTION: Louisiana, Massachusetts. Québec. Great Britain, Europe, Africa, South America.

PLATE XCIX, figs. 3–6.

39i. Micrasterias truncata var. concatenata Wolle 1884a, Desm. U.S., p. 114 Pl. 38, Fig. 8.

Cells united in chains, the apical ends mutually flattened, the lateral lobes horizontally extended, twice divided, ending in 4 short teeth. L. c. 60 μm. W. c. 62 μm. W. at apex c. 43 μm. I. c. 30 μm.

Apparently Wolle's (*l.c.*) is the only record.

DISTRIBUTION: United States.

PLATE C, fig. 1.

39j. Micrasterias truncata var. crenata (Bréb.) Reinsch 1867, Abh. Naturh. Ges. Nürnberg 3(2): 143 f. **crenata.**

Micrasterias crenata Bréb. ex Ralfs 1848, British Desm., p. 75. Pl. VII, Fig. 2; Pl. 10, Fig. 4.

Plants slightly longer than broad, polar lobe broadly cuneate, with convex apex that may be flattened in middle; lateral lobes twice divided by very shallow incisions, and wholly without teeth at their ends. L. 75–125 μm. W. 74–102 μm. W. at apex 45–65 μm. I. 15–30 μm.

This variety resembles, and may be, merely a reduced form of *M. truncata* var. *truncata*. Lowe (1924, p. 41, Pl. 3, Fig. 3) shows from central Canada a form with small papillae within the subpolar incision and the sinus (Pl. C, Fig. 3). He remarks: "Sometimes the polar lobes were more pointed than is usual, and resembled *M. truncata*."

DISTRIBUTION: Widely distributed in U.S. and Canada. Great Britain, Europe, Asia, Africa, South America.

PLATE C, figs. 2, 3.

39k. **Micrasterias truncata** var. **crenata** f. **carteri** (Krieg.) Croasdale comb. nov.

Micrasterias crenata Bréb. ex Ralfs forma Carter 1935, Linn. Soc. Jour. Bot. 50: 158. Fig. 18.
Micrasterias truncata var. *carteri* Krieger 1939, Rabenhorst's Kryptogamen-Flora 13(1:2): 28. Pl. 103, Fig. 1.

Cells longer than broad (1.17 to 1.34 times); lateral lobes only once divided, the ends typically emarginate or only slightly dentate; polar lobe broadly convex, subpolar incision and sinus linear, closed or nearly so. L. 108–116 μm. W. 84–92 μm. W. at apex 70–75 μm. I. 27–30 μm.

A variety characterized chiefly by the relatively longer cells and the relatively smooth outline. A form found by Scott in Mississippi (Pl. C, Fig. 5) shows stronger dentation on one of the two semicells.

DISTRIBUTION: Mississippi. British Columbia.

PLATE C, figs. 4–6.

39l. **Micrasterias truncata** var. **mississippiensis** Prescott & Scott 1952, Trans. Amer. Microsc. Soc. 71: 251. Pl. 5, Fig. 7.

Cells circular; polar lobe transversely elliptic, subpolar incision deep, open interiorly with rounded apex, closed outwardly, dorsal margin convex with bidentate lateral extensions; lateral lobes very slightly incised, forming 2 broad lobules, each of which bears 2 widely separated teeth; face of semicell showing a protuberance within and at the base of the polar lobe, bearing 2 horizontally directed spines; in lateral view semicells broadly oval with 2 blunt spinelike processes on each side just above the midregion; in vertical view turgid, narrowly oval with a prominent protrusion in the midregion which bears 2 spines directed parallel with the longitudinal axis of the figure. L. 53–58 μm. W. 50–58 μm. I. 9–12 μm. Th. 27–32.1 μm.

What is probably a form of this variety from Florida (Pl. CI, Fig. 8) shows a facial protuberance without the horizontally directed spines. It resembles var. *truncata* f. *convexa* in the presence of 3 teeth on the outer dorsal margin of the polar lobe, but differs from both varieties in its more dissected lateral lobes, bearing a total of 8 teeth.

DISTRIBUTION: Florida, Mississippi, Wisconsin.

PLATE CI, figs. 5–7.

39m. **Micrasterias truncata** var. **pusilla** G. S. West 1914, Mém. Soc. Neuchatel 1914: 1035. Pl. 22, Figs. 42, 43 f. **pusilla**.

A small variety; cells subcircular with relatively deep and often open incisions; amount of dissection of lateral lobes variable; sinus opening rather widely from a narrow interior; polar lobe rounded and slightly retuse in middle, subpolar incision narrow to wide but never wide-angular as in *M. decemdentata* (Näg.) Arch. L. 51–65 μm. W. 54–66 μm. W at apex 38–42 μm. I. 11–15 μm.

In North America this variety seems extremely variable; the forms with deep incisions merge into *M. decemdentata*, from which var. *pusilla* differs primarily in its relatively short polar lobe and parallel-sided polar incision, and into *M. abrupta* West & West, from which the variety differs in its differently shaped polar lobe and closed sinus. Perhaps these three could be combined into one taxon.

DISTRIBUTION: Florida, Louisiana, Massachusetts, Wisconsin. Africa. South America, West Indies.

PLATE CII, figs. 1–5.

39n. Micrasterias truncata var. **pusilla** f. **cuneata** Prescott & Scott 1943, Pap. Michigan. Acad. Sci., Arts & Lettr. 28: 80. Pl. 2, Fig. 12.

Cells small, generally broader than long; the incisions and sinus deeper, open, with rounded interior and often sinuous margins; polar lobe shallower and more laterally extended, with stouter terminal teeth which are usually somewhat upcurved. L. 55–58 μm. W. 57–66 μm. I. 10–13.5 μm.

This form, with its deep and open subpolar incisions and long, sharp teeth, might equally well be considered a form of *M. decemdentata* (Näg.) Arch.

DISTRIBUTION: Florida, Louisiana, Mississippi.

PLATE CII, figs. 6, 7.

39o. Micrasterias truncata var. **turgida** Taylor 1935, Pap. Michigan Acad. Sci., Arts & Lettr. 20: 216. Pl. 46, Figs. 6–8; Pl. 47, Figs. 2, 3 f. **turgida.**

Cells usually very slightly longer than broad; polar lobe broad, in length occupying about two-fifths of the length of the semicell, with a curved dorsal margin bearing 1 or 2 strong teeth near the lateral extremities; lateral lobes simpler than usual in this species, once narrowly and shallowly cleft, with single strong teeth at the angles; subpolar incisions and sinus nearly closed; vertical view diamond-shaped, thick and rounded at the center, tapering directly to the acute margins, the ratio of thickness to width about 4 to 7; in lateral view the semicells subcircular, centrally somewhat inflated. L. 59–75 μm. W. 58–91 μm. I. 11–20 μm. Th. 28–48 μm.

DISTRIBUTION: Florida, Massachusetts, Montana, Wisconsin, Wyoming. Newfoundland, Québec. Japan (Kuriles), Venezuela.

PLATE C, Fig. 7; PLATE CI, figs. 1, 2.

39p. Micrasterias truncata var. **turgida** f. **elevata** Prescott f. nov.

Forma valde reducta, 1.16 plo longior quam lata, lobus polaris late rotundatus, sine incisione subpolari; lobi laterales vix indentati, sine dentibus; sinus profundus, per plus quam dimidium longitudinis inapertus; semicellula ab apice visa elliptico-rhomboidea, ratio crassitudinis et latitudinis circa 4 : 6.8. Cellulae 79 μm long., 68 μm lat., 18 μm lat. ad isthmum, 38.5 μm crass.

ORIGO: E palude *Carice* et *Equiseto* pleno in loco Mission Mountains, Montana dicto.

HOLOTYPUS: G. W. Prescott Coll. 6Mon-24.

ICONOTYPUS: Plate CI, fig. 3.

A much reduced form, 1.16 times longer than broad; polar lobe broadly rounded, without a subpolar incision; lateral lobes barely indented, with no teeth; sinus deep and closed for more than half its length; semicell in ventral view elliptic-rhomboid, ratio of thickness to width about 4 to 6.8. L. 79 μm. W. 68 μm. I. 18 μm. Th. 39.5 μm.

In outline this form is reminiscent of certain *Cosmaria* such as *C. hammeri* Reinsch var. *homalodermum* (Nordst.) West & West, but it is much larger and in vertical view the semicell is more sharply tapered. Compare some of the variable or teratological forms of *M. truncata* var. *turgida* figured by Taylor (1935. Pl. 47, Fig. 3).

DISTRIBUTION: Montana.

PLATE CI, fig. 3.

39q. Micrasterias truncata var. **turgida** f. **elobata** Prescott f. nov.

Forma reducta, 1.13 plo longior quam lata; lobus polaris late convexus ad extremitates laterales vix mucronatus, sine, autem, incisione subpolari; lobi laterales omnino non divisi, late rotundati, 3 dentibus acutis praediti. Cellulae 72 μm long., 64 μm lat., 11 μm lat. ad isthmum.

ORIGO: E loco Cascade Mountains, Oregon dicto.

HOLOTYPUS: G. W. Prescott Coll. OI-509.

ICONOTYPUS: Plate CI, fig. 4.

A reduced form, 1.13 times longer than broad; the polar lobe broadly convex, very slightly mucronate at the lateral extremities but without a subpolar incision; lateral lobes completely undivided, broadly rounded and furnished with 3 sharp teeth. L. 72 μm. W. 64 μm. I. 11 μm.

Compare *"M. truncata f. abnormale"* Irénée-Marie (1958a, Fig. 9) and *"M. truncata* variety" Rozeria (1954, p. 58, Figs. 7–15).

DISTRIBUTION: Cascade Mountains, Oregon.

PLATE CI, fig. 4.

40. Micrasterias verrucosa Biss., in Roy & Bissett 1893, Ann. Scottish Nat. Hist. 1893: 174. Pl. 1(4), Fig. 2 f. **verrucosa.**

Micrasterias verrucosa Roy, in Wolle 1885a, Bull. Torrey Bot. Club 12(12): 127. Pl. 51, Fig. 10.

Cells large, 1.1 to 1.2 times longer than broad, in outline resembling *M. denticulata* Bréb., with sinus and incisions very narrow or closed, with lateral lobes nearly equal, divided usually to the fourth order, their extremities crenate or weakly bidentate; polar lobe narrowly cuneate with very slightly concave lateral margins, depressed apex with rather deep open notch and rounded angles, sometimes with mammillate swellings on either side of the notch; surface of cell with a pattern of radially arranged low circular swellings, outlined with circles of granules and decreasing in size away from the middle of the cell; semicell in vertical view fusiform with a middle swelling on each side and lateral swellings decreasing in size toward the poles. L. 210–287 μm. W. 180–251 μm. W. at apex 60–70 μm. I. c. 29 μm. Zygospore unknown.

Since Roy & Bisset (*l.c.*) cite Bissett as the authority of this species, this is probably correct, although Wolle and others cite Roy.

DISTRIBUTION: Florida, Louisiana, Massachusetts, Mississippi, Montana, Virginia. Labrador. Great Britain, Europe.

PLATE CXXXVIII, figs. 2–7.

Key to the North American Forms of *Micrasterias verrucosa*

1. Cells less than 1.3 times longer than broad; apex depressed.	40. f. *verrucosa.*
1. Cells more than 1.3 times longer than broad; apex notched.	40a. f. *elongata.*

40+. Micrasterias verrucosa forma

A small form with a very broad polar lobe with concave lateral margins; otherwise as in the typical. Two plants seen. L. 139 μm. 163 μm. W. 115 μm, 132 μm. I. 22 μm, 24 μm.

The broad polar lobe with widely spreading apex, and the smaller size remind one of *M. tetraptera* West & West and *M. conferta* Lund. The flattened, deeply notched apex with mammillate tumors on either side, and the surface ornamentation, however, place the form in *M. verrucosa.*

DISTRIBUTION: Florida, Mississippi.

PLATE CXXXIX, fig. 1.

40a. Micrasterias verrucosa f. **elongata** Prescott & Scott 1952, Trans. Amer. Microsc. Soc. 71(3): 252. Pl. 4, Fig. 5.

A form differing from the typical by being relatively longer (1.32 times) and by having a broader and shorter polar lobe which has a deeper notch, with the angles decidedly outturned. L. 148 μm. W. 112 μm. I. 27 μm.

The figure of *M. verrucosa* shown as typical by Wolle (1885a, p. 127. Pl. 51, Fig. 10) is nearly as elongate (1.26 times) but has only a very shallow apical notch. Our form listed above has an equally broad polar lobe but is not as elongate. Perhaps these two forms fall within the normal range of variation of the species.

DISTRIBUTION: Louisiana.

PLATE CXXXIX, fig. 2.

MICRASTERIAS: North American Taxa Rejected or in Synonymy

Micrasterias americana (Ehrenb.) Ralfs in, Taylor 1935, p. 211. Pl. 45, Fig. 5 = *Micrasterias americana* var. *americana* f. *calcarata* Croasdale nom. nov.

Micrasterias americana f. *ampullacea* Maskell 1888, p. 9. Pl. 2, Fig. 13 = *Micrasterias mahabuleshwarensis* var. *ampullacea* (Mask.) Nordstedt 1888a, p. 30. Pl. 2, Fig. 16.

Micrasterias americana f. *hermanniana* (Reinsch) Maskell 1888, p. 9. Pl. 2, Fig. 11 = *Micrasterias mahabuleshwarensis* var. *wallichii* (Grun.) West & West 1905a, p. 122. Pl. 54, Figs. 7, 8; Pl. 55, Figs. 1–3.

Micrasterias americana f. *taylorii* Irénée-Marie 1949, p. 18. Pl. 1, Fig. 2 = *Micrasterias americana* var. *americana* f. *calcarata* Croasdale nom. nov.

Micrasterias americana var. *boldtii* Gutwinski 1890, p. 73; 1891, p. 74. Pl. 3, Fig. 27 = *Micrasterias americana* var. *americana* f. *boldtii* (Gutw.) Croasdale stat. nov.

Micrasterias americana var. *hermanniana* Reinsch in Wolle 1892, p. 124. Pl. 36, Fig. 5 = *Micrasterias mahabuleshwarensis* var. *wallichii* (Grun.) West & West 1905a, p. 122. Pl. 54, Figs. 7, 8; Pl. 55, Figs. 1–3.

Micrasterias americana var. *recta* Wolle 1876, p. 122; 1892, p. 124. Pl. 36, Fig. 3 = *Micrasterias americana* var. *americana* f. *boldtii* (Gutw.) Croasdale comb. nov.

Micrasterias americana var. *spinosa* Turner 1885, p. 936. Pl. 15, Fig. 13 = *Micrasterias americana* var. *americana* f. *spinosa* (Turn.) Croasdale stat. nov.

Micrasterias americana var. *spinulifera* Cushman 1904a, p. 359 = *Micrasterias americana* (Ehrenb.) Ralfs 1848, pp. xix, 74. Pl. 10, Figs. 1a–d var. *americana* f. *americana.*

Micrasterias americana var. *wollei* Raciborski 1889, p. 106. Pl. 7, Fig. 3 = *Micrasterias americana* (Ehrenb.) Ralfs 1848, pp. xix, 74. Pl. 10, Figs. 1a–d var. *americana* f. *americana.*

Micrasterias angulosa Hantzsch in Rabehnorst 1862, No. 1407 = *Micrasterias denticulata* var. *angulosa* (Hantzsch) West & West 1902, p. 30.

Micrasterias apiculata var. *brachyptera* (Lund.) West & West 1905a, p. 101. Pl. 46, Fig. 5; Pl. 47, Fig. 6 = *Micrasterias brachyptera* Lundell 1871, p. 12. Pl. 1, Fig. 4.

Micrasterias apiculata var. *brachyptera* f. *rectangularis* Irénée-Marie & Hilliard 1963, p. 109. Pl. 2, Fig. 17 = *Micrasterias brachyptera* Lundell 1871, p. 12. Pl. 1, Fig. 4.

Micrasterias apiculata var. *simplex* Irénée-Marie 1952, p. 142, Pl. 13, Fig. 2 = *Micrasterias brachyptera* Lundell 1871, p. 12. Pl. 1, Fig. 4.

Micrasterias arcuata var. *expansa* f. *intermedia (p.p.)* Nordstedt 1878, p. 22. Textfig. II: 4 on p. 23 = *Micrasterias arcuata* var. *expansa* Nordstedt 1878, p. 23. Textfig. II:5, 6b on p. 23.

Micrasterias arcuata var. *expansa* f. *intermedia (p.p.)* Nordstedt 1878, p. 23. Textfig. II: 3 on p. 23 = *Micrasterias arcuata* Bailey 1851, p. 37. Pl. 1, Fig. 6 var. *arcuata.*

Micrasterias arcuata var. *robusta* f. *recurvata* Prescott & Scott 1952, p. 232, Pl. 5, Fig. 5 = *Micrasterias arcuata* var. *expansa* f. *recurvata* (Presc. & Scott) Prescott comb. nov.

Micrasterias baileyi Ralfs 1848, p. 211. Pl. XXXV, Fig. 4 = ?*Micrasterias mahabuleshwarensis* var. *ringens* (Bail.) Krieger 1939, p. 52. Pl. 110, Fig. 9; in Wolle 1892, p. 130, Pl. 41, Fig. 6 = ?*Micrasterias muricata* var. *simplex* Grönblad ex Krieger 1939.

Micrasterias brachyptera var. *americana* Wolle 1883, p. 18. Pl. 27, Fig. 19 = *Micrasterias brachyptera* Lundell 1871, p. 12. Pl. 1, Fig. 4.

Micrasterias conferta Lund. in Taylor 1935, p. 212. Pl. 46, Fig. 3 = *Micrasterias tetraptera* f. *taylorii* (Krieg.) Croasdale stat. nov.

Micrasterias conferta var. *glabra* Prescott & Scott 1943, p. 70. Textfig. 1 = *Micrasterias conferta* var. *hamata* f. *glabra* (Presc. & Scott) Prescott comb. nov.

Micrasterias conferta var. *hamata* f. *spinosa* Prescott & Scott 1943, p. 71. Textfig. 2 = *Micrasterias conferta* var. *hamata* Wolle 1883, p. 19. Pl. 27, Fig. 1 f. *hamata.*

Micrasterias conferta var. *novae-angliae* Cushman 1904, p. 583. Pl. 26, Fig. 8; and in Taylor 1935, p. 212. Pl. 43, Fig. 3; Pl. 48, Fig. 1 = *Micrasterias novae-terrae* (Cushm.) Krieger 1939, p. 91. Pl. 127, Figs. 6, 7.

Micrasterias cornuta Bennett 1886, p. 7. Pl. 1, Fig. 6 = *Micrasterias denticulata* Bréb. ex Ralfs 1848, p. 70. Pl. 7, Fig. 1 var. *denticulata*.

Micrasterias crenata Bréb. ex Ralfs 1848, p. 75. Pl. VII, Fig. 2; Pl. 10, Fig. 4 = *Micrasterias truncata* var. *crenata* (Bréb.) Reinsch 1867, p. 143; in Wailes 1931, p. 75. Pl. 2, Fig. 19 = *Micrasterias truncata* var. *truncata* f. *neodamensis* (A. Br.) Dick 1926, p. 448. Pl. 20, Fig. 8.

Micrasterias crenata forma Carter 1935, p. 158. Fig. 18 = *Micrasterias truncata* var *crenata* f. *carteri* (Krieg.) Croasdale comb. nov.

Micrasterias crux-melitensis (Ehrenb.) Ralfs 1848, p. 73. Pl. IX, Fig. 3 in Woodhead & Tweed 1960, p. 348. Pl. 2, Fig. 12 = *Micrasterias crux-melitensis* f. *evoluta* Turner 1892, p. 92.

Micrasterias crux-melitensis var. *janeira* (Racib.) Grönblad 1920, p. 35. Pl. 6, Figs. 17, 18 = *Micrasterias crux-melitensis* (Racib.) Croasdale comb. nov.

Micrasterias crux-melitenis var. *rabenhorstii* (Kirchn.) Krieger 1939, p. 66 , Pl. 115, Fig. 4 = *Micrasterias crux-melitensis* f. *rabenhorstii* (Kirchn.) Croasdale comb. nov.

Micrasterias crux-melitensis var. *simplex* Borge 1894, p. 35. Pl. 3, Fig. 40 = *Micrasterias crux-melitensis* f. *simplex* (Borge) Croasdale stat. nov.

Micrasterias crux-melitensis var. *spinosa* Roll 1925, pp. 248, 252. Pl. 14, Fig. 3 = *Micrasterias crux-melitensis* f. *spinosa* (Roll) Croasdale stat. nov.

Micrasterias crux-melitensis var. *superflua* Turner 1885, p. 936. Pl. 15, Fig. 11 = *Micrasterias crux-melitensis* f. *superflua* (Turn.) Croasdale stat. nov.

Micrasterias decemdentata (Näg.) Archer, in Miller 1936, p. 82. Pl. 5, Fig. 3 and Wolle 1884a, p. 113. Pl. 33, Figs. 5, 6 = *Micrasterias truncata* var. *truncata* f. *neodamensis* (A. Br.) Dick 1926, p. 448. Pl. 20, Fig. 8.

Micrasterias denticulata f. *mucronata* Irénée-Marie 1951, p. 182. Pl. 1, Fig. 4 =*Micrasterias rotata* var. *japonica* Fujisawa 1936, p. 14. Pl. 2, Figs. 6a, b.

Micrasterias denticulata var. *angustosinuata* Gay 1884a, p. 52. Pl. 1, Fig. 4; 1884, p. 334 = *Micrasterias denticulata* var. *angulosa* (Hantzsch) West & West 1902, p. 30.

Micrasterias denticulata var. *minnesotensis* Turner 1885, p. 937. Pl. 16, Fig. 14 = *Micrasterias rotata* var. *japonica* Fujisawa 1936, p. 14. Pl. 2, Figs. 6a, b.

Micrasterias denticulata var. *notata* Nordstedt 1888a, p. 29. Pl. 2, Fig. 13 = *Micrasterias thomasiana* var. *notata* (Nordst.) Grönblad 1920, p. 38.

Micrasterias depauperata Nordstedt 1870, p. 222. Textfig. forma West & West 1896a, p. 238. Pl. 4, Fig. 1 = *Micrasterias depauperata* var. *wollei* Cushman 1904a, p. 396. Textfig. 3 f. *wollei*.

Micrasterias depauperata var. *kitchelii* (Wolle) West & West 1896a, p. 239 in Smith 1924, p. 42. Pl. 59, Fig. 4; in Krieger 1939, p. 39. Pl. 106, Figs. 3, 4; in Croasdale & Grönblad 1964, p. 172. Pl. 9, Fig. 1; in Prescott & Scott 1943, p. 73. Pl. 3, Fig. 5 = *Micrasterias depauperata* var. *wollei* Cushman 1904a, p. 396. Textfig. 3 f. *wollei*.

Micrasterias depauperata var. *kitchelii* (Wolle) West & West f. *apiculata* (Irénée-Marie) Graffius 1963, p. 128. Pl. 7, Fig. 5 = *Micrasterias depauperata* var. *wollei* f. *apiculata* Irénée-Marie 1952, p. 146. Pl. 14, Figs. 2–5.

Micrasterias dichotoma Wolle 1884a, p. 111. Pl. 52, Fig. 2 = *Micrasterias radiata* var. *dichotoma* (Wolle) Cushman 1908, p. 108.

Micrasterias dichotoma Wolle 1884a, p. 111. Pl. 52, Fig. 2 in Lagerheim 1885, p. 231. Pl. 27, Fig. 4 = *Micrasterias radiata* var. *lagerheimii* Krieger 1939, p. 71. Pl. 118, Fig. 2.

Micrasterias expansa Bailey 1851, p. 37. Pl. 1, Fig. 7 = *Micrasterias arcuata* var. *expansa* (Bail.) Nordstedt 1878, p. 22. Textfig. II: 5, 6b on p. 23 f. *expansa*.

Micrasterias expansa var. *robusta* Borge 1903 (wrong cit.) in Irénée-Marie 1949, p. 22. Pl. 1, Fig. 5 =*Micrasterias arcuata* var. *robusta* Borge 1899, p. 27. Pl. 2, Figs. 38, 39.

Micrasterias fimbriata Ralfs 1848, p. 71. Pl. VIII, Fig. 2 = *Micrasterias apiculata* var. *fimbriata* (Ralfs) Nordstedt 1888, p. 187. Pl. 6, Figs. 1, 2 f. *fimbriata*.

Micrasterias fimbriata f. *elephanta* Wolle 1884a, p. 110. Pl. 36, Fig. 3 = *Micrasterias apiculata* var. *fimbriata* f. *elephanta* (Wolle) Croasdale comb. nov.

Micrasterias fimbriata var. *elephanta* (Wolle) Krieger 1939, p. 82. Pl. 125, Fig. 2 = *Micrasterias apiculata* var. *fimbriata* f. *elephanta* (Wolle) Croasdale comb. nov.

Micrasterias fimbriata var. *nuda* Wolle 1884a, p. 110. Pl. 36, Fig. 4 = *Micrasterias rotata* (Grev.) Ralfs 1848, p. 71. Pl. VIII, Fig. 1a var. *rotata* f. *rotata*.

Micrasterias fimbriata var. *simplex* Wolle 1892, p. 121. Pl. 40, Fig. 8 = *Micrasterias apiculata* var. *fimbriata* (Ralfs) Nordstedt 1888, p. 187. Pl. 6, Figs. 1, 2 f. *fimbriata*.

Micrasterias fimbriata var. *spinosa* Biss., in Roy & Bissett 1893, p. 173. Pl. 1(4), Fig. 3. =

Micrasterias apiculata (Ehrenb.) Menegh. var. *fimbriata* f. *spinosa* (Biss.) West & West 1905a, p. 100. Pl. 47, Fig. 5.

Micrasterias floridensis f. *spinosa* Whelden 1941, p. 269. Pl. 6, Fig. 8 = *Micrasterias floridensis* var. *spinosa* (Wheld.) Prescott & Scott 1952, p. 236. Pl. 2, Fig. 8.

Micrasterias foliacea var. *granulifera* Cushman 1908, p. 111 = *Micrasterias foliacea* Bréb. ex Ralfs 1848, p. 210. Pl. XXXV, Fig. 3 var. *foliacea*.

Micrasterias furcata Ag. ex Ralfs 1848, p. 73. Pl. IX, Fig. 2 = *Micrasterias radiata* Hassall 1845, p. 386. Pl. 90, Fig. 2 var. *radiata* f. *radiata*.

Micrasterias furcata var. *decurta* Turner 1885, p. 936. Pl. 16, Fig. 10 = *Micrasterias radiata* Hassall 1845, p. 386. Pl. 90, Fig. 2 var. *radiata* f. *radiata*.

Micrasterias furcata var. *simplex* Wolle 1885a, p. 128. Pl. 51, Figs. 6, 7 = *Micrasterias radiata* Hassall 1845, p. 386. Pl. 90, Fig. 2 var. *radiata* f. *radiata*.

Micrasterias granulata Wood 1874, p. 146. Pl. 21, Fig. 16 = *Micrasterias conferta* Lundell 1871, p. 14. Pl. 1, Fig. 5 var. *conferta*.

Micrasterias janeira Raciborski 1885, p. 97. Pl. 14, Fig. 4 = *Micrasterias crux-melitensis* f. *janeira* (Racib.) Croasdale comb. nov.

Micrasterias jenneri var. *lata* Woodhead & Tweed 1960, p. 327. Pl. 2, Fig. 16 = ?*Micrasterias tetraptera* var. *tetraptera* f. *taylorii* (Krieg.) Croasdale stat. nov.

Micrasterias johnsonii West & West 1898, p. 297. Pl. 16, Fig. 5 forma Prescott & Scott 1952, p. 236. Pl. 3, Fig. 1 = *Micrasterias johnsonii* f. *maior* Prescott f. nov.

Micrasterias johnsonii var. *novae-angliae* Whelden 1942, p. 178. Fig. 3 = *Micrasterias johnsonii* var. *ranoides* f. *novae-angliae (Wheld.) Croasdale comb. nov.*

Micrasterias johnsonii var. *papillata* f. *canadense* Irénée-Marie 1957, p. 75. Figs. 2, 3 = *Micrasterias floridensis* var. *floridensis* f. *canadensis* (Irénée-Marie) Croasdale comb. nov.

Micrasterias kitchelii Woll 1880, p. 45. Pl. 5, Fig. M = *Micrasterias depauperata* var. *wollei* Cushman 1904a, p. 396. Textfig. 3 f. *wollei*.

Micrasterias kitchelii Wolle 1884a, p. 116. Pl. 37, Figs. 1-3 = *Micrasterias depauperata* var. *kitchelii* (Wolle) West & West 1896a, p. 239.

Micrasterias kitchelii "var." Wolle 1892. Pl. 42, Fig. 2 = *Micrasterias depauperata* var. *wollei* Cushman 1904, p. 396. Textfig. 3 f. *wollei*.

Micrasterias laticeps f. *major* Nordstedt 1870, p. 220. Pl. 2, Fig. 14 = *Micrasterias laticeps* Nordstedt 1870, p. 220. Pl. 2, Fig. 14 var. *laticeps* f. *laticeps*.

Micrasterias laticeps var. *crassa* Prescott & Magnotta forma Prescott & Scott 1943; p. 75. Pl. 2, Fig. 6 = *Micrasterias laticeps* var. *crassa* f. *robustior* Förster 1964, p. 378. Pl. 14, Fig. 6; Pl. 45, Photos 3, 4.

Micrasterias mahabuleshwarensis f. *dichotoma* Smith 1922 p.p. p. 345. Pl. 9, Fig. 12; Irénée-Marie 1938, p. 231. Pl. 40, Fig. 1 = *Micrasterias mahabuleshwarensis* Hobson 1863, p. 169. Fig. on p. 168 var. *mahabuleshwarensis* f. *mahabuleshwarensis*.

Micrasterias mahabuleshwarensis var. *americana* Wolle 1881, p. 1. Pl. 6, Fig. 1 = *Micrasterias mahabuleshwarensis* Hobson 1863, p. 169. Fig. on p. 168 var. *mahabuleshwarensis* f. *mahabuleshwarensis*.

Micrasterias mahabuleshwarensis var. *compacta* f. *americana* Nordstedt 1888a, p. 31 = *Micrasterias mahabuleshwarensis* Hobson 1863, p. 169. Fig. on p. 168 var. *mahabuleshwarensis* f. *mahabuleshwarensis*.

Micrasterias mahabuleshwarensis var. *serrulata* (Wolle) Smith 1924a, p. 143. Pl. 16, Fig. 2; Irénée-Marie 1938, p. 231. Pl. 40, Fig. 1 = *Micrasterias mahabuleshwarensis* Hobson 1863, p. 169. Fig. on p. 169 var. *ringens* (Bail.) Krieger 1939, p. 52. Pl. 110, Fig. 9.

Micrasterias mamillata Turner 1885, p. 936. Pl. 16, Fig. 12 = *Micrasterias apiculata* var. *apiculata* f. *mamillata* (Turn.) Croasdale comb. nov.

Micrasterias moebii (C. Agardh) Ralfs 1848, in Meyer & Brooks 1968, p. 262 = Sp. Inquir.

Micrasterias mucronata (Dixon) Rabenhorst 1868, p. 187. Textfig. 68b = *Micrasterias oscitans* var. *mucronata* (Dixon) Wille 1880, p. 21. Pl. 1, Fig. 3.

Micrasterias mucronata f. *intermedia* Nordstedt 1873, p. 6 = *Micrasterias oscitans* Ralfs 1848, p. 76. Pl. X, Fig. 2 var. *oscitans*.

Micrasterias muricata (Bail.) Ralfs 1848, p. 210, in Grönblad 1956, p. 26. Pl. 5, Figs. 38, 39; Bourrelly 1966, p. 96. Pl. 13, Fig. 3; Förster 1972, p. 543. Pl. 11, Fig. 1 = *Micrasterias muricata* var. *muricata* f. *basituberculata* Scott ex Croasdale f. nov.

Micrasterias muricata (Bail.) Ralfs "abnormal semicells," in West & West 1896a, p. 239. Pl. 14, Figs. 5, 6 = *Micrasterias muricata* var. *laevigata* Irénée-Marie 1952, p. 148. Pl. 14, Fig. 9.

Micrasterias muricata var. *furcata* Förster 1972, p. 543. Pl. 11, Fig. 2 = *Micrasterias muricata* var. *laevigata* Irénée-Marie 1952, p. 148. Pl. 14, Fig. 9.

Micrasterias murrayi West & West 1903, p. 538. Pl. 15, Figs. 1, 2 = *Micrasterias radiosa* var. *murrayi* (West & West) Allorge & Allorge 1931, p. 352. Pl. 7, Fig. 6 f. *murrayi.*

Micrasterias murrayi var. *glabra* Irénée-Marie 1949a, p. 289. Pl. 3, Fig. 32 = *Micrasterias radiosa* var. *murrayi* f. *glabra* (Irénée-Marie) Croasdale comb. nov.

Micrasterias novae-terrae var. *spinosa* Prescott & Scott 1952, p. 242. Pl. 2, Figs. 1, 2 = *Micrasterias novae-terrae* var. *speciosa* (Wolle) Krieger & Bourrelly 1956, p. 152. Pl. 6, Figs. 56-59.

Micrasterias oscitans var. *inflata* Wolle 1876, p. 122 = *Micrasterias pinnatifida* var. *pinnatifida* f. *inflata* (Wolle) Croasdale stat. nov.

Micrasterias oscitans var. *pinnatifida* (Kütz.) Rabenhorst 1868, p. 189 = *Micrasterias pinnatifida* (Kütz.) Ralfs 1848, p. 77. Pl. X, Fig. 3 var. *pinnatifida* f. *pinnatifida.*

Micrasterias papillifera Bréb. ex Ralfs forma Taylor 1935, p. 214. Pl. 46, Fig. 2 = *Micrasterias tetraptera* var. *longesinuata* Krieger 1939, p. 93. Pl. 127, Fig. 2.

Micrasterias papillifera f. *maior* West & West 1905, p. 92. Pl. 44, Fig. 3 = *Micrasterias papillifera* Bréb. ex Ralfs 1848, p. 72. Pl. IX, Fig. 1 var. *papillifera* f. *papillifera.*

Micrasterias papillifera f. *Murrayi* (West & West) Kossinskaja 1960, p. 463. Pl. 72, Figs. 8-10 = *Micrasterias radiosa* var. *murrayi* (West & West) Allorge & Allorge 1931, p. 352. Pl. 7, Fig. 6 f. *murrayi.*

Micrasterias papillifera var. *novae-scotiae* Turner 1885, p. 937. Pl. 15, Fig. 16 = *Micrasterias papillifera* var. *papillifera* f. *novae-scotiae* (Turn.) Croasdale.

Micrasterias papillifera var. *speciosa* (Wolle) Krieger 1939, p. 90. Pl. 130, Fig. 3 = *Micrasterias novae-terrae* var. *speciosa* (Wolle) Krieger & Bourrelly 1956, p. 152. Pl. 6, Figs. 56-59.

Micrasterias papillifera var. *varvicensis* Turner 1893, p. 345. Fig. 13 = *Micrasterias papillifera* Bréb. ex Ralfs 1848, p. 72, Pl. IX, Fig. 1 var. *papillifera* f. *papillifera.*

Micrasterias papillifera var. *verrucosa* Schmidle 1896, p. 23. Textfig. on p. 23 = *Micrasterias papillifera* Bréb. ex Ralfs 1848, p. 72. Pl. IX, Fig. 1 var. *papillifera* f. *papillifera.*

Micrasterias phymatophora Prescott & Scott 1952, p. 242. Pl. 4, Fig. 3 = *Micrasterias conferta* var. *phymatophora* (Presc. & Scott) Prescott comb. nov.

Micrasterias phymatophora f. *circularis* Prescott & Scott 1952, p. 244. P. 4, Fig. 4 = *Micrasterias conferta* var. *phymatophora* (Presc. & Scott) Prescott comb. nov.

Micrasterias pinnatifida f. *rhomboidea* Brunel 1938, p. 71. Fig. 1b = *Micrasterias pinnatifida* var. *pseudoscitans* Grönblad 1920, p. 36. Pl. 6, Figs. 7, 8.

Micrasterias pinnatifida var. *divisa* West 1891, p. 354. Pl. 315, Fig. 8 = *Micrasterias pinnatifida* var. *pinnatifida* f. *furcata* (Krieg.) Croasdale stat. nov.

Micrasterias pinnatifida var. *furcata* Krieger 1939, p. 18. Pl. 99, Figs. 11, 12 = *Micrasterias pinnatifida* var. *pinnatifida* f. *furcata* (Krieg.) Croasdale stat. nov.

Micrasterias pinnatifida var. *inflata* Wolle 1881, p. 1. Pl. 6, Fig. 5 = *Micrasterias pinnatifida* var. *pinnatifida* f. *inflata* (Wolle) Croasdale stat. nov.

Micrasterias pinnatifida var. *inflata* f. *ornata* Irénée-Marie 1938, p. 220. Pl. 33, Fig. 14 = *Micrasterias pinnatifida* var. *pinnatifida* f. *inflata* (Wolle) Croasdale comb. nov.

Micrasterias pinnatifida var. *tridentata* Krieger 1939, p. 20. Pl. 100, Figs. 8, 9 = *Micrasterias pinnatifida* var. *pinnatifida* f. *tridentata* (Krieg.) Croasdale stat. nov.

Micrasterias pinnatifida var. *trigona* West 1889a, p. 206. Pl. 291, Fig. 15 = *Micrasterias pinnatifida* var. *pinnatifida* f. *trigona* (West) Croasdale stat. nov.

Micrasterias pseudofurcata Wolle 1881, p. 1. Pl. 6, Fig. 3 = *Micrasterias radiata* (Ehrenb.) Hassall 1845, p. 386. Pl. 90, Fig. 2 var. *radiata* f. *radiata.*

?*Micrasterias pseudofurcata* Wolle 1892, p. 122. Pl. 39, Fig. 4 = *Micrasterias muricata* var. *simplex* Grönblad ex Krieger 1939, p. 74. Pl. 120, Figs. 1, 2.

?*Micrasterias pseudofurcata* var. *minor* Wolle 1892, p. 122. Pl. 41, Fig. 11 = *Micrasterias muricata* var. *simplex* Grönblad ex Krieger 1939, p. 74. Pl. 120, Figs. 1, 2.

Micrasterias pseudotorreyi Wolle 1883, p. 19. Pl. 27, Fig. 2 = *Micrasterias triangularis* Wolle 1881, p. 1. Pl. 6, Fig. 4.

Micrasterias quadrata Bailey 1851, p. 37. Pl. 1, Fig. 5 = *Micrasterias pinnatifida* var. *pinnatifida* f. *quadrata* (Bail.) Croasdale comb. nov.

Micrasterias rabenhorstii Kirchner 1878, p. 163 = *Micrasterias crux-melitensis* f. *rabenhorstii* (Kirchn.) Croasdale comb. nov.

Micrasterias radians var. *ambadiensis* Grönbl. & Scott, in Grönblad, Prowse & Scott 1958, p. 21. Pl. 11, Fig. 119; Pl. 26, Fig. 363 = *Micrasterias ambadiensis* (Grönbl. & Scott) Thomasson 1960, p. 22. Pl. 4, Fig. 10; Pl. 10, Fig. 4.

Micrasterias radiata Hassall, in Brown 1930, p. 114. Pl. 11, Fig. 12. = *Micrasterias radiata* var. *gracillima* Smith, 1922, p. 244, Pl. 9, Figs. 6-11 f. *gracillima.*

Micrasterias radiata f. *deflexa* Irénée-Marie 1951, p. 188. Pl. 2, Figs. 2–4 = *Micrasterias radiata* Hassall 1845, p. 386. Pl. 90, Fig. 2 var. *radiata* f. *radiata.*

Micrasterias radiata var. *simplex* (Wolle) Smith 1922, p. 344. Pl. 9, Figs. 1–3 = *Micrasterias radiata* Hassall 1845, p. 386. Pl. 90, Fig. 2 var. *radiata* f. *radiata.*

Micrasterias radiosa Ralfs (*p.p.*) in Wolle 1892, p. 119. Pl. 35, Fig. 3 = *Micrasterias radiosa* var. *ornata* Nordstedt 1870, p. 223. Pl. 2, Fig. 11 f. *ornata.*

Micrasterias radiosa Ralfs forma West & West 1896a, p. 240. Pl. 13, Fig. 30 = *Micrasterias radiosa* var. *elegantior* (G. S. West) Croasdale comb. nov.

Micrasterias radiosa f. *papillifera* Irénée-Marie 1951, p. 190. Pl. 2, Fig. 8 = *Micrasterias radiosa* var. *ornata* Nordstedt 1870, p. 223.

Micrasterias radiosa var. *aculeata* Krieg. forma Prescott & Scott 1943, p. 77. Pl. 6, Fig. 7 = *Micrasterias radiosa* Ralfs 1848, p. 72. Pl. VIII, Fig. 3 var. *radiosa* f. *radiosa.*

Micrasterias radiosa var. *extensa* Prescott & Scott 1943, p. 77. Pl. 4, Figs. 7, 8, 10; 1952, p. 249. Pl. 1, Fig. 5 (as "forma") = *Micrasterias radiosa* var. *swainei* (Hast.) West & West 1896a, p. 240.

Micrasterias radiosa var. *ornata* f. *elegantior* (G. S. West) Smith 1924, p. 47. Pl. 60, Fig. 4 = *Micrasterias radiosa* var. *elegantior* (G.S. West) Croasdale comb. nov.

Micrasterias radiosa var. *punctata* West 1889, p. 20. Pl. 2 Fig. 12 in Irénée-Marie 1951, p. 192. Pl. 3, Fig. 1 = *Micrasterias radiosa* var. *punctata* f. *concentrica* Croasdale f. nov.

Micrasterias radiosa var. *taylorii* Irénée-Marie 1952, p. 154. Pl. 15, Fig. 8 = *Micrasterias radiosa* Ralfs 1848, p. 72. Pl. VIII, Fig. 3 var. *radiosa* f. *radiosa.*

Micrasterias radiosa var. *wollei* Cushman 1904a, p. 394 = *Micrasterias radiosa* var. *ornata* Nordstedt 1870, p. 223. Pl. 2, Fig. 11 f. *ornata.*

Micrasterias ranoides Salisbury 1936, p. 59. Pl. 1, Fig. 1 = *Micrasterias johnsonii* var. *ranoides* (Salisb.) Krieger 1939, p. 73. Pl. 119, Fig. 3 f. *ranoides.*

Micrasterias ringens Bailey 1851, p. 37. Pl. 1, Fig. 11 = *Micrasterias mahabuleshwarensis* var. *ringens* (Bail.) Krieger 1939, p. 52. Pl. 110, Fig. 9 f. *ringens.*

Micrasterias ringens var. *serrulata* Wolle 1885a, p. 128. Pl. 51, Fig. 15 = *Micrasterias mahabuleshwarensis* var. *ringens* (Bail.) Krieger 1939, p. 52. Pl. 110, Fig. 9 f. *ringens.*

Micrasterias rotata f. *inermis* Irénée-Marie 1951, p. 193. Pl. 3, Fig. 5 = *Micrasterias rotata* (Grev.) Ralfs 1848, p. 71. Pl. VIII, Fig. 1a var. *rotata* f. *rotata.*

Micrasterias rotata f. *nuda* (Wolle) Irénée-Marie 1938, p. 230. Pl. 37, Fig. 1 = *Micrasterias rotata* (Grev.) Ralfs 1848, p. 71. Pl. 8, Fig. 1a var. *rotata* f. *rotata.*

Micrasterias rotata var. *papillifera* Raciborski 1895, p. 34. Pl. 4, Fig. 17 = *Micrasterias apiculata* var. *fimbriata* f. *spinosa* (Biss.) West & West 1905a, p. 100. Pl. 47, Fig. 5.

Micrasterias sol (Ehrenb.) Kützing 1849, p. 171; West & West 1905a. p. 95. Pl. 46, Figs. 1, 2 = *Micrasterias radiosa* Ralfs 1848, p. 72. Pl. VIII, Fig. 3 var. *radiosa* f. *radiosa.*

Micrasterias sol var. *aculeata* Krieger 1939, p. 94. Pl. 131, Fig. 2 = *Micrasterias radiosa* var. *ornata* f. *aculeata* (Krieg.) Croasdale comb. nov.

Micrasterias sol var. *elegantior* f. *glabra* Bourrelly 1966, p. 98. Pl. 14, Fig. 4 = *Micrasterias radiosa* var. *elegantior* (G. S. West) Croasdale comb. nov.

Micrasterias sol var. *ornata* Nordstedt 1870, p. 223. Pl. 2, Fig. 11 in Taft 1934, p. 96. Pl. 6, Fig. 24 = *Micrasterias papillifera* Bréb. ex Ralfs 1848, p. 72. Pl. IX, Fig. 1 var. *papillifera* f. *papillifera.*

Micrasterias sol var. *ornata* f. *elegantior* G. S. West 1914, p. 1035. Pl. 22, Fig. 44 = *Micrasterias radiosa* var. *elegantior* (G. S. West) Croasdale comb. nov.

Micrasterias speciosa Wolle 1885, p. 4; 1887, p. 38. Pl. 56, Figs. 1, 2 = *Micrasterias novae-terrae* var. *speciosa* (Wolle) Krieger & Bourrelly 1956, p. 152. Pl. 6, Figs. 56–59.

Micrasterias speciosa Wolle forma Taylor 1935, p. 215. Pl. 48, Fig. 2 = *Micrasterias novae-terrae* (Cushm.) Krieger 1939, p. 91. Pl. 127, Figs. 6, 7 var. *novae-terrae.*

Micrasterias speciosa Wolle forma West & West 1896a, p. 240, Pl. 14, Fig. 10 = *Micrasterias novae-terrae* var. *speciosa* (Wolle) Krieger & Bourrelly 1956, p. 152. Pl. 6, Figs. 56–59.

Micrasterias suboblonga var. *maxima* Prescott & Scott 1943, p. 78. Pl. 1, Fig. 7 = *Micrasterias suboblonga* var. *australis* f. *maxima* (Presc. & Scott) Prescott comb. nov.

Micrasterias swainei Hastings in Wolle 1892, p. 119. Pl. 42, Fig. 1 = *Micrasterias radiosa* var. *swainei* (Hast.) West & West 1896a, p. 240.

Micrasterias tetraptera var. *alobulifera* Prescott & Scott 1952, p. 249. Pl. 5, Figs. 10 = *Micrasterias tetraptera* West & West var. *angulosa* f. *alobulifera* (Presc. & Scott) Prescott comb. nov.

Micrasterias tetraptera var. *protuberans* Prescott & Scott 1943, p. 79. Pl. 5, Figs. 5, 6 = *Micrasterias tetraptera* var. *longesinuata* f. *protuberans* (Presc. & Scott) Prescott comb. nov.

Micrasterias tetraptera var. *taylorii* Krieger 1939, p. 93. Pl. 127, Fig. 3 = *Micrasterias tetraptera* var. *tetraptera* f. *taylorii* (Krieg.) Croasdale stat. nov.

Micrasterias thomasiana Arch. var. Taylor 1935, p. 215, Pl. 45, Fig. 7 = *Micrasterias denticulata* var. *taylorii* Krieger 1939, p. 108. Pl. 137, Fig. 7 f. *taylorii.*

Micrasterias torreyi f. *punctata* Irénée-Marie 1951, p. 194. Pl. 4, Fig. 2, 3 = *Micrasterias torreyi* Bail. ex Ralfs 1848, p. 210. Pl. XXXV, Fig. 5 var. *torreyi.*

Micrasterias torreyi var. *nordstedtiana* (Hieron.) Schmidle 1898a, p. 48 forma in Prescott & Scott 1943, p. 79. Pl. 3, Fig. 2 = *Micrasterias torreyi* Bail. ex Ralfs 1848, p. 210. Pl. XXXV, Fig. 5 var. *torreyi.*

Micrasterias truncata (Corda) Bréb., Ralfs 1848, p. 75. Pl. VIII, Fig. 4; Pl. 10, Fig. 5 (*p.p.*) in Wolle 1892, p. 126. Pl. 44, Fig. 9 = *Micrasterias abrupta* West & West 1896a, p. 241. Pl. 14, Figs. 13–16.

Micrasterias truncata f. *granulata* Raciborski 1889, p. 105. Pl. 7, Fig. 1 = *Micrasterias truncata* (Corda) Bréb. ex Ralfs 1848, p. 75. Pl. VIII, Fig. 4; Pl. 10, Fig. 5 var. *truncata* f. *truncata.*

Micrasterias truncata var. *carteri* Krieger 1939, p. 28. Pl. 103, Fig. 1 = *Micrasterias truncata* var. *crenata* f. *carteri* (Krieg.). Croasdale comb. nov.

Micrasterias truncata var. *convexa* Prescott & Scott 1943, p. 80. Pl. 2, Fig. 3 = *Micrasterias truncata* var. *truncata* f. *convexa* (Presc. & Scott) Croasdale stat. nov.

Micrasterias truncata var. *mauritiana* Irénée-Marie 1949, p. 29. Pl. 3, Figs. 5, 6 = *Micrasterias truncata* var. *bahusiensis* Wittrock 1869, p. 9. Pl. 1, Fig. 2.

Micrasterias truncata var. *mauritiana* f. *triangularis* Irénée-Marie 1949, p. 30. Pl. 3, Figs. 7, 8 = *Micrasterias truncata* var. *bahusiensis* Wittrock 1869, p. 9. Pl. 1, Fig. 2.

Micrasterias truncata var. *minor* Wolle 1884a, p. 114. Pl. 38, Fig. 9 = *Micrasterias truncata* var. *truncata* f. *semiradiata* (Näg.) Cleve 1864, p. 487.

Micrasterias truncata var. *neodamensis* (A. Br.) Dick in Krieger 1939, p. 30 = *Micrasterias truncata* var. *truncata* f. *neodamensis* (A. Br.) Dick 1926. p. 448. Pl. 20, Fig. 8.

Micrasterias truncata var. *quadrata* Bulnheim 1861, p. 52. Pl. 9, Fig. 5 = *Micrasterias truncata* var. *truncata* f. *quadrata* (Bulnh.) Croasdale stat. nov.

Micrasterias truncata var. *quebecensis* Irénée-Marie 1951, p. 196. Pl. 2, Figs. 6, 7 = *Micrasterias truncata* (Corda) Bréb. ex Ralfs 1848, p. 75. Pl. VIII, Fig. 4; Pl. 10, Fig. 5 var. *truncata* f. *truncata.*

Micrasterias truncata var. *rectangularis* Irénée-Marie 1957, p. 86. Figs. 8, 9 = *Micrasterias truncata* var. *truncata* f. *neodamensis* (A. Br.) Dick 1926, p. 448. Pl. 20, Fig. 8.

Micrasterias truncata var. *tridentata* Bennett 1890, p. 7. Pl. 1, Fig. 10 = *Micrasterias truncata* var. *truncata* f. *tridentata* (Benn.) Croasdale stat. nov.

Micrasterias truncata var. *uralensis* Krieger 1939, p. 34. Pl. 104, Fig. 5 = *Micrasterias truncata* var. *truncata* f. *uralensis* (Krieg.) Croasdale stat. nov.

Micrasterias verrucosa Roy, in Wolle 1885a, p. 127. Pl. 51, Fig. 10 = *Micrasterias verrucosa* Biss., in Roy & Bissett 1893, p. 174. Pl. 1(4), Fig. 2 f. *verrucosa.*

Micrasterias wailesi in Lackey & Lackey 1967, p. 14 = Unknown species.

Micrasterias westii Roll 1925, p. 244, 251. Pl. 11, Fig. 5 = *Micrasterias americana* var. *westii* (Roll) Krieger 1939, p. 48. Pl. 109, Fig. 6.

Micrasterias multifida Wolle 1876, p. 122 = Sp. Inquir.

Plates LVIII to CXLVIII

PLATE LVIII

Plate LVIII   211

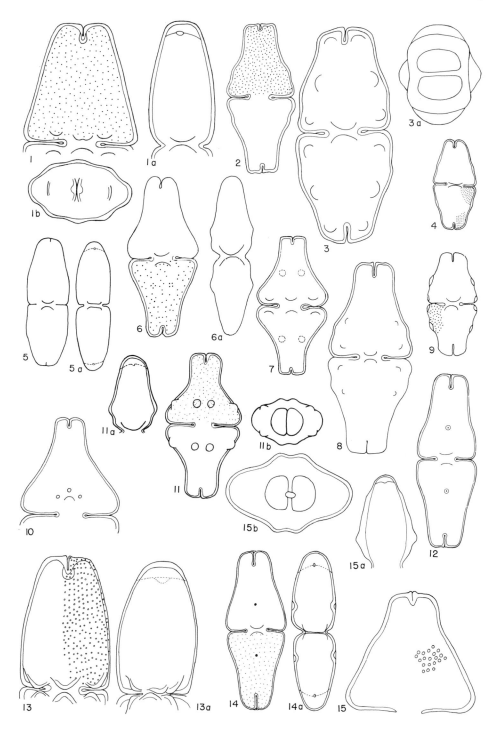

PLATE LIX

Plate LIX

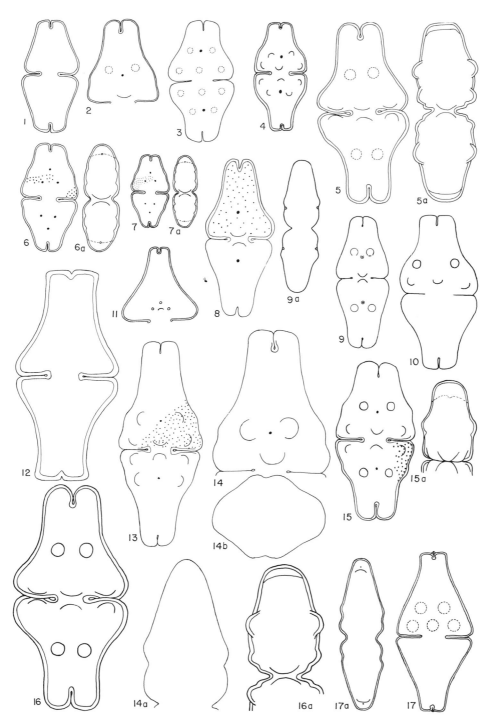

PLATE LX

Plate LX   215

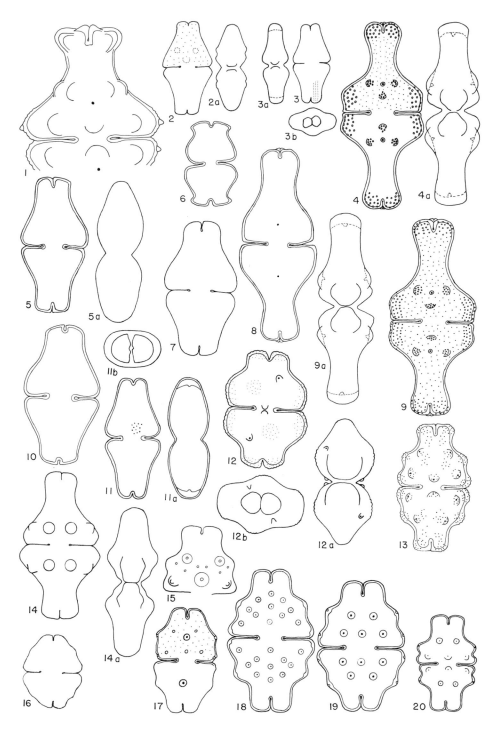

PLATE LXI

Plate LXI   217

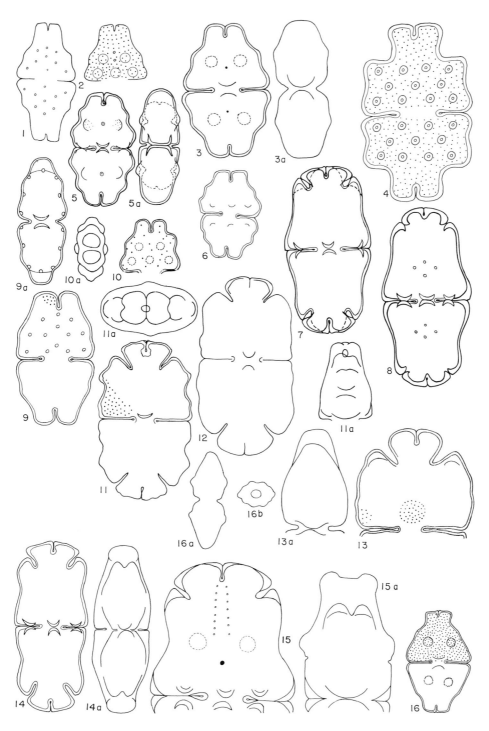

PLATE LXII

Plate LXII

219

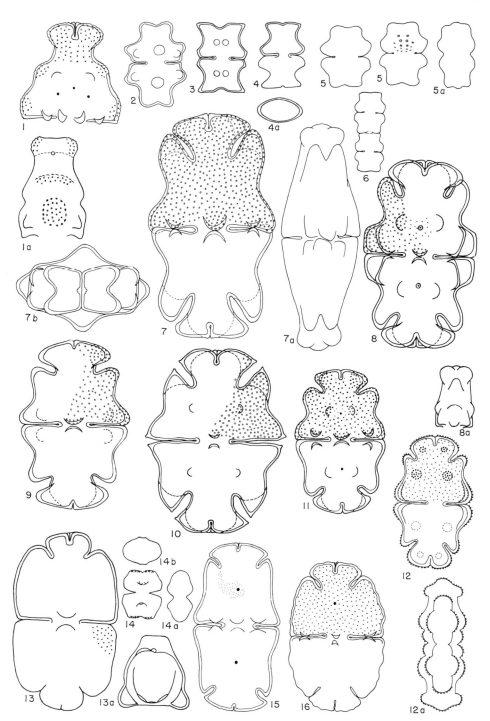

PLATE LXIII

Plate LXIII   221

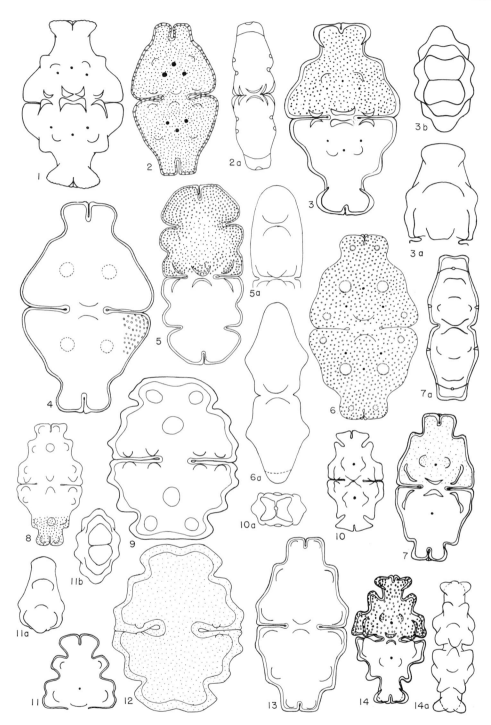

222 Legend Plate LXIV

PLATE LXIV

Plate LXIV

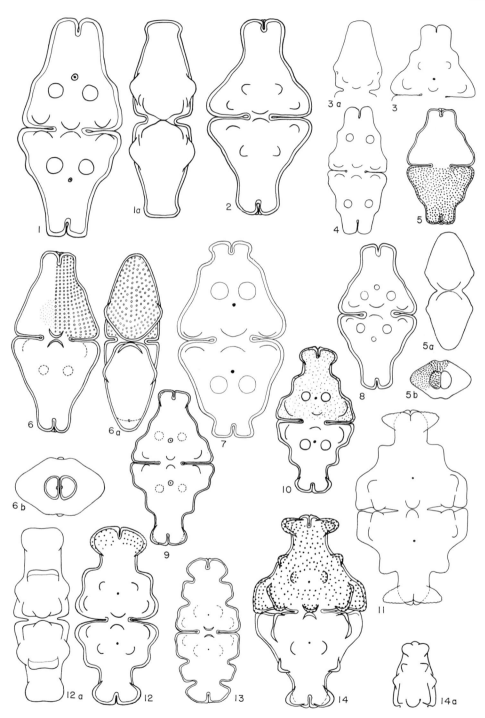

PLATE LXV

Plate LXV

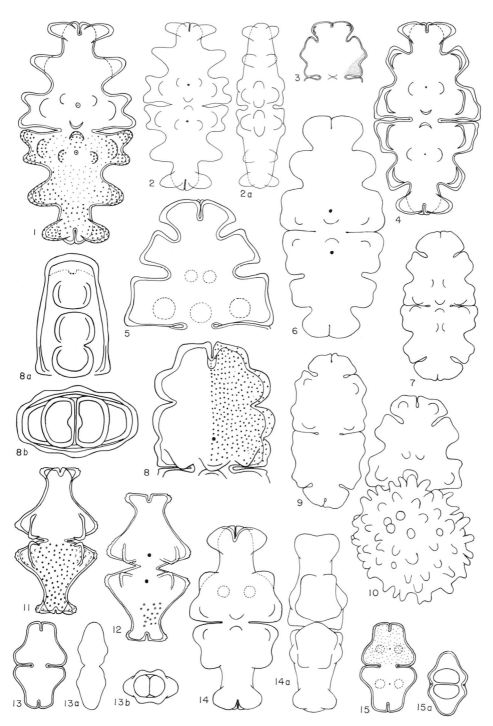

PLATE LXVI

Plate LXVI   227

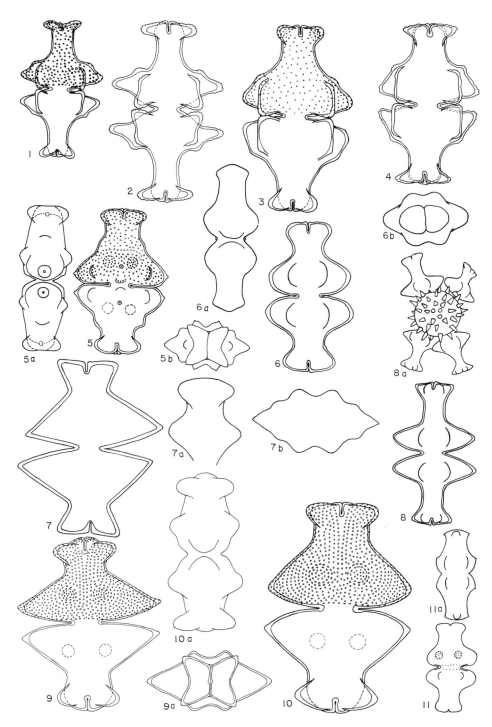

PLATE LXVII

Plate LXVII   229

PLATE LXVIII

Plate LXVIII　　　　　　　　　　231

PLATE LXIX

Plate LXIX

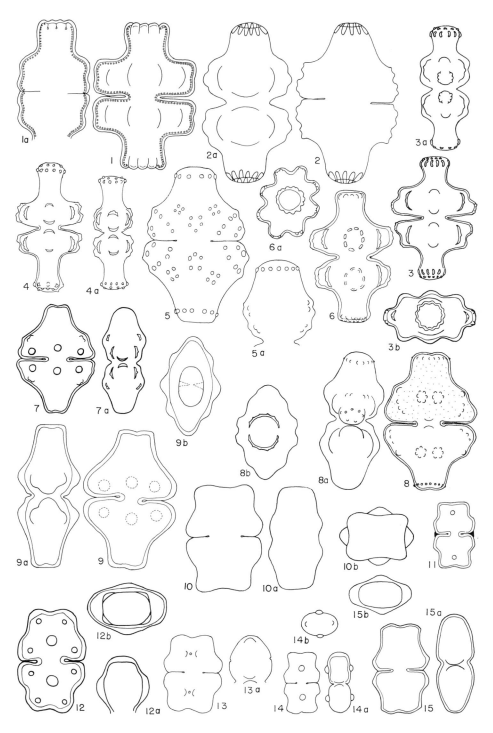

PLATE LXX

Plate LXX   235

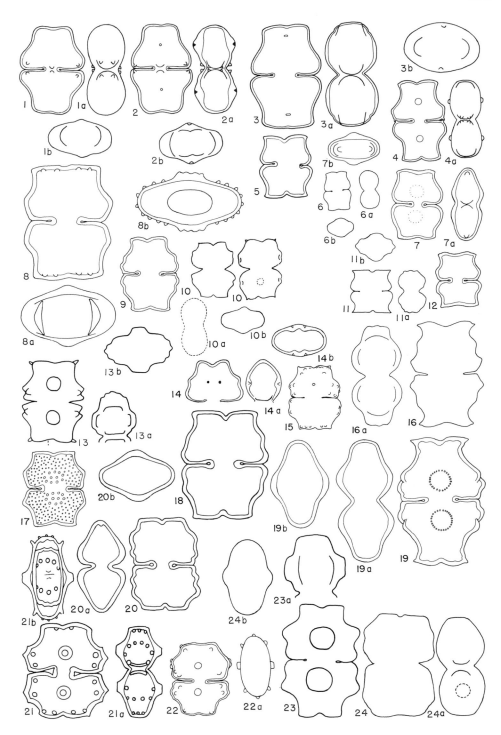

PLATE LXXI

Plate LXXI   237

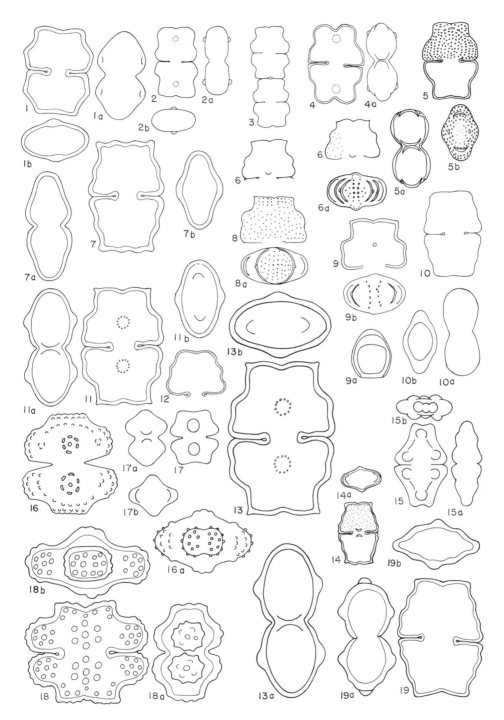

PLATE LXXII

Plate LXXII 239

PLATE LXXIII

Plate LXXIII 241

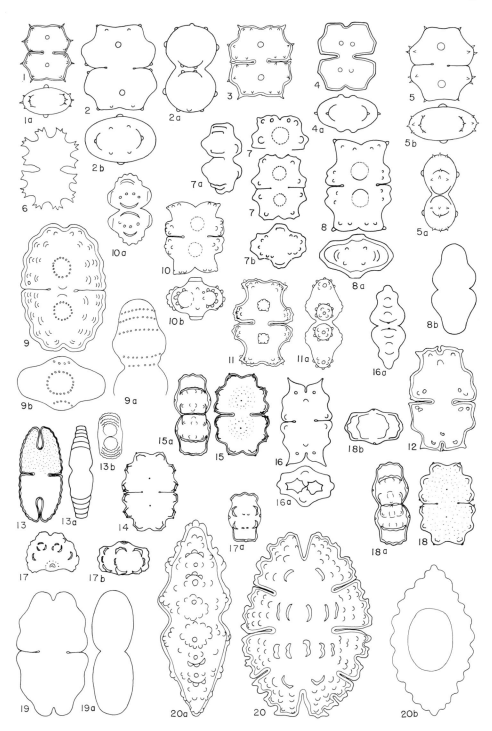

PLATE LXXIV

Plate LXXIV 243

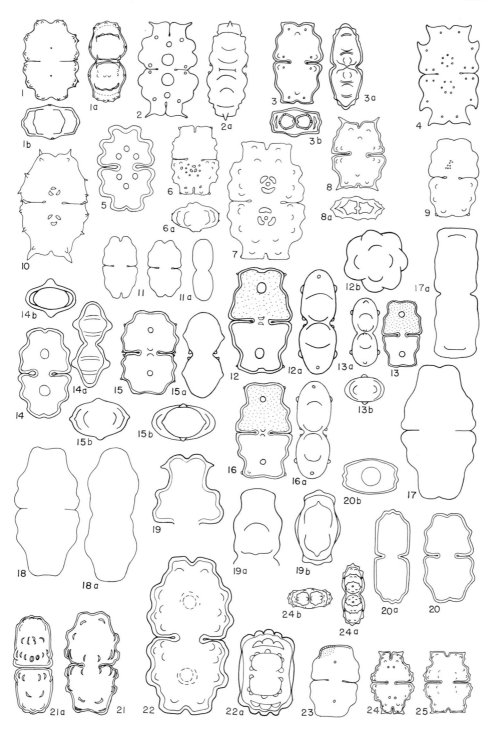

PLATE LXXV

Plate LXXV   245

PLATE LXXVI

Plate LXXVI   247

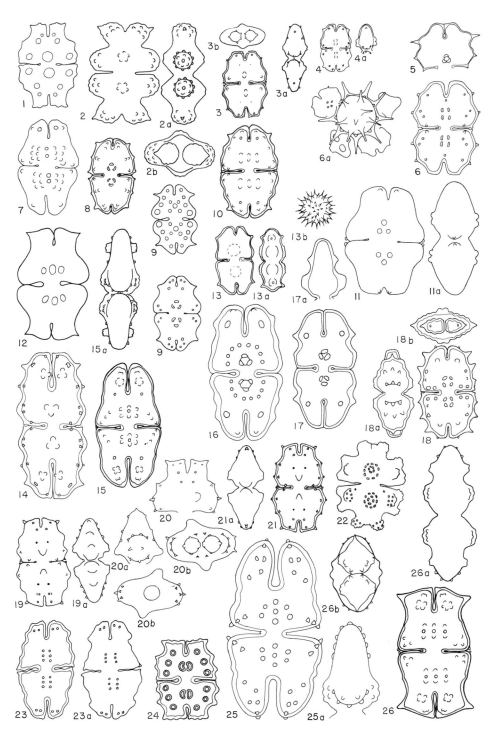

PLATE LXXVII

Plate LXXVII 249

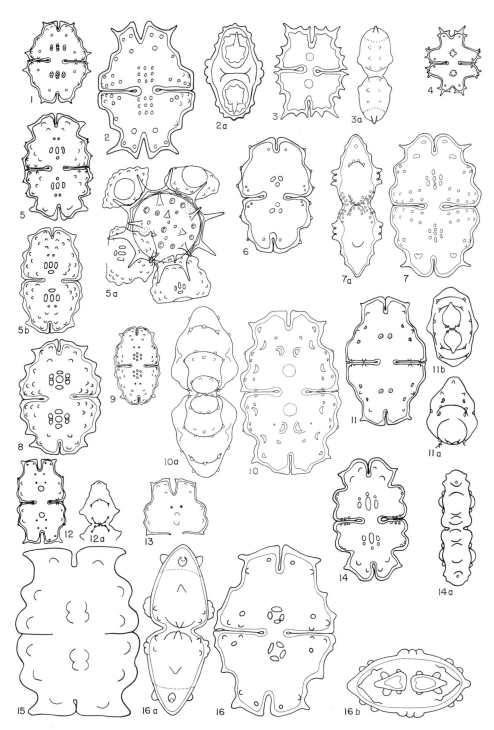

PLATE LXXVIII

Plate LXXVIII   251

PLATE LXXIX

Plate LXXIX   253

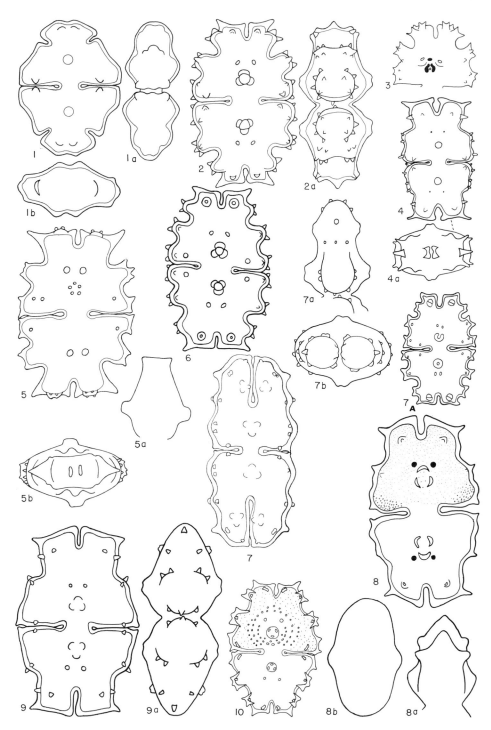

PLATE LXXX

Plate LXXX

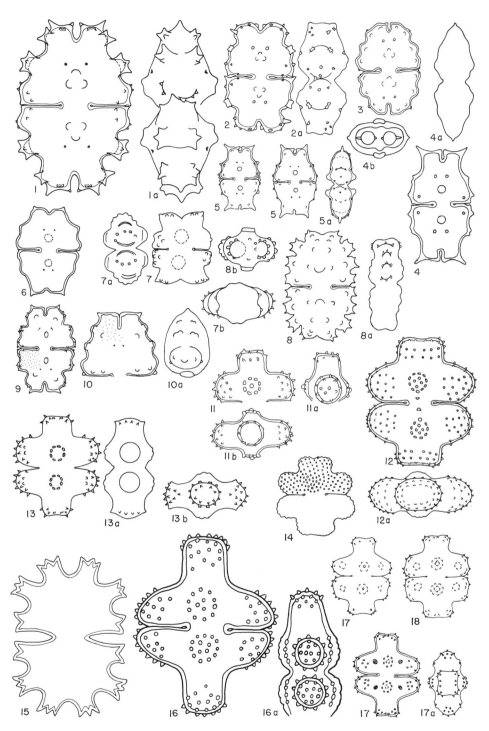

PLATE LXXXI

Plate LXXXI   257

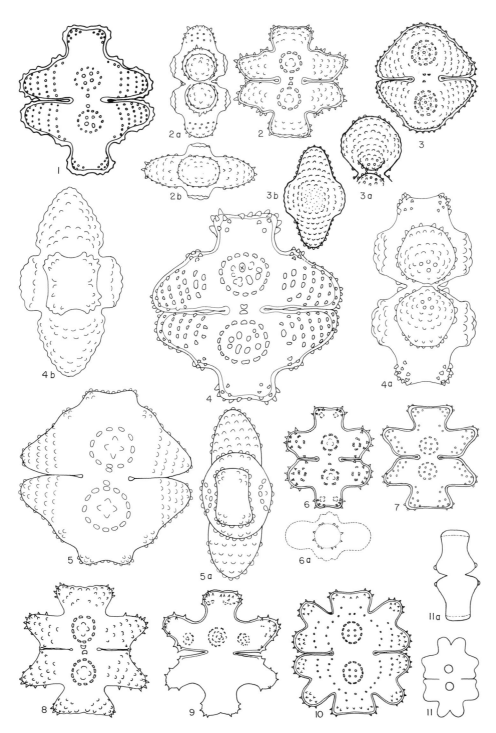

PLATE LXXXII

Plate LXXXII   259

PLATE LXXXIII

Plate LXXXIII   261

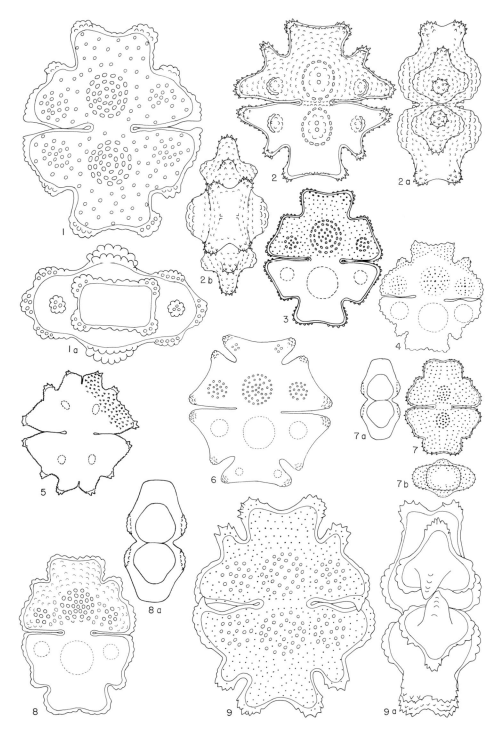

PLATE LXXXIV

Plate LXXXIV   263

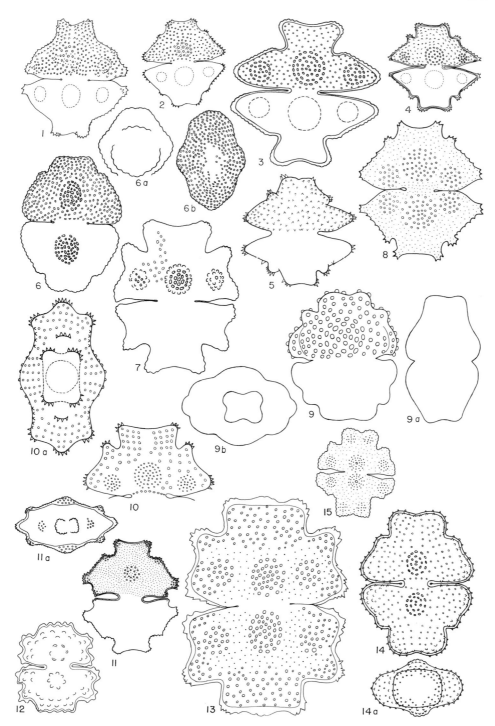

PLATE LXXXV

Plate LXXXV   265

PLATE LXXXVI

Plate LXXXVI 267

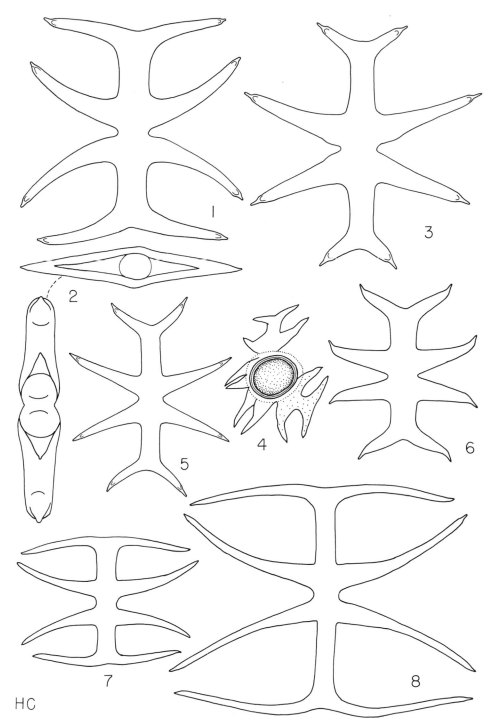

HC

PLATE LXXXVII

Plate LXXXVII   269

PLATE LXXXVIII

Plate LXXXVIII   271

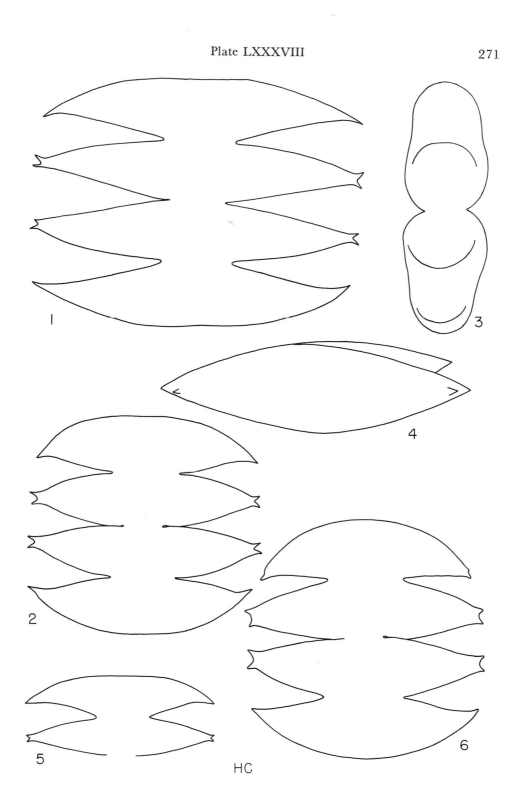

HC

PLATE LXXXIX

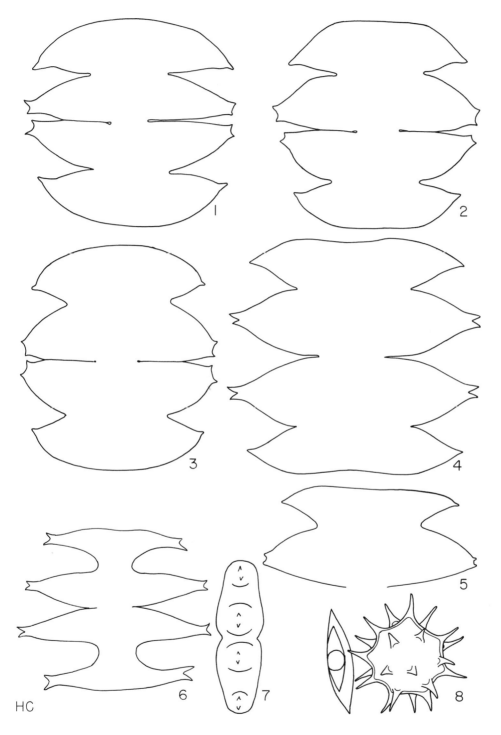

Plate LXXXIX

273

PLATE XC

Plate XC   275

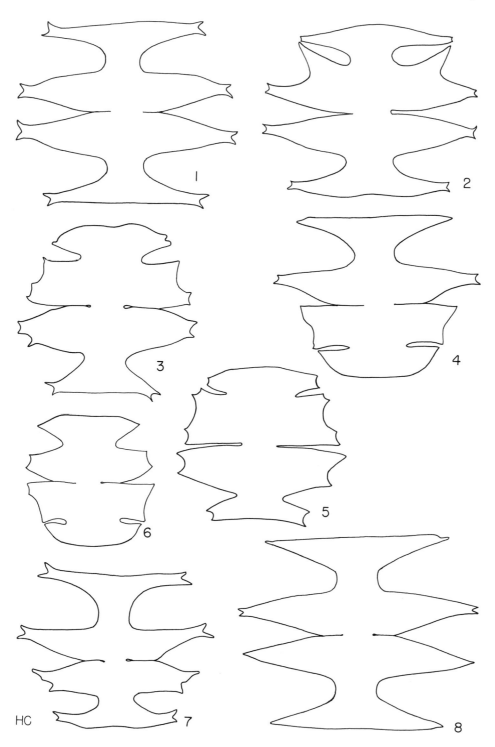

PLATE XCI

Plate XCI 277

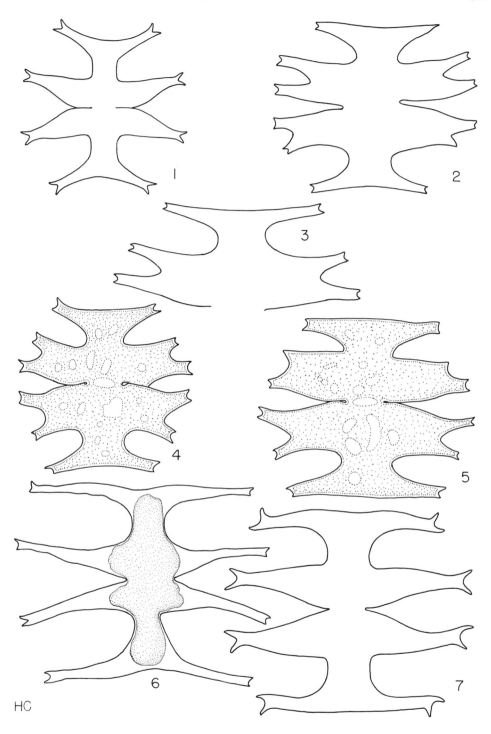

HC

PLATE XCII

Plate XCII 279

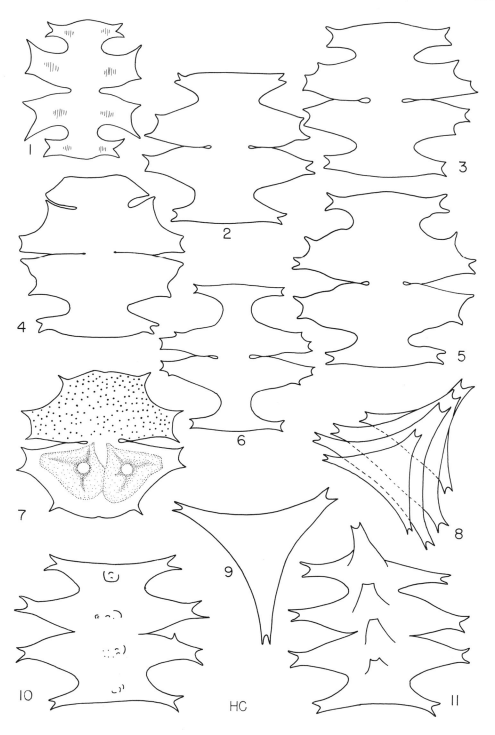

HC

PLATE XCIII

Plate XCIII 281

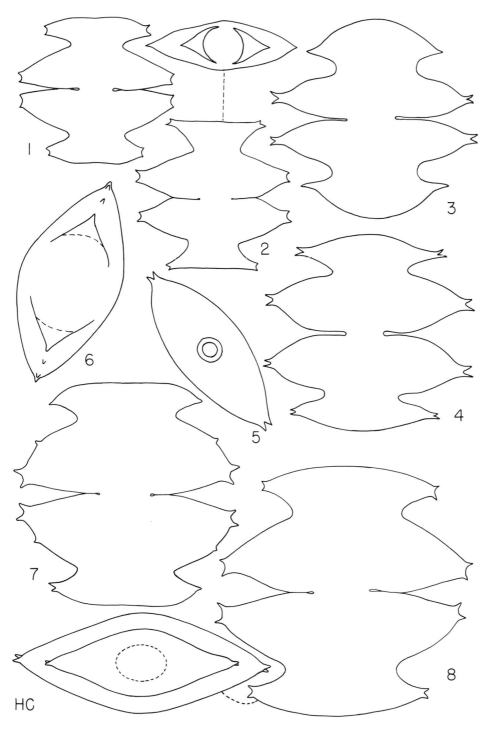

PLATE XCIV

Plate XCIV

283

PLATE XCV

Plate XCV   285

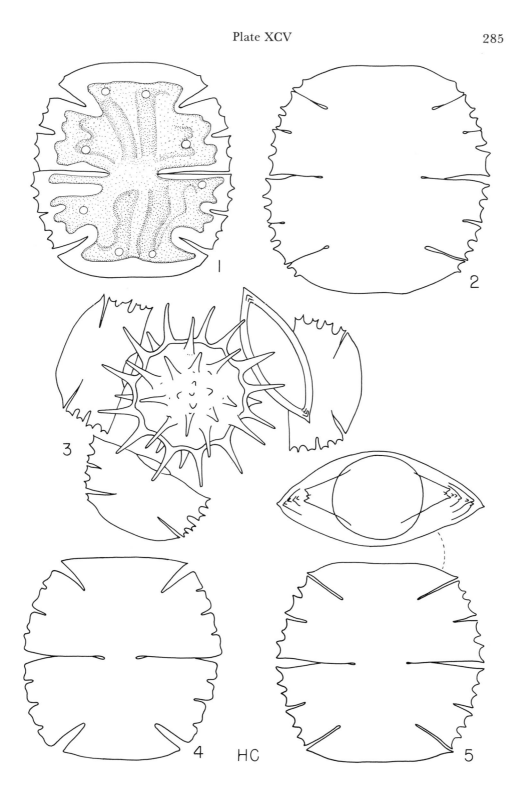

HC

PLATE XCVI

Plate XCVI 287

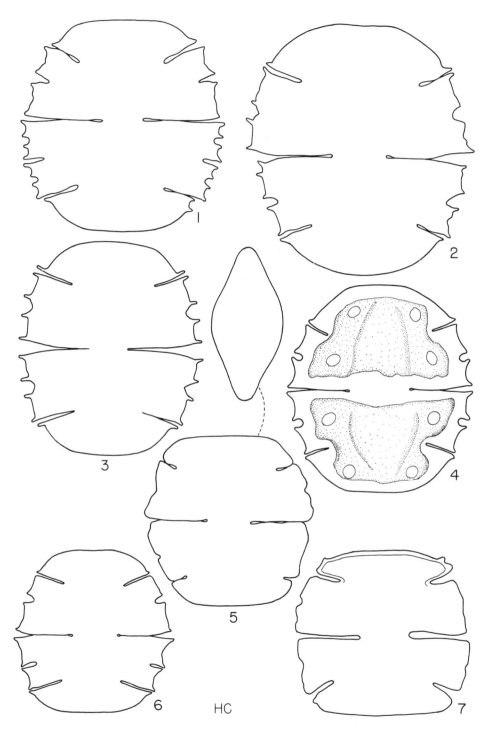

HC

PLATE XCVII

Plate XCVII   289

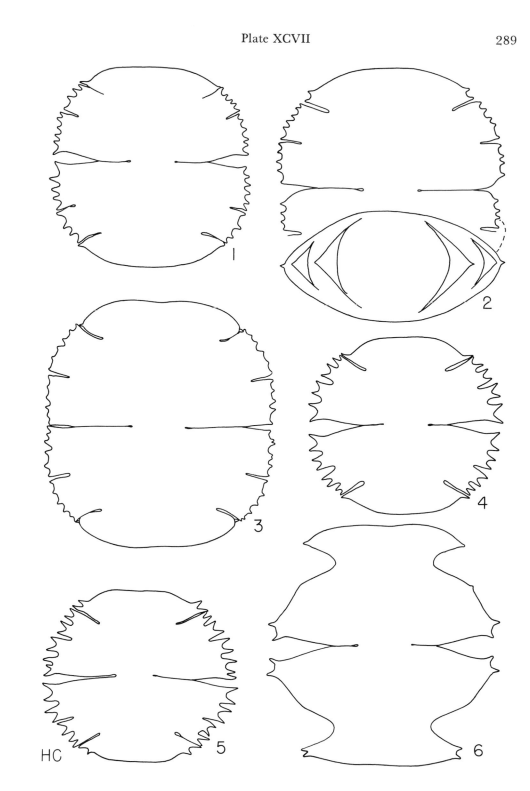

PLATE XCVIII

Plate XCVIII   291

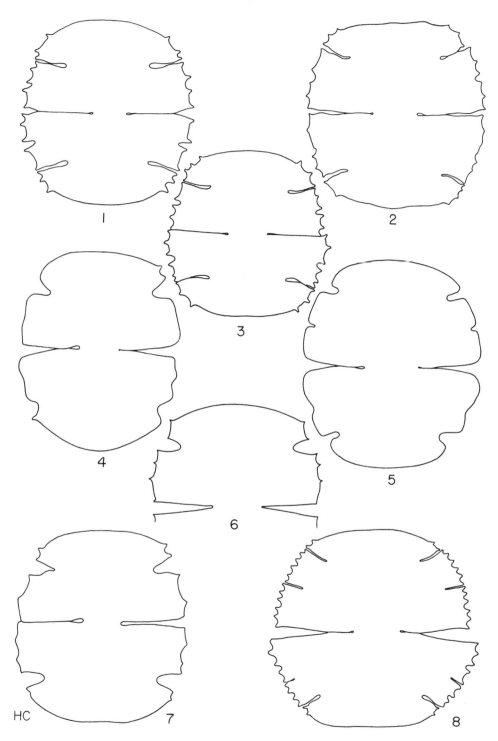

HC

PLATE XCIX

Plate XCIX 293

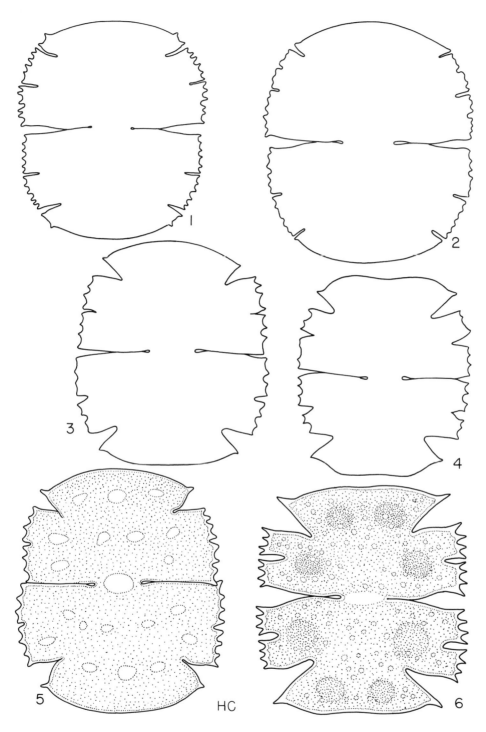

PLATE C

Plate C   295

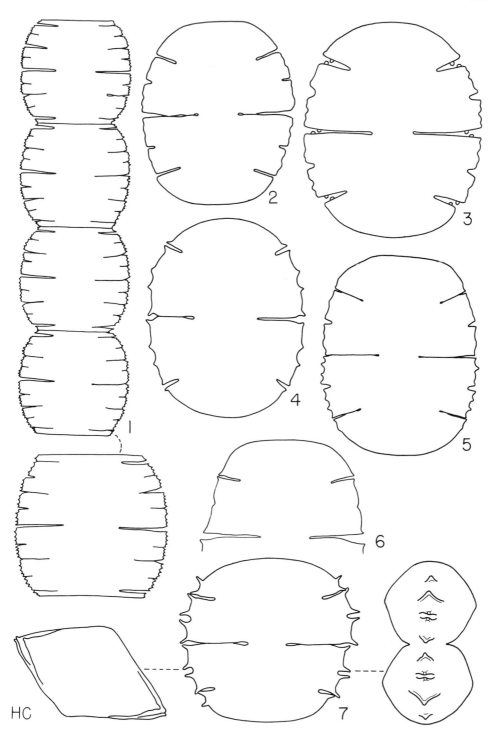

PLATE CI

Plate CI 297

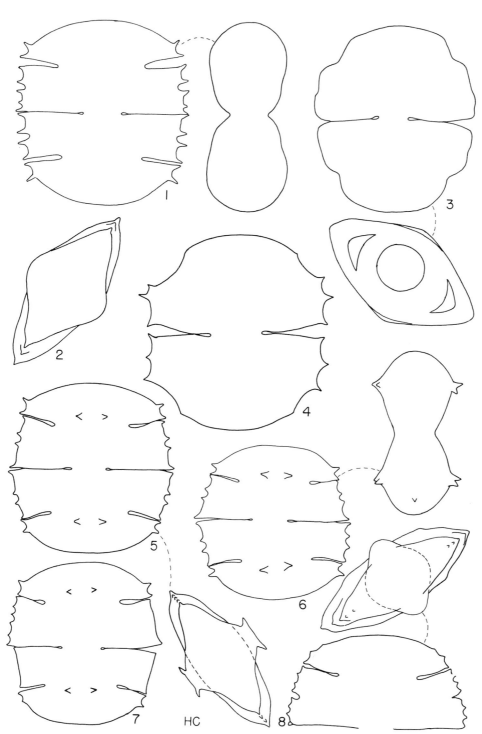

PLATE CII

Plate CII

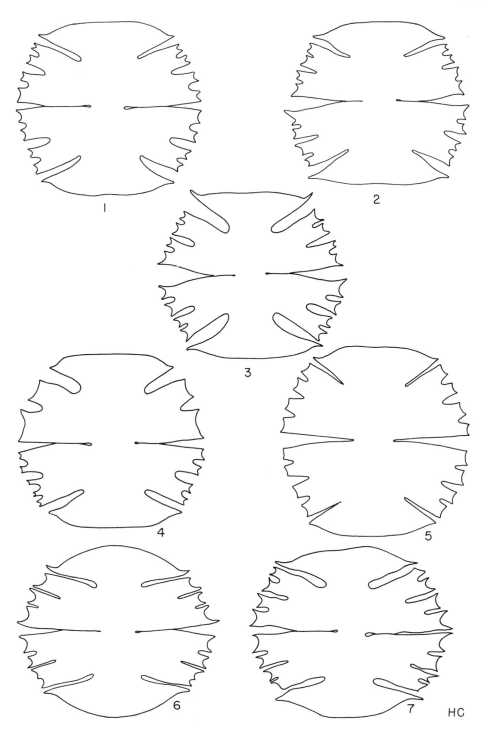

PLATE CIII

Plate CIII 301

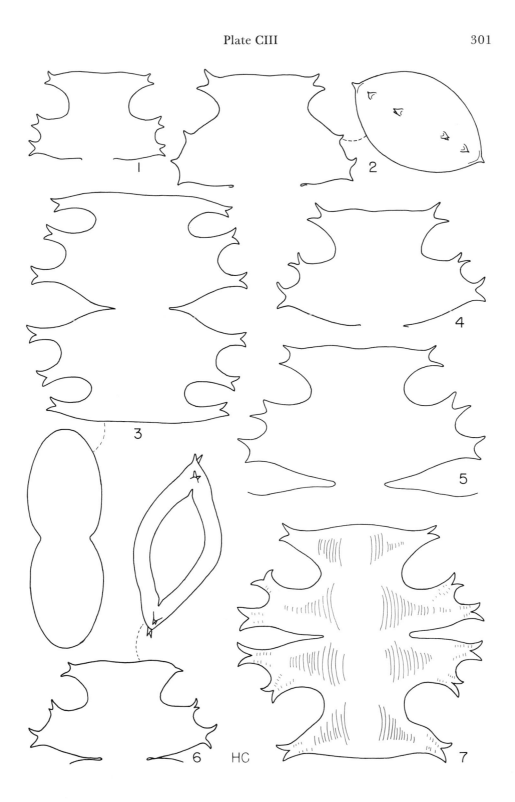

HC

PLATE CIV

Plate CIV   303

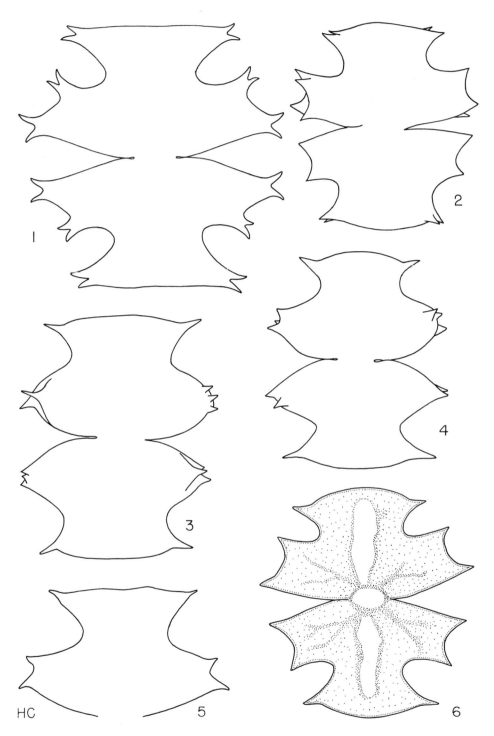

HC

PLATE CV

Plate CV   305

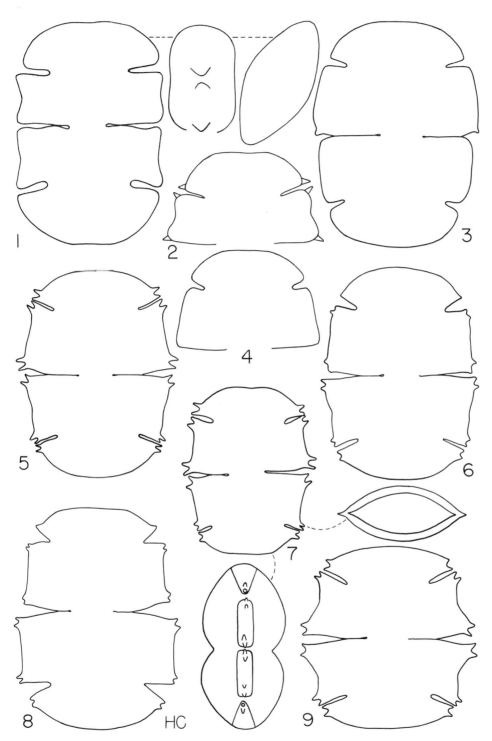

PLATE CVI

Plate CVI   307

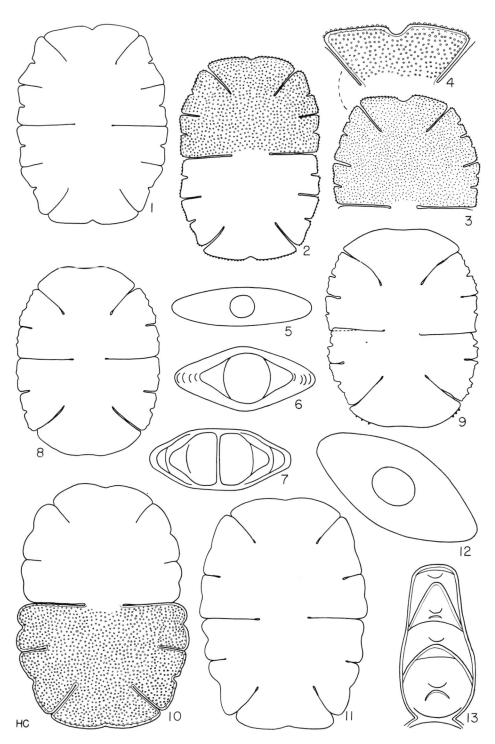

PLATE CVII

Plate CVII   309

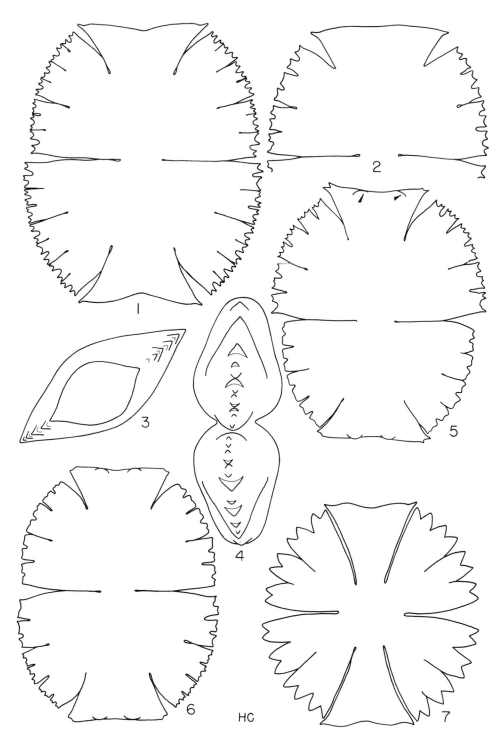

HC

PLATE CVIII

Plate CVIII   311

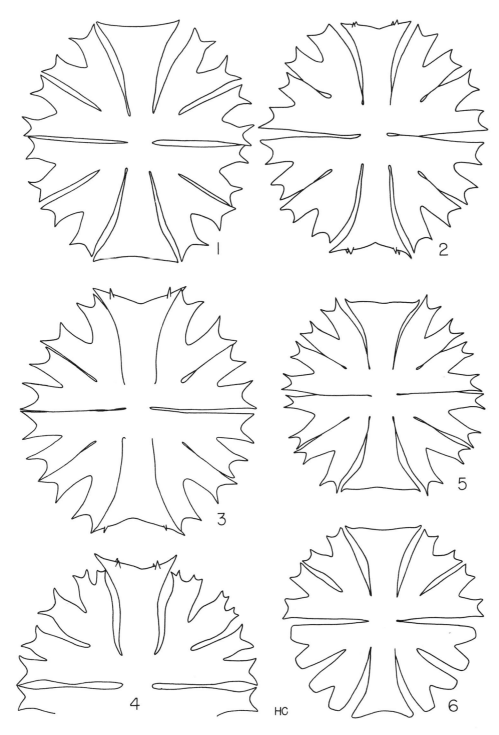

PLATE CIX

Plate CIX   313

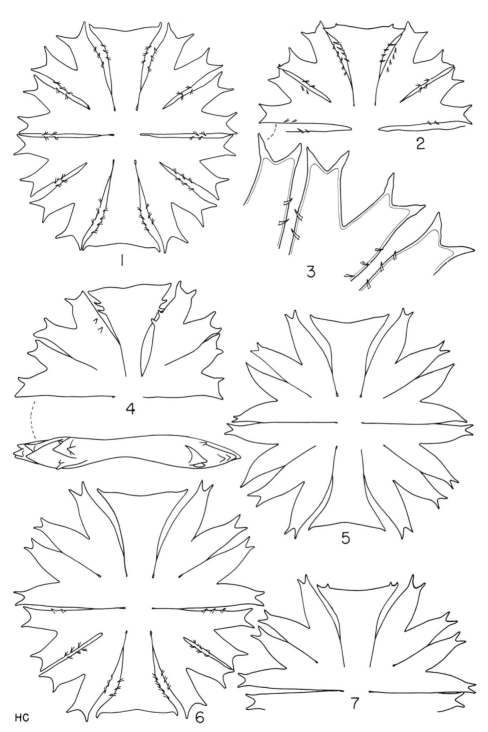

HC

PLATE CX

Plate CX
315

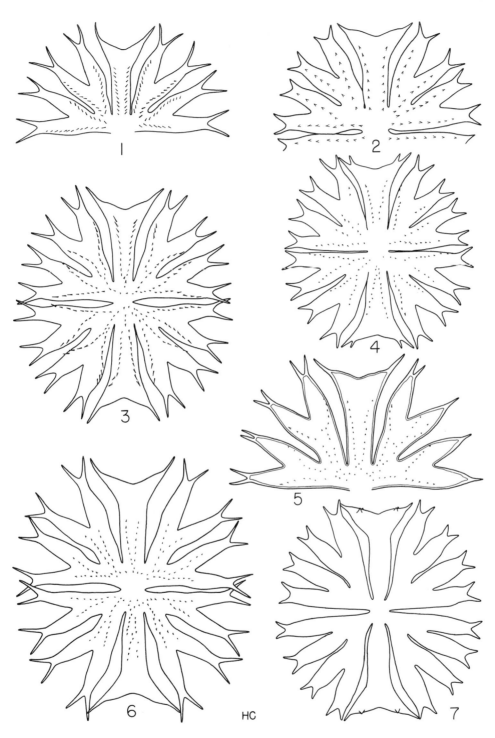

HC

PLATE CXI

Plate CXI   317

PLATE CXII

Plate CXII   319

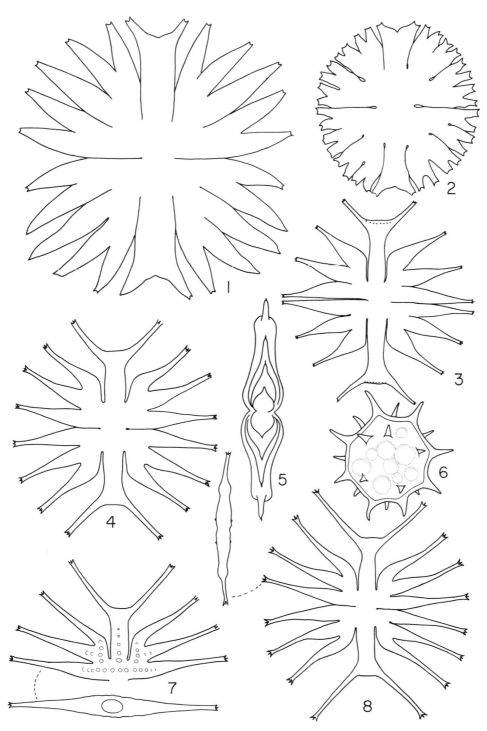

PLATE CXIII

Plate CXIII   321

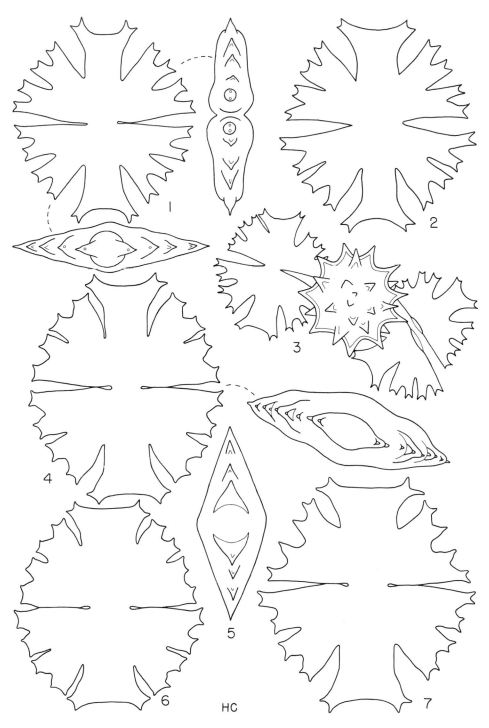

HC

PLATE CXIV

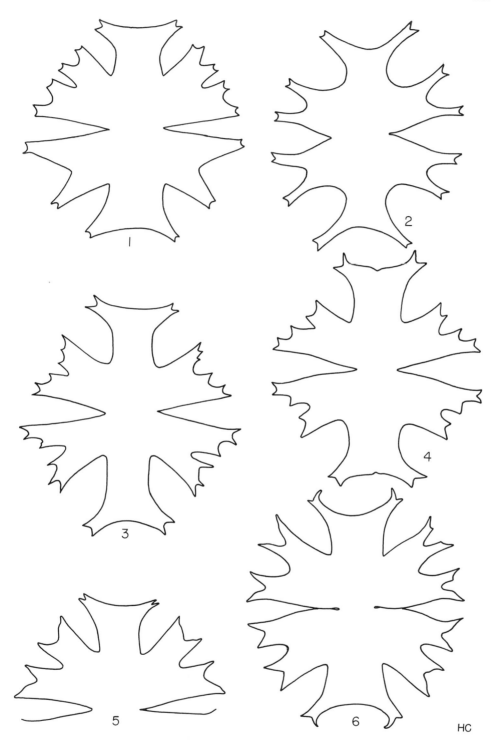

Plate CXIV   323

1

2

3

4

5

6

HC

PLATE CXV

Plate CXV

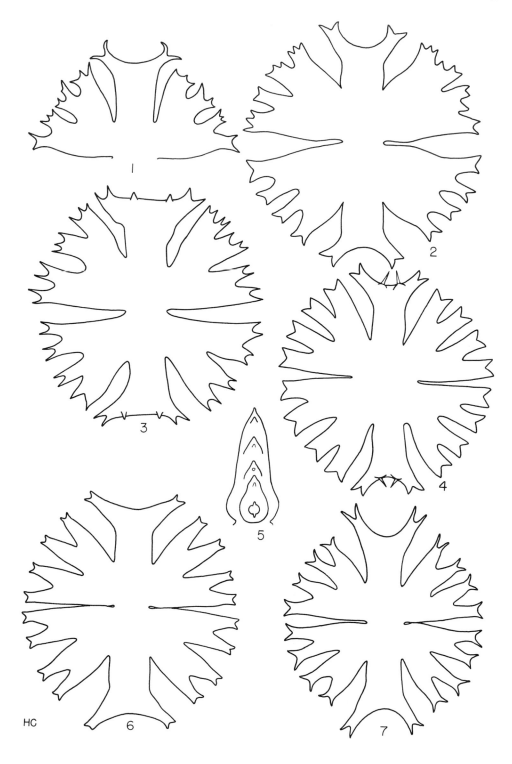

HC

PLATE CXVI

Plate CXVI 327

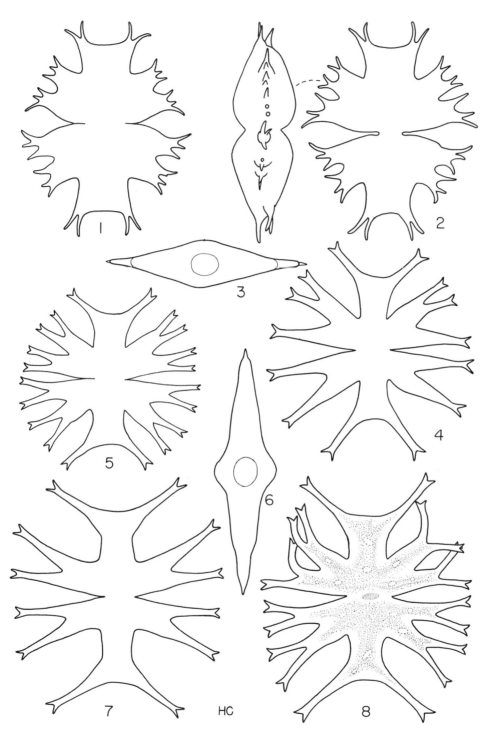

PLATE CXVII

Plate CXVII   329

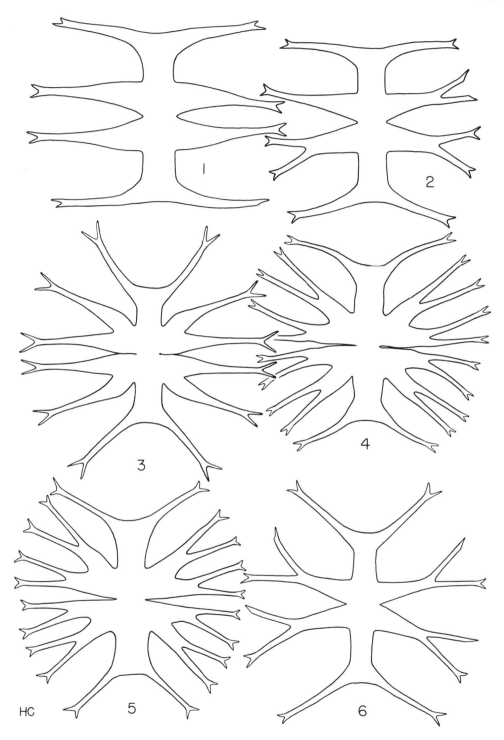

HC

PLATE CXVIII

Plate CXVIII 331

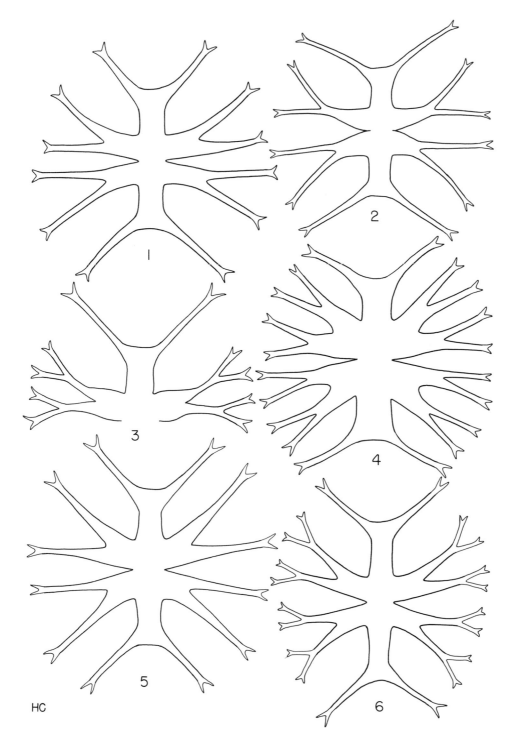

HC

PLATE CXIX

Plate CXIX 333

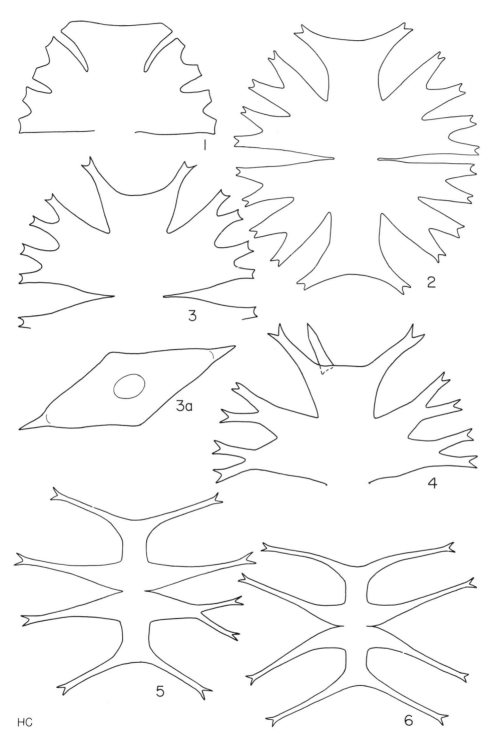

HC

PLATE CXX

Plate CXX 335

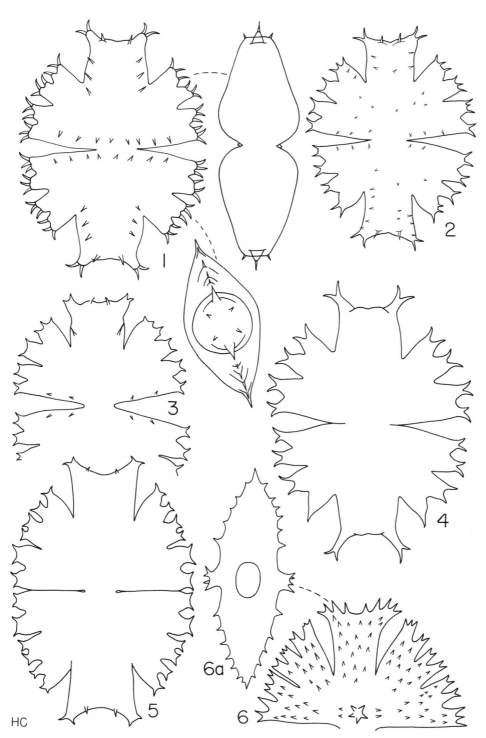

HC

PLATE CXXI

Plate CXXI   337

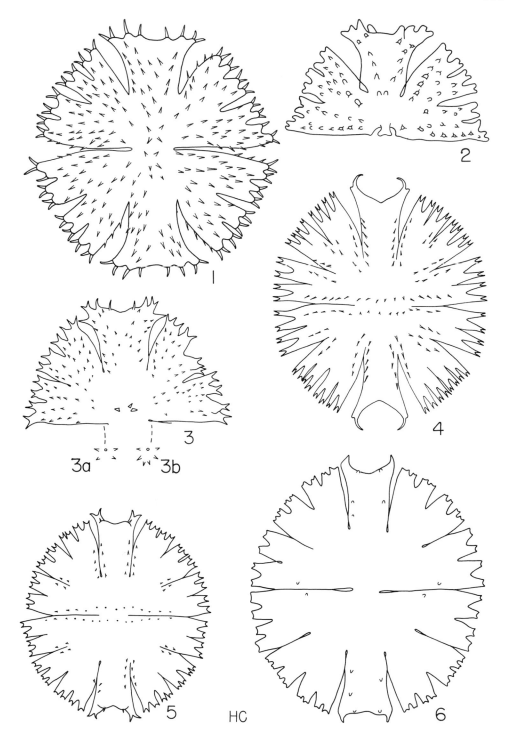

HC

PLATE CXXII

Plate CXXII   339

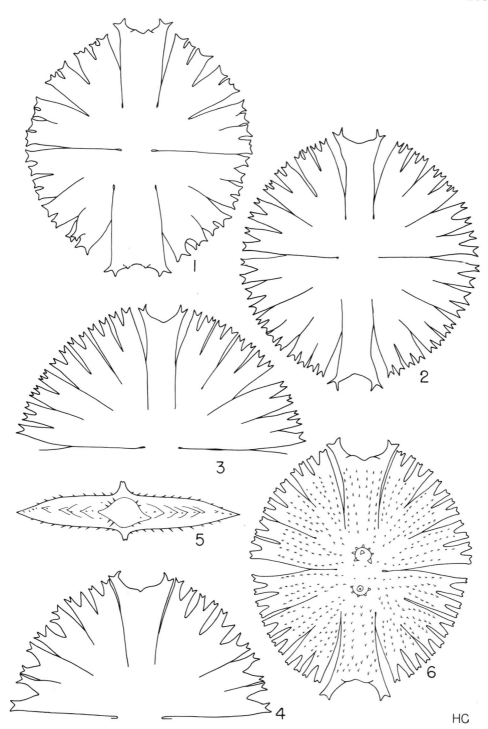

HC

PLATE CXXIII

Plate CXXIII   341

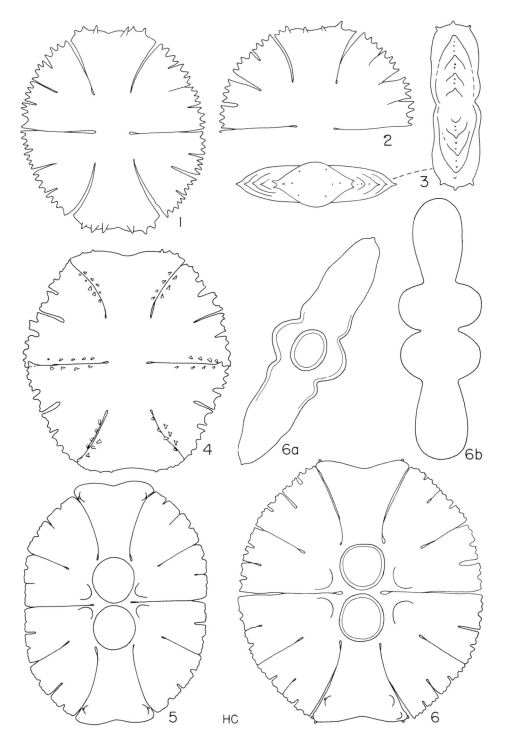

HC

PLATE CXXIV

Plate CXXIV   343

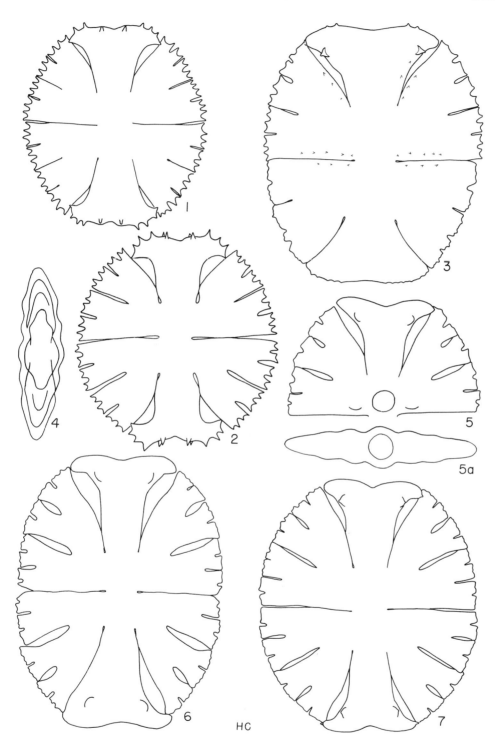

HC

PLATE CXXV

Plate CXXV 345

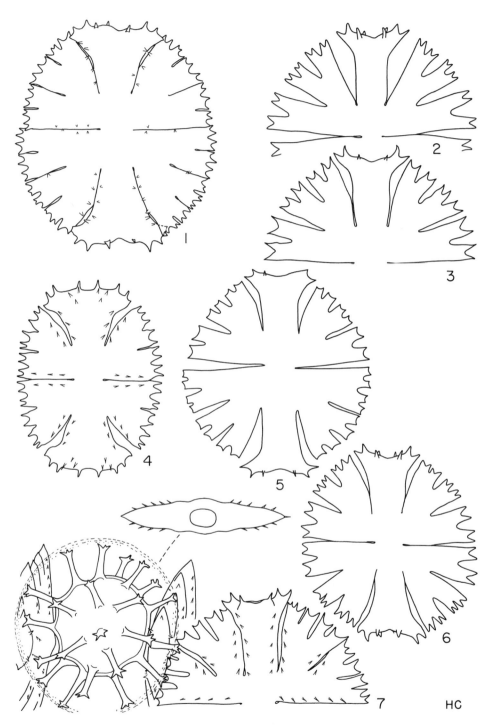

HC

PLATE CXXVI

Plate CXXVI   347

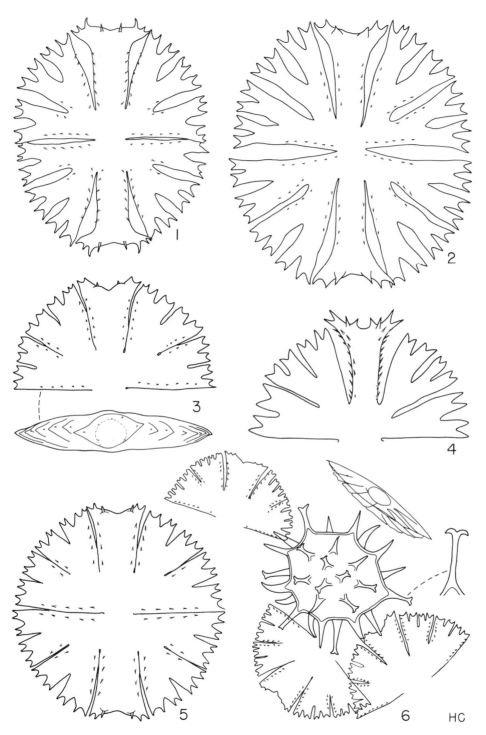

PLATE CXXVII

Plate CXXVII   349

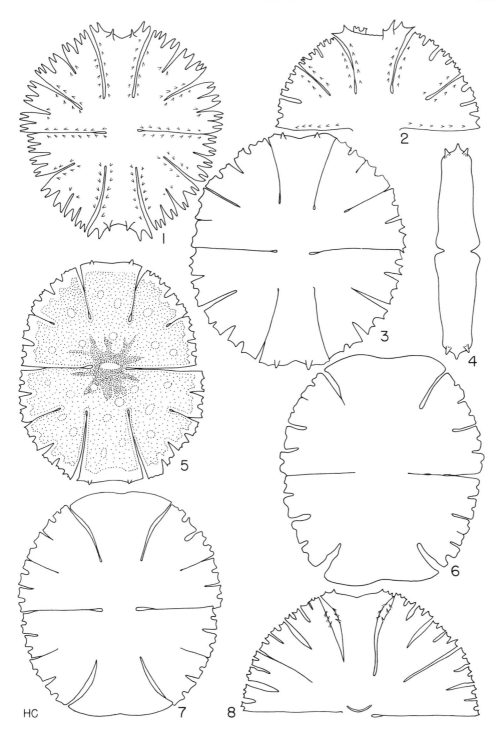

PLATE CXXVIII

Plate CXXVIII   351

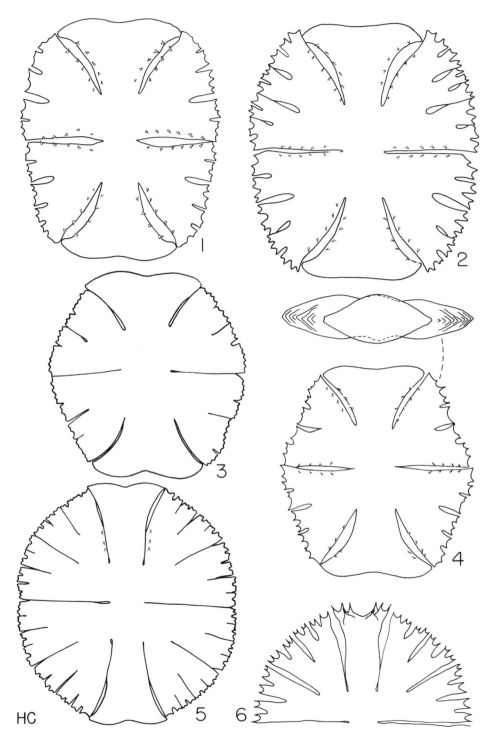

PLATE CXXIX

Plate CXXIX   353

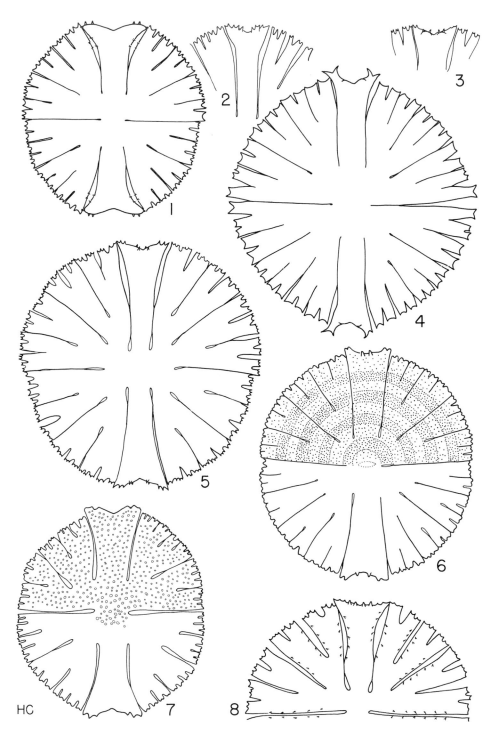

HC

PLATE CXXX

Plate CXXX   355

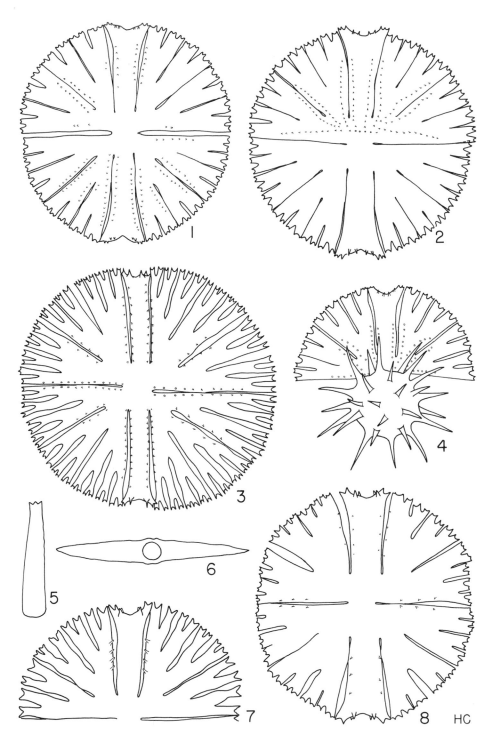

PLATE CXXXI

Plate CXXXI   357

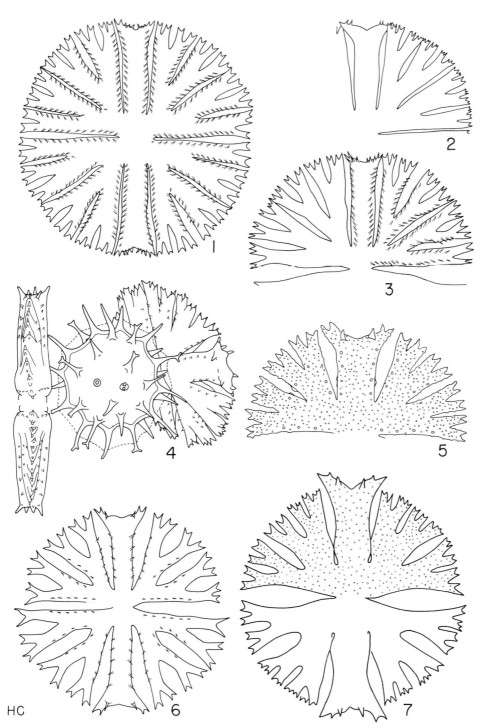

HC

PLATE CXXXII

Plate CXXXII   359

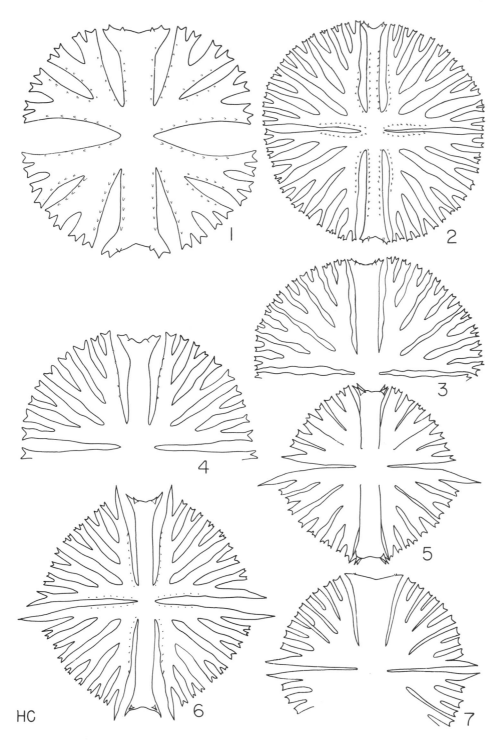

HC

PLATE CXXXIII

Plate CXXXIII   361

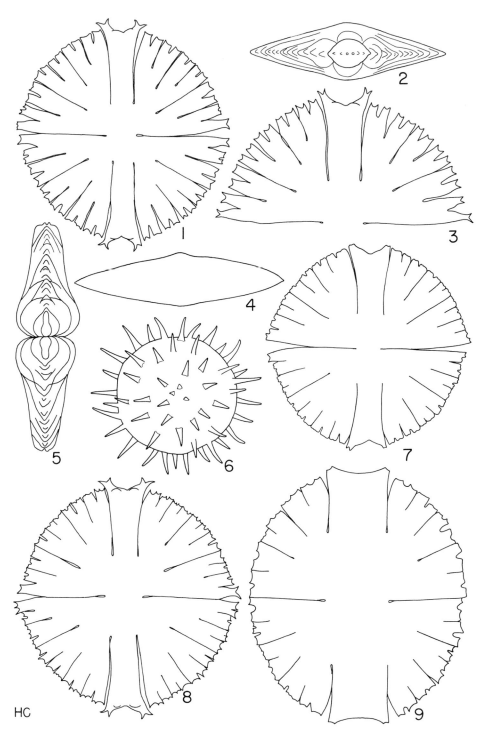

HC

PLATE CXXXIV

Plate CXXXIV   363

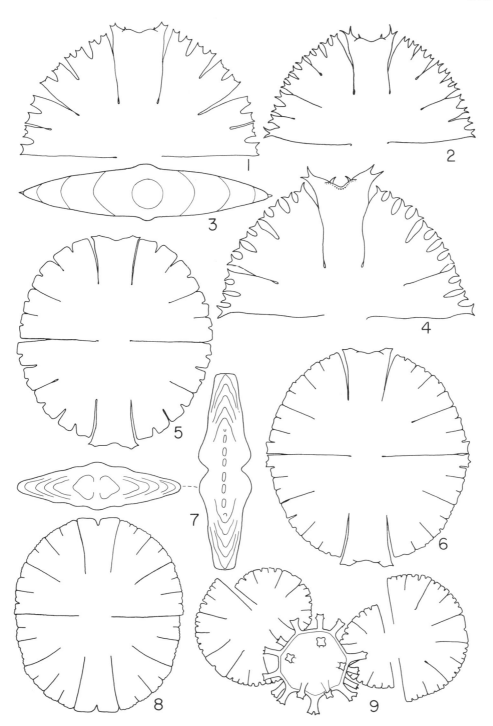

PLATE CXXXV

Plate CXXXV 365

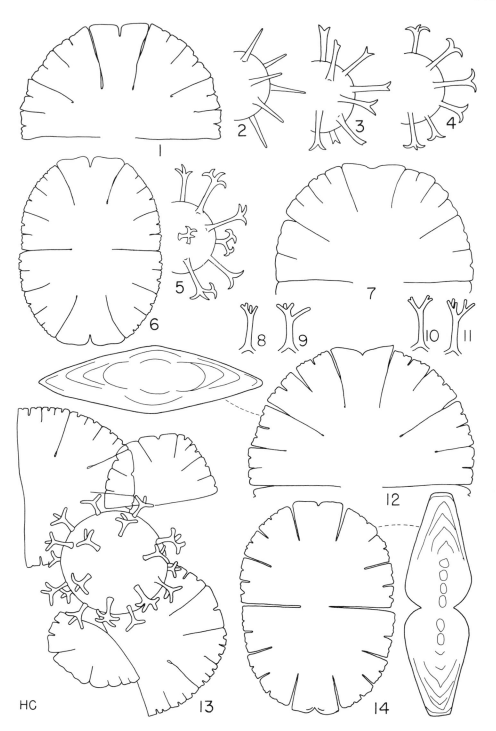

HC

PLATE CXXXVI

Plate CXXXVI   367

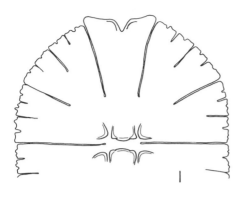

PLATE CXXXVII

Plate CXXXVII 369

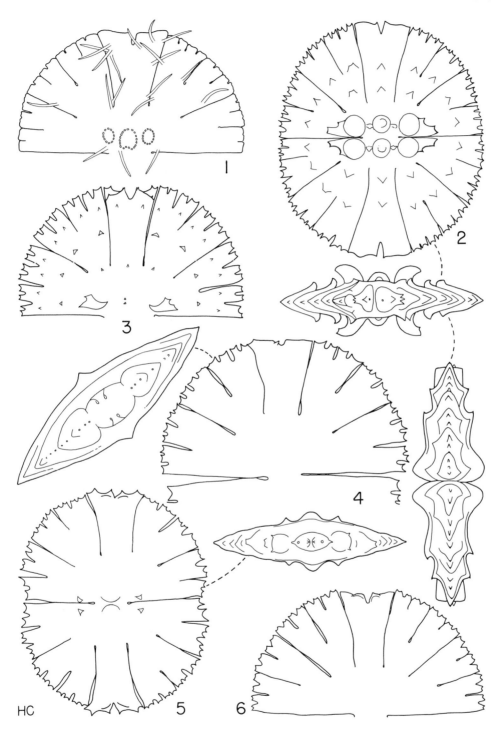

HC

PLATE CXXXVIII

Plate CXXXVIII 371

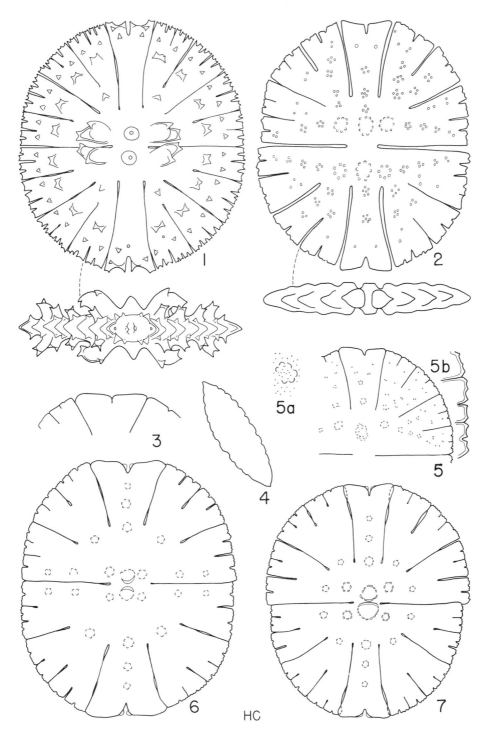

1

2

3

4

5a

5b

5

6

HC

7

PLATE CXXXIX

Plate CXXXIX 373

HC

PLATE CXL

Plate CXL   375

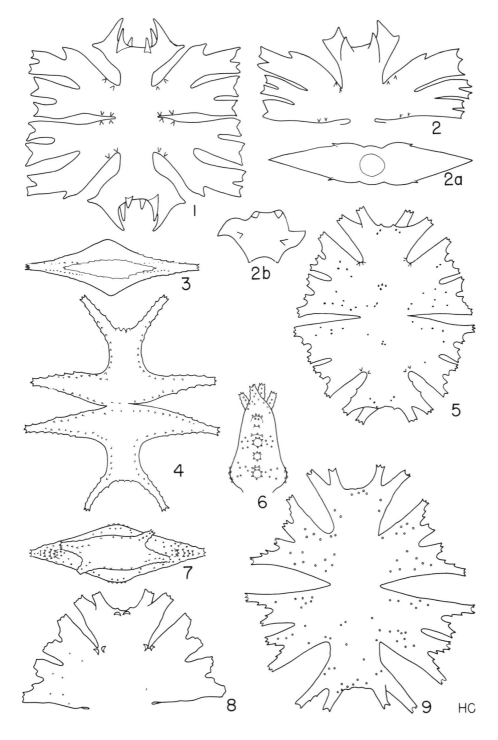

PLATE CXLI

Plate CXLI   377

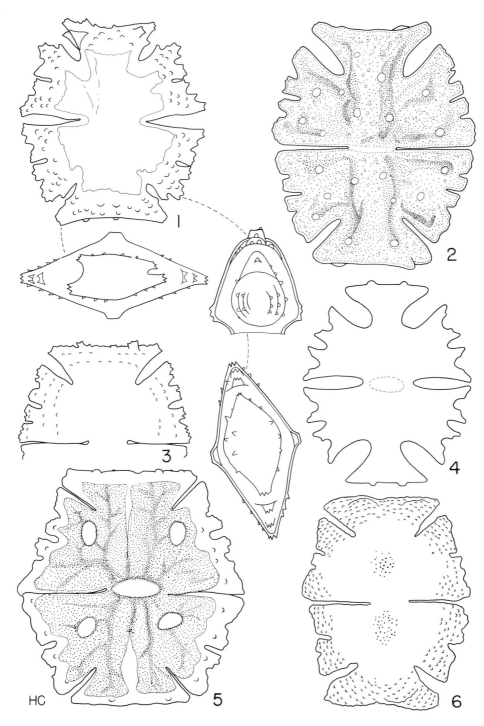

.

PLATE CXLII

Plate CXLII 379

PLATE CXLIII

Plate CXLIII   381

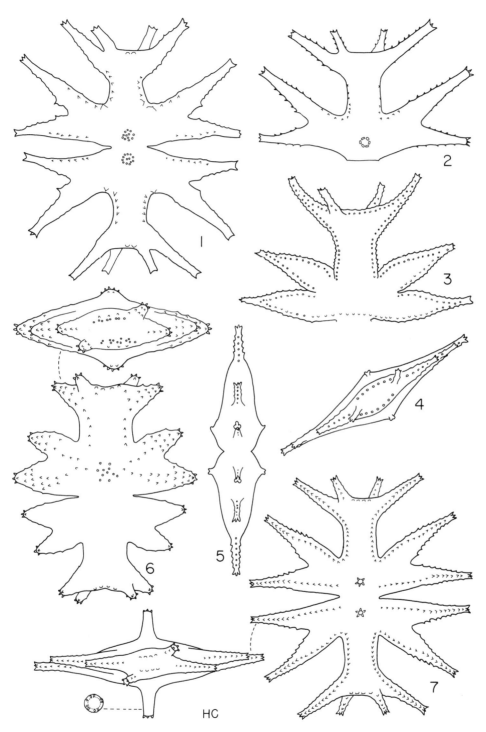

PLATE CXLIV

Plate CXLIV   383

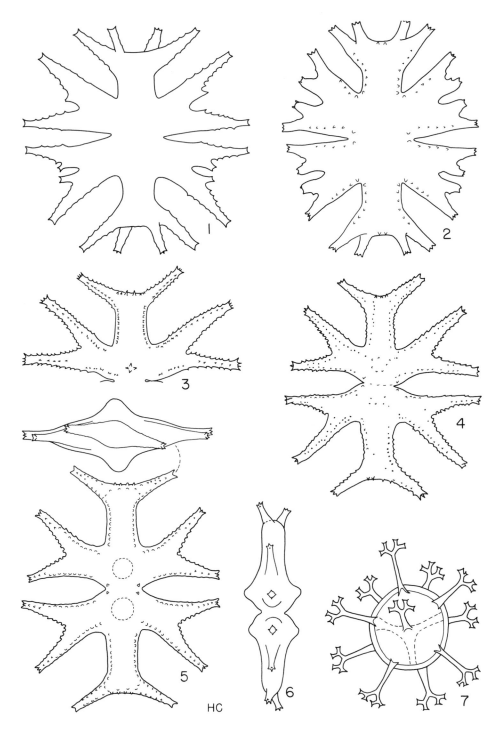

HC

PLATE CXLV

Plate CXLV   385

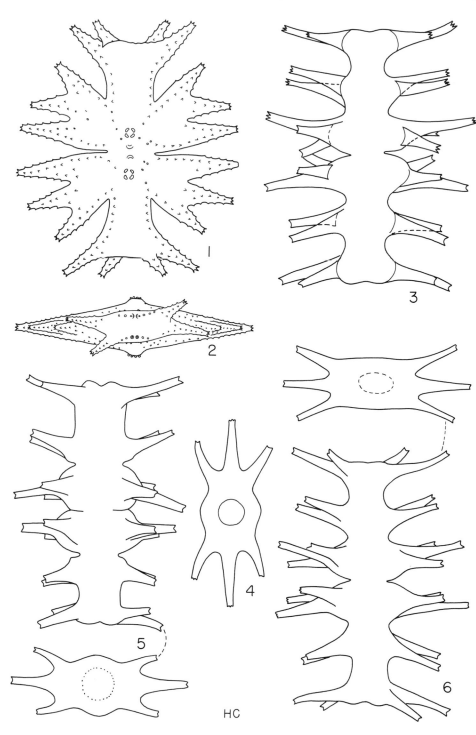

HC

PLATE CXLVI

Plate CXLVI   387

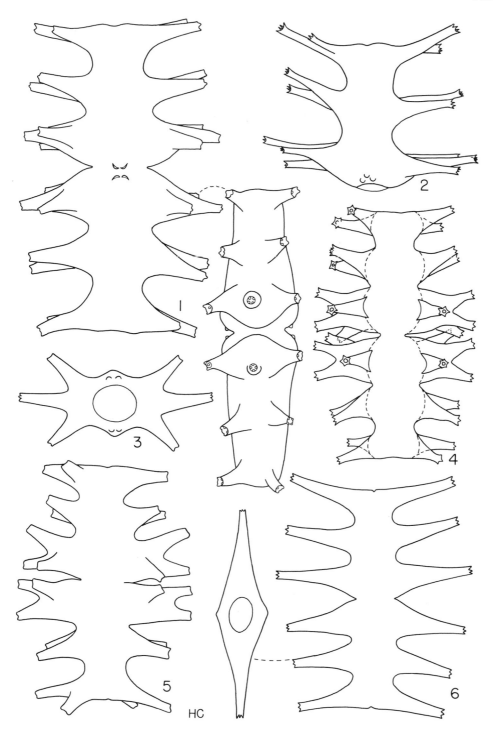

HC

PLATE CXLVII

Plate CXLVII 389

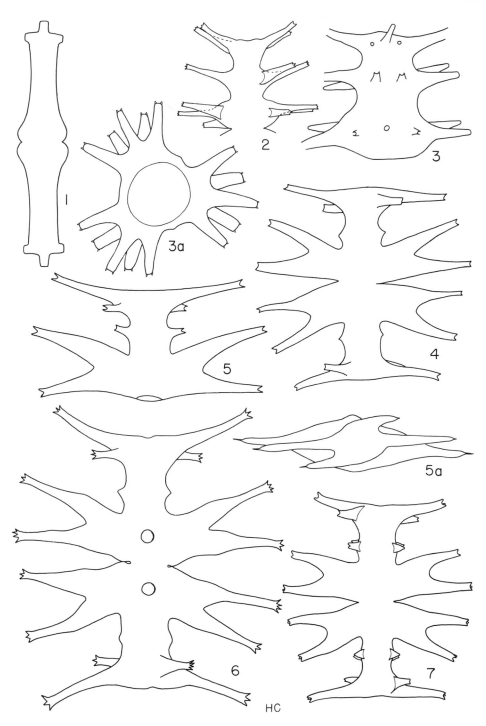

HC

Bibliography[1]

Ackley, Alma B. 1929. New species and varieties of Michigan algae. Trans. Amer. Microsc. Soc. 48(3): 302–309. Pls. 35, 36.

Agardh, C. A. 1827. Aufzählung einiger in den östreichischen Ländern gefundenen neuen Gattungen und Arten von Algen, nebst ihrer Diagnostik und beigefügten Bemerkungen. Flora 40/41: 625–641; 642–646.

Allorge, V. and Allorge, P. 1931. Hétérocontes, Euchlorophycées et Conjuguées de Galice. Matériaux pour la flore des algues d'eau douce de la Peninsule Ibérique. Rev. Algol. 5(3/4): 327–382. Pls. 17–32.

Archer, W. 1861. Sub-group Desmidieae or Desmidiaceae, in Pritchard, A. A history of Infusoria, including the Desmidiaceae and Diatomaceae British and foreign. Ed. 4, pp. 715–751. Pls. 1–17.

Archer, W. 1862. Description of a new species of *Micrasterias* (Ag. *et aliorum, non* Ehr.), with remarks on the distinction between *Micrasterias rotata* (Ralfs) and *M. denticulata* (Bréb.). Quart. Jour. Microsc. Soc., n.s., 2: 236–247. Pl. 12.

*Baglin, R. E. 1972. Surface phytoplankton and some aspects of the physical-chemical limnology of three areas on Lake Texoma. Southwest Nat. 17(1): 11–19. 1 Fig. 5 Tabls.

Bailey, J. W. 1846. On some new species of American Desmids from the Catskill Mountains. Sillim. Amer. Jour. Sci. & Arts, n.s., 1: 126–127. Figs.

Bailey, J. W. 1851. Microscopical observations made in South Carolina, Georgia and Florida. Smithson. Contrib. to Knowledge 2 (Art. 8): 1–48. Pls. 1–3.

Bennett, A. W. 1886. Freshwater algae (including Chlorophyllaceous Protophyta) of the English Lake District I, with descriptions of twelve new species. Jour. Roy. Microsc. Soc., II, 6: 1–15. Pls. 1,2.

Bennett, A. W. 1890. Freshwater algae and Schizophyceae of Hampshire and Devonshire. Jour. Roy. Microsc. Soc. 1890: 1–10. Pl. 1.

Bernard, Ch. 1909. Algues unicellulaires d'eau douce recoltées dans le Domaine Malais. Dept. Agric. Indes-Néerl. 1909. 94 pp. 6 Pls.

*Bicudo, C. E. M. 1972. Revisão do género *Arthrodesmus*, das Desmidiáceas (Chlorophyceae). Thèse. Univ. São Paulo.

Bicudo, C. E. M. and Sormus, Laine. 1972. Polymorphism in the desmid *Micrasterias laticeps* and its taxonomic implications. Jour. Phycol. 8(3): 237–242. Figs. 1–14.

*Bigelow, E. O. and Ashley, L. M. 1955. Preliminary report of investigations of the plankton of Lake Sutherland, Clallam County, Washington. Walla Walla Coll. Publ. Biol. Sci. & Biol. Station No. 14. (unpaged). 3 Pls.

Boldt, R. 1885. Studier öfver sötvattensalger och deras utbredning. I. Bidrag till kannedomen om Sibiriens Chlorophyllophycéer. Öfv. Kongl. Vet.-Akad. Förhandl. 42(2): 91–128. Pls. 5,6.

Boldt, R. 1888. Studier öfver söttvattensalger och deras utbredning. II. Desmidieer från Grönland. III. Grunddragen af Desmidieernas utbreding i norden. 154 pp. Pls. 1,2 (Repr.); Bih. Kongl. Svenska Vet.-Akad. Handl. 13, Afd. III(5/6): 1–110.

Borge, O. 1893. Übersicht der neu erscheinenden Desmidiaceen-Litteratur. Nuova Notarisia Ser. IV: 389–398.

Borge, O. 1894. Süsswasser-Chlorophyceen gesammelt von Dr. A. Osw. Kihlman in nordlichsten Russland, Gourvernement Archangel. Bih. Kongl. Svenska Vet.-Akad. Handl. 19, III(5): 1–41. Pls. 1–3.

Borge, O. 1896. Australische Süsswasserchlorophyceen. Bih. Kongl. Svenska Vet.-Akad. Handl. 22, III(9): 1–3. Pls. 1–4.

---

1. References marked with an asterisk (*) deal with North American desmids not previously appearing in Part I and II of this synopsis.

Borge, O. 1899. Über tropische und subtropische Süsswasser-Chlorophyceen. Bih. Kongl. Svenska Vet.-Akad. Handl. 24, Afd. III (12):1-33. Pls. 1, 2.

Borge, O. 1903. Die Algen der ersten Regnellschen Expedition. 2. Desmidiaceen. Ark. f. Bot. 1: 71-138. Pls. 1-5.

Borge, O. 1913. Beiträge zur Algenflora von Schweden. 2. Die Algenflora um den Torne-Tråsksee in Schwedische-Lappland. Bot. Notiser 13: 1-32; 49-109. 2 Figs. 3 Pls.

Borge, O. 1918. Die von Dr. A. Löfgren in São Paulo gesammelten Süsswasseralgen. Ark. f. Bot. 15(13): 1-108.

Borge, O. 1925. Die von Dr. F. C. Hoehne während der Expedition Roosevelt-Rondon gesammelten Süsswasseralgen. Ark. f. Bot. 19(17): 1-56. 3 Figs. Pls. 1-6.

Borge, O. 1930. Beiträge zur Algenflora von Schweden. 4. Die Algenflora am Grövelsee. Ark. f. Bot. 23A(2): 1-64. 2 Taf. 9 Figs.

Börgesen, F. 1890. Symbolae ad floram Brasiliae centralis cognoscendam. (Edit. Eug. Warming, Particula XXXIV). Vid. Medd. Naturh. Foren. Kjöbenhavn 1890: 929-958. Pls. 2-5.

Bourrelly, P. 1957. Algues d'eau douce du Soudan Français, région du Macina (A. O. F.). Bull. l'Inst. Francaise Afrique Nord 19, Ser. A(4): 1047-1102. Pls. 1-21.

Bourrelly, P. 1966. Quelques algues d'eau douce du Canada. Intern. Rev. Ges. Hydrobiol. 51(1): 45-126. Pls. 1-24.

Brandham, P. E. 1965. Some new chromosome counts in the desmids. British Phycol. Bull. 2(6): 451-455.

Brébisson, A. de. 1856. Liste des Desmidiées observées en Basse-Normandie. Mém. Soc. Impér. Sci. Nat. Cherbourg 4: 113-166. 2 Pls.

Brown, Helen J. 1930. The desmids of the southeastern coastal plain region of United States. Trans. Amer. Microsc. Soc. 49(2): 97-139. 4 Pls.

Brunel, J. 1938. Sur deux formes nouvelles de *Micrasterias*. Nat. Canadien 65: 71-74. Fig. 1.

*Buchanan, R. E. 1907. Notes on the algae of Iowa. Proc. Iowa Acad. Sci. 14: 77-84.

Bulnheim, O. 1861. Beiträge zur Flora der Desmidieen Sachsens I. Hedwigia 2(9): 50-52. Pl. 9.

*Button, K. S. and Blinn, D. W. 1973. Preliminary seasonal studies on algae from Upper Lake Mary, Arizona. Arizona Acad. Sci. 8(2): 80-83.

Carter, Nellie. 1935. Alpine desmids from British Columbia. Linn. Soc. Jour. Bot. 50(333): 151-174. 49 Figs.

Cedercreutz, C. 1932. Süsswasseralgen aus Petsamo II. Mem. Soc. Fauna Flora Fennica 7(3): 136-248. Figs. 1-17.

Cedergren, G. R. 1932. Die Algenflora der Provinz Härjedalen. Ark. f. Bot. 25A(4): 1-107. 25 Figs. 4 Pls.

Cleve, P. T. (1863)1864. Bidrag till kännedomen om Sveriges sötvattensalger af familjen Desmidieae. Öfv. Kongl. Vet.-Akad. Förhandl. 20(10): 481-497. Pl. 4.

*Cole, G. A. 1957. Studies on a Kentucky Knobs lake, III. Some quantitative aspects of the net plankton. Trans. Kentucky Acad. Sci. 18: 88-101.

Collins, F. S., Holden, I. and Setchell, W. A. 1895-1917. Phycotheca Boreali-Americana 1-45. (Exsiccatae).

Conn, H. W. and Webster, L. W. 1908. A preliminary report on the algae of the fresh waters of Connecticut. Connecticut Geol. & Nat. Hist. Surv. Bull. 10: 1-78. 44 Pls.

Cooke, M. C. 1886-1887. British Desmids. A Supplement to British Fresh-water algae. Nos. 1-6: 1-96. Pls. 1-48; Nos. 7-10: 97-205 + i-xiv. Pls. 49-66.

Croasdale, Hannah T. 1956. Freshwater algae of Alaska. I. Some desmids from the interior. Part 2. *Actinotaenium*, *Micrasterias*, and *Cosmarium*. Trans. Amer. Microsc. Soc. 75(1): 1-70. Pls. 1-17.

Croasdale, Hannah T. 1965. Desmids from Devon Island, N.W.T., Canada. Trans. Amer. Microsc. Soc. 84(3): 301-333. Pls. 1-8.

Croasdale, Hannah T. and Grönblad, R. 1964. Desmids from Labrador 1. Desmids of the southeastern coastal area. Trans. Amer. Microsc. Soc. 83(2): 142-212. Pls. 1-21.

Cushman, J. A. 1904. Desmids from Newfoundland. Bull. Torr. Bot. Club 31: 581-584. Pl. 26.

Cushman, J. A. 1904a. Notes on *Micrasterias* from southeastern Massachusetts. Bull. Torr. Bot. Club 31: 393-397. 3 Figs.

Cushman, J. A. 1905. A contribution to the desmid flora of New Hampshire. Rhodora 7: 111-119; 251-266. Pls. 61, 64.

Cushman, J. A. 1905a. Notes on the zygospores of certain New England desmids with descriptions of a few new forms. Bull Torr. Bot. Club 32(3): 223-229. Pls. 7,8.

Cushman, J. A. 1908. A synopsis of the New England species of *Micrasterias*. Rhodora 10(114): 97-111.

*Davis, C. C. 1973. A seasonal quanititative study of the plankton of Bauline Long Pond, a Newfoundland lake. Nat. Canadien 100(2): 85-105.

Delponte, J. B. 1873, 1876, 1877, 1878. Specimen Desmidiacearum subalpinum. Aug. Tauri-norum, 1873. 96 pp. Pls. 1–5. 1877. Pars altera. Ibid. pp. 97–283. Pls. 7–23. 1876. Ibid. Mem. R. Acad. sci. Torino, II, 28: 19–208. Pls. 1–5. 1878. Ibid. 30: 1–186. Pls. 7–23.

DeToni, G. B. 1889. Sylloge algarum omnium hucusque cognitarum. Vol. I. pp. 709–1315. Padua.

Dick, J. 1923. Beiträge zur Kenntnis der Desmidiaceen-Flora von Süd-Bayern II. Bot. Arch, 3: 21–236. Pls. 1–7.

Dick, J. 1926. Beiträge zur Kenntnis der Desmidiaceen-Flora von Süd-Bayern III. Folge. Oberschwaben (Bayr. Allgäu). Krypt. Forsch. Bayer. Bot. Ges. 7: 444–454. Pls. 18–21.

Dick, J.. 1930. Pfälzische Desmidiaceen. Mitt. Pfalz. ver. Naturkde. 3: 1–44. 10 Pls.

*Dillard, G. E. 1974. An annotated catalog of Kentucky algae. Ogden Coll. Sci. & Tech., Western Kentucky Univ. 135 pp.

*Dillard, G. E. and Tindall, D. R. 1973. Notes on the algal flora of Illinois, III. Additions to the Chlorophyceae. Ohio Jour. Sci. 73(4): 229–233.

Ducellier, F. 1915. Contributions à l'étude du polymorphisme et des monstruosities chez les Desmidiacées. Bull. Soc. Bot. Genève 7: 75–118. Pls. 1–3.

Ducellier, F. 1918. Étude critique sur Euastrum ansatum Ralfs et quelques-unes de ses variétiés helvétiques. Bull. Soc. Bot. Genève, II, 10: 35–46. Figs. 1–29.

Ducellier, F. 1918a. Contribution a l'étude de la flore-desmidiologique de la Suisse II. Bull. Soc. Bot. Genève, II, 10: 85–154. Figs. 62–134.

*Duthie, H. C., Ostrofsky, M. L., Sreenivasa, M. R. and Brown, D. J. 1974. Freshwater algae from western Labrador. Dept. Biol. Univ. Waterloo, Ontario. Multilithed.

Ehrenberg, C. G. (1831)1832. Über die Entwicklung und Lebensdauer der Infusionsthiere; nebst ferneren Beiträgen zu einer Vergleichung ihrer verschiedenen Systeme. Phys. Abh. d. K. Acad. Wiss. Berlin 1831: 1–154. 4 Pls.

Ehrenberg, C. G. 1835. Dritter Beitrag zur Erkenntniss grosser Organisation in der Richtung des kleinsten Raumes. Phys. Abh. d. K. Acad. Wiss. Berlin 1833: 145–336. Pls. 1–6.

Ehrenberg, C. G. 1838. Die Infusionsthierchen als vollkommene Organismen. 548 pp. 64 Pls. Leipzig.

Ehrenberg, C. G. (1840)1841. Charakteristik von 274 neuen Arten von Infusorien. Monats. Berlin Akad. 1840: 197–219.

Eichler, B. and Gutwinski, R. 1894. De nonnullis speciebus algarum novarum Krakowie. Rozpr. Wydz. matem.-przy. Akad. umiej. w Krakowie 28: 162–178. Pls. 4, 5.

Elfving, F. 1881. Anteckningar om Finska Desmidiéer. Acta Soc. Fauna Flora Fennica 2(2): 1–18. 1 Pl.

*Flensburg, T. F. and Sparling, J. H. 1973. The algal microflora of a string mire in relation to the chemical composition of water. Canadian Jour. Bot. 51(4): 743–749. 3 Tabls.

Förster, K. 1964. Desmidiaceen aus Brasilien. Hydrobiologia 23(3/4): 321–505. Pls. 1– 51.

Förster, K. 1968. Beitrag zur Desmidiaceenflora des Ost-Allgäus 1. Pfronter-Reid (3 Teil). Ber. Bayer. Bot. Ges. 40: 17–30. Pls. 5–8.

Förster, K. 1969. Amazonische Desmidieen 1. Areal Santarém. Amazonia 11(1/2): 5–116. Pls. 1–56.

Förster, K. 1970. Beitrag zur Desmidieenflora von Süd-Holstein und der Hansestadt Hamburg. Nova Hedwigia 20: 253–411. Pls. 1–29.

*Förster, K. 1972. Desmidieen aus dem Südosten der Vereinigten Staaten von Amerika. Nova Hedwigia 23(2/3): 515–644. Pls. 1–29.

Förster, K. 1974. Amazonische Desmidieen. 2 Teil: Areal Maures-Abacaxis. Amazonia 5(2): 135–242. Pls. 1–36.

Fritsch, F. E. 1918. Contributions to our knowledge of the freshwater algae of Africa. A first report on the freshwater algae mostly from the Cape Peninsula in the herbarium of the South African Museum. Ann. South African Mus. 9: 483–611. Figs. 1–43.

Fritsch, F. E. and Rich, Florence. 1924. Freshwater and subaerial algae from Natal. Trans. Roy. Soc. South Africa 11: 297–398. Figs. 1–31.

Fritsch, F. E. and Rich, Florence. 1937. Contributions to our knowledge of the freshwater algae of Africa. Algae from the Belfast Pan, Transvaal. Trans. Roy. Soc. South Africa 25(2): 153–228. 31 Figs.

Fujisawa, R. 1936. On the genus Micrasterias Ag. 1827 found in Japan. Nat. Hist. Mag. 34(58): 12–17. 2 Pls.

Gay, F. 1884. Note sur les Conjuguées du midi de la France. Bull. Soc. Bot. France 31: 331–342.

Gay, F. 1884a. Essai d'une monographie locale des Conjuguées. Thèse, Montpellier. 112 pp. 4 Pls. (Rev. Sci. Nat., III, 3(1): 187–228; 285–335. Pls. 5–6a.).

*Glooschenko, W. A. and Alvis, Carolyn. 1973. Changes in species composition of phytoplankton due to enrichment by N, P, and Si of water from a north Florida lake. Hydrobiologia 42(2/3): 285–294.

Graffius, J. H. 1963. A comparison of algal floras in two lake types, Barry County, Michigan. Dissert., Michigan State Univ.

Grönblad, R. 1920. Finnländische Desmidiaceen aus Keuru. Acta Soc. Fauna Flora Fennica 47(4): 1–98. Pls. 1–6.

Grönblad, R. 1921. New desmids from Finland and northern Russia with critical remarks on some known species. Acta Soc. Fauna Flora Fennica 49(7): 1–78. Pls. 1–7.

Grönblad, R. 1924. Observations on some desmids I. Some remarks on the genus *Pleurotaenium* Naeg. II. Critical remarks on some less known desmids and descriptions of a few new forms. Acta Soc. Fauna Flora Fennica 55(3): 3–18. Pls. 1, 2.

Grönblad, R. 1926. Beitrag zur Kenntnis der Desmidiaceen Schlesiens. Soc. Sci. Fennica. Commen. Biol. 2(5): 1–39. 9 Figs. 3 Pls.

Grönblad, R. 1936. Desmids from northern Russia (Karelia) collected 1918 at Uhtua (Ukhtins-skaya) and Hirvisalmi. Soc. Sci. Fennica. Commen. Biol. 5(6): 1–12. 2 pls.

Grönblad, R. 1942. Algae, hauptsächlich Desmidiaceen, aus dem finnischen, norwegischen und schwedischen Lappland, mit Berücksichtigung der Organismen des Phytoplanktons und des Zooplanktons. Acta Soc. Sci. Fennicae, n.s., B, II(5): 1–46. Pls. 1–5.

Grönblad, R. 1945. De algis Brasiliensibus, praecipue Desmidiaceis in regione inferiore fluminis Amazonas a Professore August Ginzberger (Wien) anno 1927 collectis. Acta Soc. Sci. Fennicae, n.s., B, II(6): 1–43. 6 Textfigs. 3 photos. 1 Map. Pls. 1–16.

Grönblad, R. 1945a. A further report of freshwater-algae recorded from the neighbourhood of the Zoological Station at Tvärminne. Mem. Soc. Fauna Flora Fennica 21: 48–53. 5 Textfigs. 1 Pl.

Grönblad, R. 1956. Desmids from the United States, collected in 1947–1949 by Dr. Hannah Croasdale and Dr. Edwin T. Moul. Soc. Sci. Fennica Commen. Biol. 15(12): 1–38. Pls. 1–12.

Grönblad, R. 1956a. A contribution to the knowledge of the algae of brackish water in some ponds in the Woods Hole region, U.S.A. Mem. Soc. Fauna Flora Fennica 31: 63–69. 24 Figs.

Grönblad, R. 1963. Desmids from Jamtland, Sweden and adjacent Norway. Soc. Sci. Fennica Commen. Biol. 26(1): 1–43. 1 Textfig. 5 Pls.

Grönblad, R., Prowse, G. A. and Scott, A. M. 1958. Sudanese desmids. Acta Bot. Fennica 58: 1–82. Pls. 1–29.

Grönblad, R. and Scott, A. M. 1959. Concerning later starting points in algae. Taxon 8(3): 88–90.

Grönblad, R., Scott, A. M. and Croasdale, Hannah. 1964. Desmids from Uganda and Lake Victoria collected by Dr. Edna M. Lind. Acta Bot. Fennica 66: 1–57. Pls. 1–12.

Gutwinski, R. 1890. Zur Wahrung der Priorität, Vorläufige Mittheilungen über einige neue Algen-Species und Varietäten aus der Umgebung von Lemburg. Bot. Centralbl. 43: 65–73.

Gutwinski R. (1891)1892. Flora glonów okolic Lwowa (Flora algarum agri Leopoliensis) Kraków 1891. Spraw. kom. Fizyj. Akad. Umiej. 27 Materyal. fizyjogr. Kraków. Czesc. II: (1)–(124). Pls. 1–3.

Gutwinski, R. 1897. Über die bis jetzt in Bosnien und der Hercegovina entdeckten Algen (mit Ausschluss der Diatomen). Wiss. Mitt. aus Bosnien u. Hercegovina 5: 453–463. Pl. 77.

Gutwinski, R. 1902. De algis a Dre M. Raciborski anno 1899 in insula Java collectis. Bull. Intern. l'Acad. Sci. d. Cracovie, Cl. Sci. Mat. et Nat. 1902(9): 575–617. Pls. 36–40.

Gutwinski, R. 1909. Flora glonów Tatranskich: Flora algarum montium Tatrensium. Bull. Acad. Sci. Cracovie, Cl. Sci. Mat. et Nat. 1909: 415-560. 2 Pls.

*Hale, F. E. 1950. The use of copper sulphate in control of microscopic organisms. Phelps Dodge Refin. Corp. 44 pp. 6 Pls. New York.

Hansgirg, A. 1888. Prodromus der Algenflora von Böhmen. Erster Theil. 2: 1–288. Figs. 46–124.

Hassall, A. H. 1845. A history of the British freshwater algae. Vol. I. 8 + 462 pp. Vol. II. 24 pp. 103 Pls.

Hirano, M. 1967. Freshwater algae collected by the joint Thai-Japanese Biological Expedition to South Asia 1961–62. Nature & Life in Southeast Asia 5: 1–71. Pls. 1–16.

Hobson, J. 1963. Notes on Indian Desmidieae. Quart. Jour. Microsc. Sci., n.s., 3: 168–170. 2 Figs.

Homfeld, H. 1929. Beitrag zur Kenntnis der Desmidiaceen Nordwestdeutschlands, besondere ihrer Zygoten. Pflanzenfam. 12: 1–96. 1 Map. 9 Pls.

Hooker, W. J. 1830. The English Flora of Sir James Edward Smith. Vol. II, Part 1. i–viii + 432 pp.

Huber-Pestalozzi, G. 1931. Der formkreis von *Euastrum verrucosum* Ehrenberg. Arch. f. Hydrobiol. 22: 415–459. 50 Figs.

Hughes, E. O. 1950. Fresh-water algae of the Maritime Provinces. Proc. Nova Scotia Inst. Sci. 22(2): 1–63. Pls. I–IV.

*Hutchinson, G. E. 1967. A treatise on limnology. Vol. 2. Introduction to lake biology and the limnoplankton. 1115 pp. New York.

*Hutchinson, G. E. 1973. Eutrophication. Amer. Scientist 61(3): 269–279. 4 Figs.

Huzel, C. 1937. Beitrag zur Kenntnis der mikroskopischen Pflanzenwelt der Rauhen Wiese bei Böhmenkirch. Ver. Württemberg Landes. f. Natur. 13: 1–148. Pls. 1–15.

Hylander, C. J. 1928. The algae of Connecticut. Connecticut Geol. & Nat. Hist. Surv. 42: 1–245. Pls. 1–28.

Irénée-Marie, Fr. (1938)1939. Flore desmidiale de la région de Montréal. 547 pp. 69 Pls. La Prairie, Canada.

Irénée-Marie, Fr. 1947. Contribution à la connaissance des Desmidiées de la région des Trois-Rivières. Nat. Canadien 74(3): 102–124. Pls. 1, 2.

Irénée-Marie, Fr. 1949. Contribution à la connaissance des Desmidiées de la région des Trois-Rivières. IVe Partie. Nat. Canadien 76(1/2): 16–42. Pls. 1–4.

Irénée-Marie, Fr. 1949a. Quelques Desmidiées du Lac Mistassini. Nat. Canadien 76(8/10): 242–261; 265–316. Pls. 1–6.

Irénée-Marie, Fr. 1951. Desmidiées de la région de Québec. 2e Partie. Les *Micrasterias*, les *Euastrum* et les *Closterium*. Nat. Canadien 78(7/8): 177–221. Pls. 1–4.

Irénée-Marie, Fr. 1952. Contribution à la connaissance des Desmidiées de la région du Lac-St.-Jean. Hydrobiologia 4(1/2): 1–208. Pls. 1–19.

Irénée-Marie, Fr. 1956. *Les Euastrum* du Lac-Saint-Jean et de Parc des Laurentides de la Province de Québec. Rev. Algol. 2(1/2): 112–121. 1 Pl.

Irénée-Marie, Fr. 1957. Les *Micrasterias* de la région des Trois-Rivières. Hydrobiologia 9(1): 66–88. 1 Pl.

Irénée-Marie, Fr. 1958. Contribution à la connaissance des Desmidiées de la région des Trois-Rivières. Nat. Canadien 85(5): 105–147. 1 Pl.

Irénée-Marie, Fr. 1958a. Contribution à la connaissance des Desmidiées du Sud-est de la Province de Québec et de la Gaspésie. Hydrobiologia 12–(2/3): 107–128. 15 Figs.

Irénée-Marie, Fr. 1958b. Contribution à la connaissance des Desmidiées de la région des Trois-Rivières. Rev. Algol. 4(2): 94–124. 2 Pls.

Irénée-Marie, Fr. 1959. Expedition algologique dans la nord de la Maurice, bassin de la Mattawin. Hydrobiologia 13(4): 319–381. Pls. 1–8.

Irénée-Marie, Fr. and Hilliard, D. K. 1963. Desmids from southcentral Alaska. Hydrobiologia 21(1/2): 90–124. Pls. 1, 2.

Istavànffi, G. 1887. Diagnose praeviae algarum novarum in Hungaria observatarum, etc. Notarisia 1887(5): 234–242.

Istvànffi, G. 1888. Jelentes a felsö-magyarországi tözegképletek algologiai megviszsgálásáról. Math.-term.-tud. Közl. 23: 205–262. 2 Pls.

Jackson, D. C. 1971. A study of selected genera of the families Gonatozygaceae, Mesotaeniaceae and Desmidiaceae in Montana. Dissert., Michigan State Univ.

Jacobsen, J. P. 1875. Aperçu systématique et critique sur les Desmidiacées du Danemark. Bot. Tidsk. 8: 143–215. Pls. 7, 8.

Johnson, L. N. 1894. On some species of *Micrasterias*. Bot. Gaz. 19: 56–60. Pl. 6.

Johnson, L. N. 1895. Some new and rare desmids of the United States, II. Bull. Torr. Bot. Club 22(7): 289–298. Pls. 232, 233.

Joshua, W. 1885. On some new and rare Desmidieae. III. Jour. Bot. 23: 33–35. Pl. 254.

Joshua, W. 1886. Burmese Desmidieae, with descriptions of new species occurring in the neighbourhood of Rangoon. Linn. Soc. Jour. Bot. 21(40): 634–655. Pls. 23–25.

Kaiser, P. E. 1919. Desmidiaceen des Berechtesgadener Landes. Krypt. Forsch. 1(4): 216–230. 34 Textfigs.

Kallio, P. 1951. The significance of nuclear quantity in the genus *Micrasterias*. Ann. Bot. Soc. Zool.-Bot Fennicae 'Vanamo' 24(2): 1–122. Illus.

Kallio, P. 1968. On the morphogenetic system of *Micrasterias sol*. Ann. Acad. Sci. Fennicae, Ser. A, IV, Biol. 124: 1–23. 7 Figs.

Kiermayer, O. 1968. The distribution of microtubules in differentiating cells of *Micrasterias denticulata* Bréb. Planta 83: 223–236. 10 Figs.

Kiermayer, O. 1970. Causal aspects of cytomorphogenesis in *Micrasterias*. Ann. New York Acad. Sci. 175(2): 686–701. 11 Figs.

Kiermayer, O. and Dobberstein, B. 1973. Membrankomplexe dictyosomalen Herkunft als

"Matrizen" fur die extraplasmatische Synthese und Orientierung von Mikrofibrillen. Proto-plasm 77(4): 437–451. 12 Figs.

Kirchner, O. 1878. Algen. In: Cohn, F. Kryptogamen-Flora von Schlesien 2(1): 1–284.

Kossinskaja, C. C. 1936. Sur la flore des Demidiées du lac Montsché. Acta Inst. Bot. Acad. Sci. USSR (Plantae Crypt.) 3: 451–467. Pls. 1, 2.

Kossinskaja, C. C. 1936a. (Desmidien der Arktis). Acta Inst. Bot. Acad. Sci. USSR (Plantae Crypt.), II, 3: 401–440. 5 Pls.

Kossinskaja, C. C. 1936b. *Euastrum verrocosum* Ehr. var. *alatum* Wolle f. *minor* (Lobik) Kossinskaja comb. nov. Sovietskaja Bot. 1936: 109–110. 1 Pl.

Kossinskaja, C. C. 1949. Desmidiaceae novae et rariores. Not. Syst. e Sect. Crypt. Inst. Bot. nomine V. L. Komarovii Acad. Sci. URSS 6(1/6): 42–50. 2 Pls.

Kossinskaja, C. C. 1960. Flora plantarum cryptogamarum URSS. Vol. 5, Conjugatae (II), Desmidiales Fasc. 1: 1–706. Pls. 1–87. Acad. Sci. URSS, Leningrad.

Krieger, W. 1932. Die Desmidiaceen der Deutschen Limnologischen Sunda Expedition. Arch. f. Hydrobiol. Suppl. 11: 129–230. Pls. 3–26.

Krieger, W. 1937. Die Desmidiaceen Europas mit Berüchtigung der aussereuropaischen Arten. Rabenhorst's Kryptogamen-Flora 13(1:3): 1–712. Pls. 1–96. (Printing of Paul Dünnhaupt, Köthen).

Krieger, W. 1939. Die Desmidaceen Europas mit Berüchtigun der aussereuropaischen Arten. Rabenhorst's Kryptogamen-Flora 13(1:2): 1–117. Pls. 97–142.

Krieger, W. and Bourrelly, P. 1956. Desmidiacées des Andes du Venezuela. Ergebn. deutsch. limnolog. Venezuela-Expedition 1932. 1: 141–195. Pls. 1–12.

Kützing, F. T. 1845. Phycologia germanica, d. i. Deutschlands Algen in bündigen Besch-reibungen. 10 + 340 pp. Nordhausen.

Kützing, F. T. 1849. Species algarum. 6 + 922 pp. Leipzig.

Lacalli, T. C. and Acton, A. B. 1974. Tip growth in *Micrasterias*. Science 183: 665–666.

*****Lackey, J. B. and Lackey, E. W.** 1967. A partial check-list of Florida fresh-water algae and Protozoa, with reference to McCloud and Cue Lakes. Florida Eng. & Indust. Exper. Sta. Bull. Ser. 131, iii, iv: 1–28.

Lagerheim, G. 1885. Bidrag till Amerikas Desmidié-Flora. Öfv. Kongl. Vet. Akad. Förhandl. 42(7): 225–255. Pl. 27.

Lagerheim, G. 1888. Über Desmidiaceen aus Bengalen nebst Bemerkungen über die geo-graphische Verbreitung des Desmidiaceen in Asien. Bih. Kongl. Svenska Vet.-Akad. Handl. 13, III(9): 1–12. Pl. 1.

Laporte, L. J. 1931. Recherches sur la biologie et la systématique des Desmidiées. Encyclop. Biol. 9: 1–147. Pls. 1–22.

Larsen, E. 1904. Botanical exploration of the east coast of Greenland between 65 deg., 35 min.–74 deg., 30 min. lat. N. II. The fresh-water algae of East Greenland. Medd. om Grönland 30: 75–100. 10 Figs.

*****Larson, D. W. and Donaldson, J. R.** (not dated). Waldo Lake, Oregon; an ultra-oligotrophic environment and some implications concerning recreational development. Mimeo. Rep. Dept. Fish. & Wildlife, Oregon State Univ. 21 pp.

*****Ling, H. U. and Tyler, P. A.** 1974. Interspecific hybridity in the desmid genus *Pleurotaenium*. Jour. Phycol. 10(2): 225–230. 31 Figs.

Lowe, C. W. 1924. The freshwater algae of central Canada. Trans. Roy. Soc. Canada, III, 18(5): 19–50. 4 Pls.

Lundell, P. M. 1871. Des Desmidiaceis, quae in Suecia inventae sunt, observationes criticae. Nova Acta Reg. Soc. Sci. Upsaliensis, III, 8(2): 1–100. Pls. 1–5.

Lütkemüller, J. 1894. Die Poren der Desmidiaceengattung *Closterium* Nitzsch. Osterr. Bot. Zeit. 44: 11–14; 49–53.

Lütkemüller, J. 1900. Desmidiaceen aus der Umgebung des Millstättersees in Kärnten. Verh. k. k. Zool.-Bot. Ges. Wien 50: 60–84. 1 Pl.

Lütkemüller, J. 1900a. Desmidiaceen aus den Ningpo-Mountains in Centralchina. Ann. des Nat. Hofmuseums 15: 115–126. 6 Pls.

Lütkemüller, J. 1902. Die Zellmembran der Desmidiaceen. Beitr. Biol. Pflanzen. 8: 347–414. Pls. 18–20.

Lütkemüller, J. 1910. Zur Kenntnis der Desmidiaceen Böhmens. Verh. k. k. Zool.-Bot. Ges. Wien 60: 478–503. 3 Figs. 2 Pls.

*****Macfie, M. F. and Swails, L. F.** 1957. An ecological study of the fauna and flora of the Savannah River area. 7. The algae. A new distribution record of a rare variety of *Micrasterias*. Univ. South Carolina Publ., III, Biol. 2(2): 61–62.

Maskell, W. M. 1888. Note on *Micrasterias americana* Ralfs and its varieties. Jour. Roy. Microsc. Soc. 1888: 7–10. Pl. 2.

*Masters, M. J. 1971. The occurrence of *Phlyctidium bumillerias* on two growth forms of *Staurastrum pingue* and other *Staurastrum* spp. in Lake Manitoba. Canadian Jour. Bot. 49: 1637–1641. 1 Pl.

*Mathews, G. W. 1974. A study of the taxonomy and ecology of southeast Kansas desmids (Chlorophyta: Conjugatae). M. S. Thesis, Kansas State Coll.

*McInteer, B. B. 1944. Algae on wet rocks at Cumber and Falls state Park. Castanea 9: 115–116.

Meneghini, G. 1840. Synopsis Desmidiarum hucusque cognitarum. Linnaea 14: 201–240.

Messikommer, E. 1927. Biologische Studien im Torfmoor von Robenhausen unter besonderer Berucksichtigung der Algenvegetation. Inaug. Dissert. Univ. Zürich. (Mitt. aus dem Bot. Mus. der Univ. Zürich 122: 1–171. Pls. 1–6.).

Messikommer, E. 1935. Algen aus den Obertoggenburg. Mitt. Bot. Mus. Univ. Zürich 148: 95–130. Pls. 1, 2. (Jahrb. St. Gallischen Naturw. Ges. 67(1933–1934): 95–130. Pls. 1, 2.).

Messikommer, E. 1938. Beitrag zur Kenntnis der fossilen und subfossilen Desmidiaceen. Hedwigia 78: 107–201. 1 Fig. Pls. II–X.

Messikommer, E. 1942. Beiträge zur Kenntnis der Algenvegetation des Hochgebirges um Davos. Beit. Geobot. Landes. der Schweiz 24: 1–452.

Meyer, R. L. and Brook, A. J. 1968. Freshwater algae from the Itasca State Park, Minnesota. Nova Hedwigia 16: 251–266. 1 Map.

Miller, L. R. 1936. Desmids of the Medicine Bow Forest of Wyoming. Univ. Wyoming Publ. Sci. Bot. 2(6): 67–120. Pls. 1 11.

Mix, Marianne. 1965. Zur Variationsbreite von *Micrasterias swainei* Hastings und *Staurastrum leptocladum* Nordst. sowie über die Bedeutung von Kulturversuchen für die Taxonomie der Desmidiaceen. Arch. f. Mikrobiol. 51: 168–178. 14 Textfigs. 1 Tabl.

Moore, Caroline and Moore, Laura. 1930. Some desmids of the San Juan Islands. Publ. Puget Sound Biol. Sta. 7: 289–335. 232 Figs.

*Moore, J. W. 1974. The benthic algae of southern Baffin Island. I. Epipelic communities in rivers. Jour. Phycol. 10(1): 50–57.

*Mulligan, H. F. and Baranowski, A. 1969. Growth of Phytoplankton and vascular aquatic plants at different nutrient levels. Verh. Intern. Ver. Limnol. 17: 802–810.

*Mutchler, F. M. 1934. The algae of a transient lake in Kentucky. M. A. Thesis, Western Kentucky State Coll.

Nägeli, C. 1849. Gattungen einzelliger Algen physiologisch und systematisch Bearbeitet. 8 + 139 pp. 8 Pls.

Nordstedt, C. F. O. (1869)1870. Symbolae ad floram Brasiliae centralis cognoscendam. Edit. Eug. Warming. Particula quinta. 18 Fam. Desmidiaceae. Vid. Medd. Naturh. Forening Kjöbenhavn 1869(14/15): 195–234, Pls. 2–4.

Nordstedt, C. F. O. 1873. Bidrag till kännedomen om sydligare Norges Desmidieer. Acta Univ. Lund 9: 1–51. 1 Pl.

Nordstedt, C. F. O. 1875. Desmidieae arctoae. Öfv. Kongl. Vet.-Akad. Förhandl. 1875(6): 13–43. Pls. 6–8.

Nordstedt, C. F. O. (1877)1878. Nonnullae aquae dulcis brasilienses. Öfv. Kongl. Vet.-Akad. Förhandl. 1877(3): 15–28. Textfigs. I–VI. Pl. 2.

Nordstedt, C. F. O. 1880. De algis et Characeis. 1. De algis nonnullis praecipue Desmidieis, inter Utricularias Musei Lugdono-Batavi. Acta Univ. Lund 16: 1–14. 1 Pl.

Nordstedt, C. F. O. 1887. Algologiska smasaker. 4. Utrag ur ett arbete öfver de af Dr. S. Berggren på Nya Seland och i Australien samlade sötvattensalgerna. Bot. Notiser 1887: 153–164.

Nordstedt, C. F. O. 1888. Desmidieer från Bornholm, samlade och delvis bestämda af R. T. Hoff, granskade af O. Nordsteat. Vidensk. Medd. Naturh. Foren. Kjöbenhavn 1888: 182–213. Pl. 6.

Nordstedt, C. F. O. 1888a. Freshwater algae collected by Dr. S. Berggren in New Zealand and Australia. Kongl. Svenska Vet.-Akad. Handl. 22(8): 1–198. 7 Pls.

Okada. Y. 1936. Notes on Japanese desmids, with special reference to the newly found species. I–IV. Bot. Mag. Tokyo 1: 79–85. Pl. 1 Textfigs. 1, 2; 255–259. Pl. 3; 313–317. Pl. 4; 430–434. 1 Textfig. Pl. 7.

Okada, Y. 1939. Desmids from the Sinsiru Island in the Middle Kuriles. Jour. Imper. Fish. Inst. (Tokyo) 33(2): 107–121. 1 Map. Pls. 1–6.

Okada, Y. 1953. Taxonomical studies on genus *Euastrum*, with special reference to the species of Japan and its surrounding areas. Mem. Fac. Fish. Kagoshima Univ. 3(1): 193–244. Pls. 1–3.

Okada, Y. 1953a. A new classification of Conjugatae, with special reference to desmids. Mem. Fac. Fish. Kagoshima Univ. 3(1): 165–192. 2 Figs. 6 Tabls.

*Patrick, Ruth. 1961. A study of the numbers and kinds of species found in rivers in eastern United States. Proc. Acad. Nat. Sci. Philadelphia 113: 215–258.

Pevalek, I. 1925. Prilog poznavanju alga jezera I Poljana kod Denog Polha u Julskim Alpama. Nuova Notarisia 36: 282–295.

*Pickett-Heaps, J. D. 1974. Scanning electron microscopy of some cultured desmids. Trans. Amer. Microsc. Soc. 93(1): 1–23, Figs. 1–30.

*Pickett-Heaps, J. D. 1975. Green algae: Structure, reproduction and evolution in selected genera. Sinauer Assoc., Inc. Sunderland, Mass.

Pickett-Heaps, J. D. and Marchant, H. J. 1972. The phylogeny of the green algae. A new proposal. Cytobios 6: 255–264.

Prescott, G. W. 1931. Iowa algae. Univ. Iowa Stud. Nat. Hist. 13(6): 1–235. Pls. 1–39.

Prescott, G. W. 1935. Notes on the desmid flora of New England. I. The genus *Euastrum* in Massachusetts. Rhodora 37: 22–30. Pl. 235.

Prescott, G. W. 1935a. Notes on the desmid flora of New England. II. Desmids from Cape Cod and the Elizabeth Islands. Rhodora 37: 113–131. Pls. 329, 330.

Prescott, G. W. 1940. Desmids of Isle Royale, Michigan. The genera *Staurastrum, Micrasterias, Xanthidium,* and *Euastrum,* with a note on *Spinoclosterium.* Pap. Michigan Acad. Sci., Arts & Lettr. 25: 89–100. Pls. 1–4.

Prescott, G. W. 1942. The fresh-water algae of southern U.S. II. The algae of Louisiana, with descriptions of some new forms and notes on distribution. Trans. Amer. Microsc. Soc. 61(2): 109–119. 1 Pl.

Prescott, G. W. and Magnotta, Angelina. 1935. Notes on Michigan desmids, with descriptions of some species and varieties new to science. Pap. Michigan Acad. Sci., Arts & Lettr. 20: 157–170. Pls. 25–27.

Prescott, G. W. and Scott, A. M. 1942. The freshwater algae of southern United States. I. Desmids from Mississippi, with descriptions of some new species and varieties. Trans. Amer. Microsc. Soc. 61(1): 1–29. 4 Pls.

Prescott, G. W. and Scott, A. M. 1943. The desmid genus *Micrasterias* Agardh in southeastern United States. Pap. Michigan Acad. Sci., Arts & Lettr. 28: 67–82. Pls. 1–6.

Prescott, G. W. and Scott, A. M. 1945. The freshwater algae of southern United States. III. The genus *Euastrum* with descriptions of some new varieties. Amer. Mid. Nat. 34: 231–257. Pls. 1–6.

Prescott, G. W. and Scott, A. M. 1952. The algal flora of southeastern United States. V. Additions to our knowledge of the desmid genus *Micrasterias* 2. Trans. Amer. Microsc. Soc. 71(3): 229–252. Pls. 1–8.

Pritchard, A. 1861. A history of Infusoria, including the Desmidiaceae and Diatomaceae, British and foreign. Ed. 4. London.

Rabenhorst, L. 1850–1879. Die Algen Sachsens. Desmidiaceae. Dresden.

Rabenhorst, L. 1868. Flora Europas algarum aquae dulcis et submarinae. Sec. 3. Algae Chlorophyllophyceas, Melanophyceas et Rhodophyceas complectens. 461 pp. Figs. Leipzig.

Raciborski, M. 1885. De nonnullis Desmidiaceis novis vel minus cognitis, quae in Polonia inventae sunt. Pamiet. Wydz. III Akad. Umiej. w Krakowie 10: 57–100. Pls. 10–14.

Raciborski, M. 1889. Desmidye nowe (Desmidiaceae novae). Pamiet. Wydz. Mat.-Przyr. Akad. Umiej. w Krakowie 17: 73–113. Pls. 5–7.

Raciborski, M. 1892. Desmidyja zebrane przez Dr. E. Ciastonia s podrozy na okolo ziemi. Rozpr. Akad. Umiej. w Krakowie 22: 361–392. Pls. 6, 7.

Raciborski, M. 1895. Die Desmidieenflora des Tapakoomasees. Flora 81: 30–35. Pls. 3, 4.

Ralfs, J. 1844. On the British Desmidieae. Ann. & Mag. Nat. Hist. 13: 375–380; 14: 187–194, 256–261, 391–396.

Ralfs, J. 1848. The British Desmidieae. i–xxii + 226 pp. 35 Pls. London.

Ramanathan, K. R. 1962. Zygospore formation in some South Indian Desmids. Phykos 1(1): 38–43 Figs. 1–20.

Reinsch, P. 1867. Die Algenflora des mittleren Theiles von Franken enthaltend die vom Autor bis jezt in diesen Gebieten beobachteten Süsswasseralgen..... Abh. Naturh. Ges. z. Nürnberg 3(2): 1–238. 13 Pls.

Reinsch, P. 1875. Contributiones ad algologiam et fungologiam I. 12 + 103 pp. 9 Pls.

Rich, Florence. 1935. Contributions to our knowledge of the freshwater algae of Africa. 11. Algae from a Pan in Southern Rhodesia, Trans. Roy. Soc. South Africa 23: 107–160. Map. 24 Figs.

Rino, J. A. 1972. Contribuicao para o conhecimento das algas de água doce de Mocambique III. Rev. Cienc. Biol. Ser. 5, A: 121–164. Pls. 1–32.

Roll, J. 1925. Contribution à l'étude de la flore des algues de l'URSS. Genre *Micrasterias* Ag. Arch. Russ. Protistol. 4(3/4): 235–253. Pls. 10–14.

Roll, J. 1928. On new or noteworthy forms of desmid algae. III. Arch. Russ. Protistol. 7: 131−138. 1 Pl.

Roy, J. 1877. Contributions to the desmid flora of Perthshire. Scottish Nat. 1877: 68−74.

Roy, J. and Bissett, P. (1893)1894. On Scottish Desmidieae. Ann. Scottish Nat. Hist. 1893(6): 106−111; 1893(7): 170−190. Pl. 1; 1893(8): 237−245; 1894(9): 40−46. Pl. 2; 1894(10): 100−105. Pl. 3; 1894(11): 167−178; 1894(12): 241−256. Pl. 4.

Rozeira, A. 1954. Manípulo de Desmídias da Guiné Portuguesa. Publ. Inst. Bot. Dr. Gonçalo Sampaio Fac. Cienc. Univ. Porto, Ser. 2(22): 56−61. 18 Figs.

Růžička, J. J. 1955. Zajímavé Krásivky (Desmidiales) horni Oravy. Biologia 10: 590−604. 41 Figs. Pls. 1−3.

Salisbury, R. K. 1936. The desmids of Florida. Ohio Jour. Sci. 36(1): 55−61. Pl. 1.

Saunders, de A. 1894. Flora of Nebraska. Introduction. Part I. Protophyta-Phycophyta. Part 2. Coleochaetaceae, Characeae. Bot. Seminar. Univ. Nebraska. 128 pp. 36 Pls.

Saunders, de A. 1901. The Algae. Pap. Harriman Alaska Exped. No. 25. Proc. Washington Acad. Sci. 3: 391−486. Pls. 43−62.

Schaarschmidt, J. (1882)1883. Tanulmányok a Magyarhoni Desmidiaceákról. Mag. Tudom. Akad. Math. e Termész. Köz. 18: 259−280. 1 Pl.

*Scheffer, V. and Robinson, R. H. 1939. A limnological study of Lake Washington. Ecol. Monogr. 9: 95−143.

Schmidle, W. 1893. Algen aus dem Gebiete des Oberrheins. Ber. d. Deutsch. Bot. Ges. 11(10): 544−555. Pl. 28.

Schmidle, W. 1893a. Beiträge zur Algenflora des Schwarzwaldes und der Rheinebene. Ber. Naturf. Ges. Freiburg i Br. 7(1): 68−112. Pls. 2−6.

Schmidle, W. (1895)1896. Beiträge zur alpinen Algenflora. Österr. Bot. Zeit. 1895: 249−253; 305−311; 316−350; 387−389; 454−459. Pls. 14−16; 1896: 20−25; 59−65; 91−94.

Schmidle, W. 1896a. Süsswasseralgen aus Australien. Flora 82: 297−313. Pl. 9.

Schmidle, W. 1897. Zur Kritik einiger Süsswasseralgen. Nuova Notarisia 1897(8): 63−70.

Schmidle, W. 1898. Über einige von Knut Bohlin in Pite Lappmark und Vesterbotten gesammelte Süsswasseralgen. Bih. Kongl. Svenska Vet.-Akad. Handl. 24, III(8): 2−69. Pls. 1−3.

Schmidle, W. 1898a. Die von Professor Dr. Volkens und Dr. Stuhlmann in Ost-Afrika gesammelten Desmidiaceen, bearbeitet unter Benützung der Vorarbeiten von Prof. G. Hieronymus. In: Engler, A. Beiträge zur Flora von Afrika. XVI. Engler's Bot. Jahrb. 26: 1−59. Pls. 1−4.

Scott, A. M. and Grönblad, R. 1957. New and interesting desmids from the southeastern United States. Acta Soc. Sci. Fennicae, II, B, II(8): 1−62. Pls. 1−25.

Scott, A. M. and Prescott, G. W. 1952. The algal flora of southeastern U.S. VI. Additions to our knowledge of the desmid genus Euastrum 2. Hydrobiologia 4(4): 377−398. Pls. 1−3.

Scott, A. M. and Prescott, G. W. 1958. Some freshwater algae from Arnhem Land in the northern territory of Australia. Rec. America-Australia Sci. Exped. Arnhem Land 3: 8−155. 1 Textfig. Pls. 1−28.

Scott, A. M. and Prescott, G. W. 1960. Notes on Indonesian freshwater algae. IV. Concerning Euastrum moebii (Borge) Scott & Prescott comb. nov. and Euastrum turgidum Wallich. Reinwardtia 5(3): 323−340. Figs. 1−5.

Scott, A. M. and Prescott, G. W. 1961. Indonesia Desmids. Hydrobiologia 17(1/2): 1−132. Pls. 1−63.

Selman, G. G. 1966. Experimental evidence for the nuclear control of differntiation in Micrasterias. Jour. Embryol. Exper. Morph. 16(3): 469−485. 1 Textfig. 2 Pls.

Sieminska, Jadwiga. 1965. Algae from Mission Wells pond, Montana. Trans. Amer. Microsc. Soc. 84(1): 98−126. Pls. 1−8.

Silva, H. 1951. Algae of the Tennessee Valley region: a manual for the identification of species. Dissert., Michigan State Coll.

Silva, H. 1953. Fifteen new algae. Bull. Torr. Bot. Club 80(4): 342−348. 15 Figs.

Skuja, H. 1934. Beitrag zur Algenflora Lettlands. I. Acta Horti Bot. Univ. Latvia 7: 25−85. 119 Figs.

Smith, G. M. (1921)1922. The phytoplankton of the Muskoka region, Ontario, Canada. Trans. Wisconsin Acad. Sci., Arts & Lettr. 20: 323−364. Pls. 8−13.

Smith, G. M. 1924. Wisconsin phytoplankton. II. Desmidiaceae. Wisconsin Geol. & Nat. Hist. Surv. 57(2): 1−227. 17 Textfigs. Pls. 52−88.

Smith, G. M. 1924a. Ecology of the plankton algae in the Palisades Interstate Park, including the relation of control methods to fish culture. Roosevelt Wild Life Bull. 2(2): 88−195. Pls. 3−24.

Sormus, Laine and Bicudo, C. deM. 1974. Polymorphism in the desmid Micrasterias pinnatifida and its taxonomic implications. Jour. Phycol. 10(3): 274−279. 28 Figs.

*Sparling, J. H. and Nalewajko, C. 1970. The chemical composition and phytoplankton in southern Ontario. Bull Fish. Res. Bd. Canada 27: 1405-

Taft, C. E. 1934. Desmids of Oklahoma II. Trans. Amer. Microsc. Soc. 53(2): 95-101. Pl. 6.

Taft, C. E. 1945. The desmids of the west end of Lake Erie. Ohio Jour. Sci. 45(5): 180-205. Pls. 1-5.

Taft, C. E. 1947. Some algae, including new species from New Mexico. Ohio Jour. Sci. 47(2): 85-87. Figs. 1-4.

Taft, C. E. 1949. New, rare, or otherwise interesting algae. Trans. Amer. Microsc. Soc. 68(3): 208-216. Pls. 1, 2.

*Taylor, W. D. 1974. freshwater algae of the Rae Lakes Basin, Kings Canyon National Park (Exclusive of Diatoms). M.A. Thesis, Humboldt State Univ.

*Taylor, W. R. 1933. Distribution of the freshwater algae in Newfoundland. Collecting Net 8(3): 68.

Taylor, W. R. 1934. The freshwster algae of Newfoundland. Part 1. Pap. Michigan Acad. Sci., Arts & Lettr. 19(1933): 217-278. Pls. 45-57.

Taylor, W. R. 1935. The freshwater algae of Newfoundland. Part 2. Pap. Michigan Acad. Sci., Arts & Lettr. 20(1934): 185-230. Pls. 33-49.

Taylor. W. R. 1935a. Alpine algae from the Santa Marta Mountains, Colombia. Amer. Jour. Bot. 22: 763-781. Pls. 1-3.

Teiling, E. 1957. Morphological investigations of asymmetry in desmids. Bot. Notiser 110(1): 49-82. 39 Figs.

Thomasson, K. 1960. Notes on the plankton of Lake Bangweulu. Part 2. Nova Acta Reg. Soc. Sci. Upsaliensis, IV, 17(12): 1-43. Figs. 1-14.

Thomasson, K. 1963. Araucanian Lakes. Plankton studies in North Patagonia with notes on terrestrial vegetation. Acta Phytogeogr. Suecia 47: 1-139. Figs. 1-47.

Thomasson, K. 1971. Amazonian algae. Mém. Inst. Roy. Sci. Nat. de Belgique, Fasc. 86: 1-57. 24 Pls.

Thomaszewicz, G. H. 1973. The typical variety and developmental stages of *Micrasterias truncata* (Corda) Bréb. Acta Soc. Bot. Poloniae 42(4): 591-598. Figs. 1-4.

Tippitt, D. H. and Pickett-Heaps, J. D. 1974. Experimental investigations into morphogenesis in *Micrasterias*. Protoplasma 81: 271-296. 43 Figs.

Turner, W. B. 1885. On some new and rare desmids. Jour. Roy. Microsc. Soc., II, 5(6): 933-940. Pls. 15, 16.

Turner, W. B. 1892. Algae aquae dulcis Indiae orientalis. The freshwater algae (principally Desmidieae) of East India. Kongl. Svenska Vet.-Akad. Handl. 25(5): 1-187. Pls. 1-23.

Turner, W. B. 1893. Desmidiological notes. Naturalist 18: 343-347. 1 Pl.

Turpin, P. J. F. 1820(1816-1829). In: Dictionnaire des Sci. Nat., Planches: Vegetaux acotyledons. Vol. 10.

Turpin, P. J. F. 1828. Aperçu organographique sur le nombre deux. Mém. Mus. Hist. Nat. Paris 16: 296-344. Pl. 13.

Vinyard, W. C. 1951. Distribution of alpine and subalpine algae in western United States. M.S. Thesis, Michigan State Coll.

Viret, L. 1909. Desmidiacées de la Vallée du Trient. (Valaise, Suisse). Bull. Soc. Bot. Genève 2(1): 251-268. 1 Pl.

*Voth, H. M. 1951. Seasonal variation in the net plankton of three north Florida lakes. M.S. Thesis, Florida State Univ.

Waddington, C. H. 1963. Ultrastructure aspects of cellular differentiation. Symposium. Soc. Exper. Biol. 17: 85-97. 3 Pls.

*Wade, W. E. 1952. A study of the taxonomy and ecology of Michigan desmids. Dissert., Michigan State Coll.

Wade, W. E. 1957. Studies on the distribution of desmids in Michigan. Trans. Amer. Microsc. Soc. 76(1): 80-86. 1 Map.

*Wade, W. E. 1957a. Additions to our knowledge of the desmid flora of Michigan. Rev. Algol. 2(4): 249-273. Pls. 1, 2.

Wailes, G. H. 1930. Protozoa and algae, Mount Ferguson, B.C. Vancouver Mus. & Art Notes 5: 160-165. 1 Pl.

Wailes, G. H. 1931. Protozoa and algae, Mount Ferguson, B.C. Part 2. Vancouver Mus. & Art Notes 6(2): 72-75. Pl. 2.

Wailes, G. H. 1933. Freshwater algae and Protozoa from Alaska. Art., Hist. & Sci. Assoc. Vancouver City Mus. 1933(March): 1-3. 25 Figs. Mimeogr.

Wallich, G. C. 1860. Descriptions of Desmidiaceae from Lower Bengal. Ann. & Mag. Nat. Hist., III, 5: 273-285. Pls. 13, 14.

Waris, H. and Kallio, P. 1964. Morphogenesis in *Micrasterias*. Advan. Morphogenesis 4: 45-80.

West, G. S. 1909. The algae of the Yan Yean Reservoir, Victoria: a biological and ecological study. Linn. Soc. Jour. Bot. 39: 1–88. Pls. 1–6.

West, G. S. 1914. A contribution to our knowledge of the freshwater algae of Columbia. In: Fuhrmann, O. and Mayor, E. Voyage d'exploration scientifique en Colombie. Mém. Soc. Neuchatel 1914, 5: 1013–1051. Pls. 21–23.

West, W. 1888. The desmids of Maine. Jour. Bot. 26: 339–340.

West, W. 1889. List of desmids from Massachusetts, U.S.A. Jour. Roy. Microsc. Soc. 1889: 16–21. Pls. 2, 3.

West, W. 1889a. The freshwater algae of Maine. Jour. Bot. 27: 205–207. Pl. 291.

West, W. 1890. Contribution to the freshwater algae of North Wales. Jour. Roy. Microsc. Soc. 1890: 227–306. Pls. 5, 6.

West, W. 1891. The freshwater algae of Maine. Jour. Bot. 29: 353–357. Pl. 315.

West, W. 1892. A contribution to the freshwater algae of West Ireland. Linn. Soc. Jour. Bot. 29: 103–216. Pls. 18–24.

West, W. 1892a. Algae of the English Lake District. Jour. Roy. Microsc. Soc. 1892: 713–748. Pls. 9, 10.

West, W. and West, G. S. 1894. New British freshwater algae. Jour. Roy. Microsc. Soc. 1894: 1–17. Pls. 1, 2.

West, W. and West, G. S. 1895. A contribution to our knowledge of the freshwater algae of Madagascar. Trans. Linn. Soc. London, Bot., II, 5: 41–90. Pls. 5–9.

West, W. and West, G. S. 1896. On some new and interesting freshwater algae. Jour. Roy. Microsc. Soc. 1896: 149–165. Pls. 3, 4.

West, W. and West, G. S. 1896a. On some North American Desmidiaceae. Trans. Linn. Soc. London, Bot., II, 5: 229–274. Pls. 12–18.

West, W. and West, G. S. 1897. Welwitsch's African freshwater algae. Jour. Bot. 35: 1–7; 33–42; 77–89; 113–183; 235–243; 264–272; 297–304. Pls. 365–370.

West, W. and West, G. S. 1897a. Desmids from Singapore. Linn. Soc. Jour. Bot. 33: 156–167. Pls. 8, 9.

West, W. and West, G. S. 1898. On some desmids of the United States. Linn. Soc. Jour. Bot. 33: 279–332. 7 Textfigs. Pls. 16–18.

West, W. and West, G. S. 1902. A contribution to the freshwater algae of the north of Ireland. Trans. Roy. Irish Acad. 32B(1): 1–100. Pls. 1–3.

West, W. and West, G. S. 1902a. A contribution to the freshwater algae of Ceylon. Trans. Linn. Soc. London, Bot., II, 6: 123–215. Pls. 17–22.

West, W. and West, G. S. 1903. Scottish freshwater plankton. No. 1. Linn. Soc. Jour. Bot. 35: 519–556. Pls. 14–18.

West, W. and West, G. S. 1905. Freshwater algae from the Orkneys and Shetlands. Trans. & Proc. Bot. Soc. Edinburgh 23: 3–41. Pls. 1, 2.

West, W. and West, G. S. 1905a. Monograph of the British Desmidiaceae, II: 1–206. Pls. 33–44. Ray Soc., London.

West, W. and West, G. S. 1907. Freshwater algae from Burma—including a few from Bengal and Madras. Ann. Roy. Bot. Gard. Calcutta 6(2): 175–260. Pls. 10–16.

Whelden, R. M. 1941. Some observations on freshwater algae in Florida. Jour. Elisha Mitchell Sci. Soc. 57(2): 261–272. Pls. 5, 6.

Whelden, R. M. 1942. Notes on New England algae. II. Some interesting New Hampshire algae. Rhodora 44: 175–187. 2 Pls.

Whelden, R. M. 1943. Notes on New England algae III. Some interesting algae from Maine. Farlowia 1(1): 9–23. 18 Figs.

*White, M. C. 1862. Note. In: Discovery of microscopic organisms in the siliceous nodules of the Paleozoic rocks of New York. Amer. Jour. Sci. & Arts, II, 33(99): 385–386.

*Whitford, L. A. and Schumacher, G. J. 1973. A manual of fresh-water algae. 324 pp. 77 Pls. Sparks Press, Raleigh, N.C.

Wille, N. 1879. Ferskvandsalger fra Novaja Semlja samlede of Dr. F. Kjellman paa Nordenskiöld's Expedition 1875. Öfv. Kongl. Vet.-Akad. Förhandl. 1879(5): 13–74. Pls. 12–14.

Wille, N. 1880. Bidrag till Kundskaben om Norges Ferskvandsalger 1. Smaalenes Chlorophyllophyceer. Christiania Vidensk.-Selsk. Förhandl. 1880(11): 1–72. Pls. 1, 2.

Wittrock, V. B. 1869. Anteckningar om Skandinaviens Desmidiacéer. Nova Acta Reg. Soc. Upsaliensis, III, 7(3): 1–28. 1 Pl.

Wittrock, V. B. 1872. Om Gotlands och Ölands sötvattensalger. Bih. Kongl. Vet.-Akad. Handl. 1(1): 1–72. Pls. 1–4.

Wittrock V. B. and Nordstedt, C. F. O. 1880. Algae aquae dulcis exsiccatae praecipue scandinavicae, quas adjectis algis marinis chlorophyllaceis et phycochromaceis distribuerunt Wittrock & Nordstedt. Bot. Notiser 4: 113–122.

Wittrock, V. B. and Nordstedt. C. F. O. 1882. Algae aquae dulcis exsiccatae..... Bot. Notiser 1882: 51–61.

Wittrock, V. B. and Nordstedt, C. F. O. 1886. Algae aquae dulcis exsiccatae.... Bot. Notiser 1886: 157–168.

Wittrock, V. B. and Nordstedt, C. F. O. 1889. Algae aquae dulcis exsiccatae..... Fasc. 21: 1–92. Figs.

Wolle, F. 1876. Freshwater algae I. Bull. Torr. Bot. Club 6(23): 121–123.

Wolle, F. 1880. Freshwater algae IV. Bull. Torr. Bot. Club 7: 43–48. Pl. 5.

Wolle, F. 1881. American fresh-water algae. Species and varieties of desmids new to science. Bull. Torr. Bot. Club 8: 1–4. Pl. 6.

Wolle, F. 1883. Freshwater algae VII. Bull. Torr. Bot. Club 10: 13–21. Pl. 27.

Wolle, F. 1884. Fresh-water algae. VIII. Bull. Torr. Bot. Club 11: 13–17. Pl. 44.

Wolle, F. 1884a. Desmids of the United States and list of American Pediastrums. xiv + 168 pp. Pls. 1–53. Bethlehem, Pa.

*Wolle, F. 1884b. First contribution to the knowledge of Kansas algae. Bull. Washburn Lab. Nat. Hist., 1(1): 17–18.

Wolle, F. 1885. Fresh-water algae. IX. Bull. Torr. Bot. Club 12(1): 1–6. Pl. 47.

Wolle, F. 1885a. Fresh-water algae. X. Bull. Torr. Bot. Club 12(12): 125–129. Pl. 51.

*Wolle, F. 1885b. Second contribution to the knowledge of Kansas algae. Bull. Washburn Lab. Nat. Hist. 1(2): 62–64.

*Wolle, F. 1886. Third contribution to the knowledge of Kansas algae. Bull. Washburn Lab. Nat. Hist. 1(6): 174–175.

Wolle, F. 1887. Fresh-water algae of the United States (exclusive of the Diatomaceae) complemental to Desmids of the United States. xi + 364 pp. Pls. 54–210. Bethlehem, Pa.

*Wolle, F. 1889. Fourth contribution to the knowledge of Kansas algae. Bull. Washburn Lab. Nat. Hist. 2(9): 64.

Wolle, F. 1892. Desmids of the United States and list of American Pediastrums. viii + 182 pp. Pls. 1–64. Bethlehem, Pa.

Woloszynska, Jadwiga. 1919. Przyczynek do znajomosci glonów Litwy. Rozpr. Sprawozd. Posied. Wydz. Mat.-Przyr. Akad. Umiej. w Krakowie, Ser. B, 57: 1–65. Pl. III.

Wood, H. C. (1869)1870. Colored drawings and mounted slides of a variety of desmids, and remarks. Biol. & Microsc. Dept. Acad. Nat. Sci. pp. 15–19, in: Academy of Nat. Sci. Philadelphia 1869. Philadelphia, Pa. 1870.

Wood, H. C. (1872)1874. A contribution to the history of fresh-water algae of North America. Smithson. Contrib. Knowledge No. 241, 19(Art. 3): ix + 262 pp. 21 Pls.

Woodhead, N. and Tweed, R. D. 1960. Additions to the algal flora of Newfoundland. Part I: New and interesting algae in the Avalon Peninsula and central Newfoundland. Hydrobiologia 15(4): 309–362. Textfigs. 81, 82. Pls. 1–8.

Woronichin, N. N. 1930. Algen des Polar- und des Nordural. Trav. Soc. Nat. Leningrad 60: 3–80. 3 Pls.

*Wunderlin, T. F. 1971. A survey of the freshwater algae of Union County, Illinois. Castanea 36(1): 1–53.

*Wunderlin, T. F. and Wunderlin, R. P. 1968. A preliminary survey of the algal flora of Horseshoe Lake, Alexander County, Illinois. Amer. Mid. Nat. 79(2): 534–539.

New Taxa and New Combinations

INDEX

Page numbers in boldface refer to descriptions of accepted taxa. Synonymous names occur on pages indicated by italics. Incidentally mentioned taxa and terms appear in lightface roman. Illustrations of taxa are given as plate numbers and figures.